RICHARD K. SPIELMAN
19200 S. MAIN STREET
GARDENA, CA. 90248

Encyclopedia
of
Aviation

Encyclopedia
of
Aviation

Charles Scribner's Sons
New York

© 1977 by Reference International
Publishers Limited

First USA publication 1977 by
Charles Scribner's Sons,
597 Fifth Avenue, New York.

Copyright under the Berne Convention

All rights reserved. No part of this
publication may be reproduced in any
form whatever without the written
permission of Reference International.

ISBN 0–684–14840–4

Library of Congress Catalog Card
Number 77–72699

1 3 5 7 9 11 13 15 17 19 I/C 20 18
16 14 12 10 8 6 4 2

Printed in the United States of America

RICHARD K. SPIELMAN
19200 S. MAIN STREET
GARDENA, CA. 90248

Contributors

Jean Alexander
A leading authority on Russian aircraft who makes frequent visits to the Soviet Union researching her specialized field, Jean Alexander is the author of *Russian Aircraft since 1940*, a major reference source on Russian aviation, and a regular contributor to aviation publications.

Chaz Bowyer
After serving in the RAF for 26 years, Chaz Bowyer became a full-time aviation writer, producing a number of outstanding reference works and contributing to many aviation magazines. He is literary editor of the British branch of the "Cross and Cockade Society," the international organization of World War I aviation historians.

Roger Freeman
Roger Freeman grew up in Suffolk, England, during the second world war and witnessed at first hand the growth of American air power in Europe during 1942–45. Subsequently he became an eminent authority on U.S. military aviation and has published a number of important books, among which is his definitive history of the 8th Air Force, *The Mighty Eighth*.

Bill Gunston
A former RAF pilot, Bill Gunston has been technical editor of the magazine *Flight International* and later technology editor of *Science Journal*. He is a noted authority on modern combat aircraft and the author of numerous books.

A. J. Jackson
A pilot and one of the world's leading authorities on civil aviation, he has published many books on aircraft, establishing himself as an outstanding aviation research worker.

Bruce Robertson
Author and editor of over twenty major works on aeronautics, Bruce Robertson's unique aeronautical archives represent thirty years of dedicated research. He is a specialist on the World War I period and the RAF.

Rodney Steel
Rodney Steel is a pilot and a journalist specializing in aviation and other technical fields. He has held editorial posts with the Chambers' Encyclopedia, the British Standards Institution, Penguin Books and the Royal Air Force, as well as contributing aviation features for Marshall Cavendish.

A

A-7. See CORSAIR

A-26. See INVADER

A.3000B. See AIRBUS

AD/A-1. See SKYRAIDER

Aerobatics

The performance of planned maneuvers involving the assumption of unorthodox flight attitudes has grown from the stunt flyer's personal exhibition of skill and daring into an internationally recognized sport, supervised by the Fédération Aéronautique Internationale and requiring specially designed aircraft.

Even stalling and spinning, two of the most basic aerobatic exercises, may be accomplished in different ways. Spins can be inverted or upright, varying from vertical to flat, or can be incorporated in a falling leaf. Simple stalls are employed in making stall turns.

Basic exercises to help teach a pilot coordination include the chandelle (180° climbing turn with maximum height gain), the "wing-over" and the "lazy-8." Rolls, on the other hand, are an essential feature of advanced aerobatics. A slow roll requires about 12 seconds for 360° rotation; a flick roll is a horizontal autorotational figure entered from a high-speed stall (it can also be executed in a vertical climb or dive); in a barrel roll the airplane follows a

The RAF's Red Arrows demonstrating high-speed formation aerobatics in their famous "bomb burst" routine

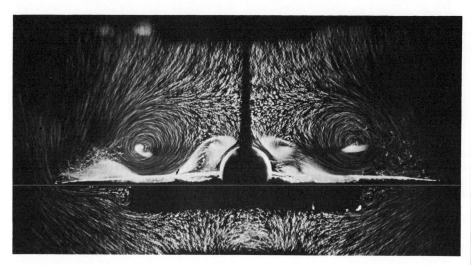

An essential tool of research in aerodynamics is Schlieren photography, which records the airflow around a model in a wind tunnel (in this case showing large vortices forming above the wings)

corkscrew path. While flying a loop the aircraft's wings remain level, and control is exercised principally by the elevators. Variations on the simple loop include the roll off the top, the avalanche (with a flick roll off the top), square and octagonal loops, diamond-shaped loops, and rolling loops.

Maneuvers involving negative *G* forces place considerable demands on both the pilot and the airplane. They include outside loops, outside stall turns, and negative flick rolls. In the 1950s Czechoslovakian pilots devised *lomcováks*, which embody rapid negative rolls with the propeller's torque assisting rotation.

Holding a vertical bank while flying straight is known as knife flight (maintainable only for a limited distance), while tail slides must be undertaken with care since there is a danger of damaging the aircraft (a torque roll is a rolling tail slide).

Some maneuvers have found a ready application in air combat. Among these is the split-S (rolling inverted and pulling through into the second half of a loop), while other now-standard aerobatic maneuvers, such as the IMMELMANN turn, had their origin in wartime dogfights.

Competitive aerobatics, with its precisely defined maneuvers and complicated scoring, is far removed from the world of stunt displays. For such an exacting sport, specially designed aircraft (such as the Zlin) must be able to withstand high accelerative and decelerative forces, and possess engine, oil, and fuel systems that will function when inverted. The maneuvers themselves are written down with a system of shorthand symbols (devised by the Spaniard J. L. Aresti) in the "Aresti Dictionary," a uniform code of maneuvers and their ratings.

Aerodynamics

The behavior of air flowing around an airplane in flight is evaluated by the science of aerodynamics.

When an airplane moves through the air it generates pressure and friction. Pressure is greater below the machine than above it, providing LIFT through the medium of downwash. The total area of the AIRFRAME over which air passes (and thus produces friction) is known as the *wetted area.*

The corridor of disturbed air trailed along behind an airplane after its passage includes not only the lift-producing downwash, but also the SLIPSTREAM or jet efflux from the propulsive source, VORTICES emanating from the wing tips where air circulates from the high pressure area below the wing to the low pressure area above it, and a wake of air caught up by the airframe and dragged along by it causing drag. The latter may vary from perhaps one-thirtieth of the lift of a high-performance sailplane to one-fifth of the lift of a supersonic transport.

The aeronautical engineer seeks to ensure that every component part of an airframe is as aerodynamically efficient as possible without disrupting the airflow over adjacent parts of the machine. Hence fairings and fillets are employed at the junction of AIRFOIL surfaces with the fuselage to minimize the loss of lift that must inevitably take place at these points, although their presence increases the wetted area and hence skin friction drag.

The lift generated by a WING depends on the amount of air the airfoil displaces downward. This is determined by the product of the distance flown and the span: a short-span wing has to travel further to generate a given amount of lift than a long-span wing.

In subsonic airplanes the fuselage in side elevation can be considered as a very low aspect ratio airfoil—its shape is in fact comparable (on a much larger scale) to the wing sections used for subsonic flight. The wings do not usually form an integral aerodynamic entity with the fuselage and the resultant junctions between the components yield a large wetted area, but this configuration nonetheless provides efficient load-carrying characteristics providing the speed of sound is not approached.

For SUPERSONIC FLIGHT, however, the drag due to shock wave formation must be minimized. Fuselage shapes tend to resemble supersonic airfoil sections (long, thin and pointed) and may even incorporate camber to minimize loss of lift at the junctions with lifting surfaces. The amount of wetted area relative to the total wing area is reduced as much as possible. Since the circulation of air from below the wing to its upper surface is only influenced by the wing- and tail-tips in the disturbed air caused by *Mach cones* shed from these points, the inefficient portions of the airfoil surfaces are often cropped. In subsonic flight the flow of air around the tips influences the pressure distribution of the entire wing, and an elliptical planform is the most efficient layout for this performance envelope since it incurs minimum vortex drag and so gives a constant downwash across the span. Elliptical wings involve manufacturing complications, however.

The achievement of supersonic speed in level flight required a revised aerodynamic approach. Swept-wing aircraft that retained the orthodox relationship of wings and fuselage merely deferred the onset of compressibility. To surpass the speed of sound and cruise at supersonic velocities it was necessary to evolve new design concepts (e.g. AREA RULE), and to re-proportion the airplane so that the entire airframe was kept within the Mach cone shed from the nose. As a result short-span, low aspect ratio wings were adopted, and the advantages of deltas became apparent: with sharply swept-back wings the lift can be maintained further forward in a DELTA than is possible on a more conventional machine, unless a special layout is used or the fuselage itself can be designed to provide lift. Swept wings of large span also encounter aero-elastic problems at high MACH NUMBERS, distortion by the disturbed airflow leading to instability.

At speeds of around Mach 3 the pre-dominant feature of the airflow is shock wave propagation, and the designer seeks to make use of the relatively high pressure regions that exist behind these waves (compression lift).

Providing a means of changing the characteristics of wings to suit different flight conditions offers many advantages at the expense of technical complexity (see SWING WING). Simple slats that open automatically at low speed can be incorporated in the leading edges; STALLING speed is reduced, but there may also be a slight loss in maximum speed due to increased drag even when the slats are closed. Trailing edge FLAPS alter the camber of a wing and may be of considerable complexity. The Boeing 727, for example, uses triple trailing edge flaps allied with a spoiler (or lift dumper, to assist braking after landing), and a leading-edge flap in combination with another spoiler. Control of the BOUNDARY LAYER can be achieved by employing suction over the surfaces of the airfoil or by generating supercirculation with compressed air discharged over the flaps, ailerons and tailplane.

Handling qualities are substantially improved (particularly at the onset of compressibility) if every section of a wing operates at the same lift coefficient. To achieve this, a negative camber may be adopted at the wing root. This causes the inboard part of the wing to stall at the same time as the tip. Other means of improving the local airflow over a wing by preventing span-wise drift of the boundary layer toward the tip and a tendency for it to separate include notched leading edges and boundary layer fences; a dog-tooth or cambered leading edge reduces the peak pressure and again serves to inhibit boundary layer separation.

Aeroflot

Aeroflot, the Soviet state airline, has operated under its present name since 1932. It was originally created as Dobrolet—together with two regional airlines, Zakavia in the Caucasus and Ukrvozdukhput in the Ukraine—in 1923. By 1928 all Russian operators except the Russo-German company Deruluft had been combined in Dobrolet. The name was changed to Aeroflot when civil aviation ceased to be administered directly by the military authorities, but even today Aeroflot can still be regarded as the transport service of the air force, with some (mostly specialized) aircraft permanently in military colors.

In the immediate postwar years Aeroflot relied heavily on the Li-2 (Russian DC-3/C-47), with the Il-12 entering service in 1947, followed by the Il-14 (1954), the Il-18 (1959) and the Tu-114 (1961). The first Soviet jet passenger service was inaugurated by the Tu-104 on September 15, 1956.

Aeroflot is the largest airline in the world and is responsible for all aspects of Soviet civil aviation, including survey work, forestry and ice patrols, ambulance services and agricultural aviation. The aircraft fleet in 1976 was estimated at some 3,500 aircraft. Aeroflot uses a number of turboprop types, including the An-3 (adapted from the An-2 biplane), the medium-range An-12, the 15-passenger An-14, the 50-passenger An-24 and the giant, long-range An-22. Among jet aircraft are the Tu-124 short-range turbofan, the Tu-134 (equivalent to the BAC-1-11), the supersonic TU-144, the Tu-154 (inspired by the TRIDENT and the BOEING 727), and the short-range Yak-40. Helicopters include the giant Mi-6 (up to 120 passengers), the smaller

Aeroflot's first jet airliner the Tu-104

Mi-4, Mi-8, and Mi-10 types, and the Mi-12, the world's largest. Passengers carried in 1977 totaled nearly 100 million, and the airline served some 2,000 places.

The Soviet Union became a member of the ICAO in November 1970, and the Aeroflot pilots' association joined IFALPA in 1972.

Aeronca

Aeronca, one of the pioneer manufacturers of light aircraft, was established as the Aeronautical Corporation of America in 1928; it assumed its present name in 1941. The firm marketed the first commercially successful light airplanes, the C-2/C-3 series of two-seat high-wing monoplanes, which continued to be built until the mid-1930s. Later in that decade the 40-hp Scout and 50-hp Chief light aircraft came into production, while during World War II the L-3 military observation plane was produced, together with license-built PT-19 and PT-23 trainers.

Over 10,000 Model 7 Champions and 600 military L-16s were built between 1946 and 1950; in 1954 the manufacturing rights for these models were sold to Flyers Services of St. Paul, Minn. Aeronca abandoned all light-airplane construction during the 1950s to undertake subcontracting work for larger manufacturers.

Aérospatiale

On January 1, 1970, the French government formed the state-owned Société Nationale Industrielle Aérospatiale by merging Sud-Aviation (originator of the CARAVELLE jet airliner and producer of numerous helicopters and light aircraft), Nord-Aviation (manufacturer of a number of military transports such as the Noratlas and Transall), and SEREB. One of the largest of European aerospace companies, Aérospatiale is involved in production of CONCORDE; the Frégate and Mohawk twin-turboprop light transports; the twin-turbofan Corvette multipurpose aircraft; the Alouette, Lama, Super Frelon, Puma, Gazelle, and Dauphin helicopters; the A.300 AIRBUS; and various tactical and ballistic missiles and research rockets.

SOCATA (Société de Construction d'Avions de Tourisme et d'Affaires) is a subsidiary which manufactures the Rallye and Diplomate light aircraft.

AEW

The initials AEW stand for Airborne Early Warning. Aircraft equipped with complex electronic and radar systems

Boeing E-3A AWACS aircraft, with its disk-like radar rotodome mounted above the fuselage

for detecting incoming hostile strike planes can patrol far out to sea from their shore bases or aircraft carriers to give advance warning of an impending attack. They can also detect low-flying aircraft invisible to ground radar.

The U.S. Navy modified two L-747 Constellations for AEW work in 1949 (designated PO-1W, later WV-1); subsequently L-1049 Constellations were used by both the Air Force (RC-121, later EC-121) and Navy (Warning Star). An equivalent Russian AEW aircraft is the Tupolev Tu-126 ("Moss"), which was operational in 1968 and is based on the Tu-114 airliner ("Cleat"). Carrierborne AEW machines include the Grumman E-1 Tracer, which entered service in 1958, and its replacement, the Grumman E-2 Hawkeye (maiden flight April 19, 1961).

A still more advanced concept is the AWACS aircraft (Airborne Warning And Control System), which has equipment enabling it to control combat aircraft in either nuclear or conventional operations.

The Air Force has proposed the use of a modified Boeing 707-320B for this task, designated the E-3 and carrying a radar rotodome 360 ft. (109.7 m) in diameter providing a surveillance radius of approximately 230 mi. (368 km) from

30,000 ft. (9,144 m), together with nine radar consoles. A crew of 17 is envisaged, with flight endurance of $11\frac{1}{2}$ hours (without in-flight refueling).

Afterburner

A device located between the turbine and the nozzle of some high-performance JET ENGINES, which provides greater thrust by burning additional fuel. Exhaust gases, at a temperature of about 1,500°F, may not be enough to produce spontaneous combustion, and an additional ignition source may be needed in the afterburner itself. The augmented thrust—an improvement of as much as 40 percent can be achieved—is the result of an increase in the exhaust velocity brought about by combustion. Afterburners are used only for brief periods during takeoff and climbing, or during combat. Heavy fuel consumption makes extended use impractical.

Agricultural aircraft

Light airplanes or helicopters built or adapted for special tasks to aid cultivation in any way. The first agricultural aircraft flew in 1921, in a test by the Ohio State Experimental Station in which lead arsenate was sprayed from

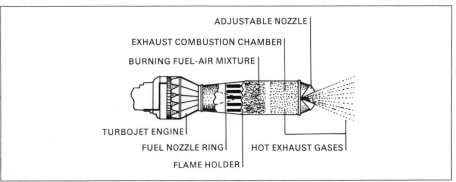

Details of an afterburner, an auxiliary source of thrust located at the rear of a turbojet engine that burns additional fuel in the engine's exhaust gases

Cessna AGtruck, a new specially designed agricultural aircraft

an airplane over a catalpa grove to poison insects destroying the trees. During the 1930s the regular crop-dusting of cotton fields established the utility of agricultural aircraft. Their scope increased from pest-control to seeding and fertilization, and to combating weeds, fungus, and compaction. In recent years over 1,000 Cessna AGwagon and some 5,000 Piper Pawnee agricultural airplanes have been built, and many helicopters have been adapted to specialized uses. Fittings include chemical tanks and spraying rigs, hoppers and dispensers for solids, and special ventilation for pilot and operators.

Ailerons

The outboard trailing edges of an airplane's wings include a pair of hinged surfaces called ailerons, which are operated to impart roll. In the early years of aviation, control of an aircraft's banking and rolling movements was often by wing-warping, a means of twisting the wings to change their curvature and angle of incidence. By the time of World War I, wing-warping, devised by the Wright Brothers, had given way to the use of ailerons.

When one aileron is depressed it increases the lift supplied by its parent wing and lifts the wing; the opposite aileron is interconnected and simultaneously comes up to reduce the lift of its own wing, which consequently drops. Ailerons are principally used to produce banking as an aircraft makes a turn.

Airacobra

One of the mainstays of American fighter squadrons at the beginning of World War II, the Bell P-39 Airacobra had its V-12 Allison engine located behind the pilot, so that its weight was concentrated around the airplane's center of gravity (thus enhancing the plane's maneuverability). The unusual layout also allowed a cannon to be installed in the propeller shaft and a tricycle undercarriage to be used.

The prototype (XP-39) first flew in April 1939, and with a turbosupercharger and a gross weight of 5,500 lb. (2,500 kg) attained 390 mph (630 km/h) at 20,000 ft. (6,100 m). The first variant to enter service in any numbers was the P-39D (February 1941), with six machine guns and a 37-mm cannon. Increased weight and deletion of the supercharger resulted in a disappointing performance; the plane was tried and rejected by the RAF, and when flown by the U.S. in the Pacific and North Africa it proved suitable only for ground strafing. Later models served with the Red Air Force and in the Mediterranean, but though remaining popular as a ground-attack plane with the Russians, the Airacobra was phased out of first-line service with U.S. forces by August 1944.

The P-63 Kingcobra, a further development of the mid-engined fighter, made its maiden flight on December 7, 1942. It was never used operationally by the United States, but 2,421 machines were supplied to Russia under Lend-Lease for ground strafing.

Airacomet

The first American jet fighter was the Bell P-59 Airacomet. The first jet flown in the United States was the XP-59 prototype on its maiden flight, October 1, 1942, at Muroc Dry Lake (later Edwards Air Force Base), Cal., powered by two General Electric 1-A turbojets of 1,400 lb. (635 kg) thrust. These were closely patterned on the jet engines designed by Sir Frank WHITTLE and brought to the United States on the initiative of General Hap Arnold. Thirteen YP-59s were ordered for service evaluation, and 20 production P-59A aircraft were delivered to the 412th Fighter Group at Santa Maria Field, Cal., by June 1945.

The 30 P-59Bs had J-31 turbojets of 2,000 lb. (900 kg) thrust, as did the last few P-59As, and also possessed increased fuel capacity. Both the A and the B variants had a 37-mm cannon in the nose, along with three .50 machine guns, but the aircraft was directionally unstable and thus proved a poor gun platform. The P-59 had a wingspan of 45 ft. 6 in. (13.86 m) and length of 38 ft. 10 in. (11.83 m). With a maximum speed of 413 mph (662 km/h) at 30,000 ft. (9,150 m), and a ceiling of 46,200 ft. (14,100 m), the performance of the P-59 was hardly an improvement on that of an advanced piston-engined aircraft like the P-51 MUSTANG, and the Airacomet was never regarded as even a potentially front-line aircraft. The last Airacomet left the Air Force inventory in October 1948.

P-39Q Airacobra

Airbus

A number of recent wide-bodied short-haul aircraft with high passenger capacities are known as airbuses. They include the McDonnell Douglas DC-10, the Lockheed TRISTAR, and the A.300B of Airbus Industrie. The A.300B is a 260–281 seat airliner powered by two wing-mounted General Electric CF6 turbofans, with a wingspan of 155 ft. 4 in. (47.34 m) and length of 181 ft. 5 in. (55.30 m). It entered service in May 1974. Airbus Industrie was formed as a multinational attempt to create a European aircraft capable of competing successfully against products of American manufacturers. Members of the Airbus Industrie consortium include AÉROSPATIALE of France, Messerschmitt-Bölkow-Blohm, VFW-Fokker of West Germany and Holland, and CASA of Spain.

A.300B Airbus

Air Canada

Air Canada was founded as Trans-Canada Air Lines on April 10, 1937. A publicly-owned company, the airline's shares were held by the government-owned Canadian National Railway. TCA's task was the establishment of a transcontinental network, but its first service, begun on September 1, 1937, only ran from Vancouver to Seattle. By the end of 1938 the entire Vancouver–Montreal route was open for mail services; passengers were carried from April 1, 1939. During the latter part of World War II TCA operated transatlantic services for the Canadian government and in September 1946 began a commercial route from Montreal to London.

In March 1964 TCA changed its name to Air Canada, and it now ranks as one of the world's major air carriers. It is seventh among world airlines (excluding Aeroflot) in terms of passengers carried, and possesses the eighth-largest fleet. The airline operates an extensive network in Canada, and services to numerous European destinations, to Bermuda and the Caribbean, and from Toronto and Montreal to Moscow. The present fleet consists of 37 DC-8s, 53 DC-9s, 12 TriStars, 14 Boeing 727s, and 6 Boeing 747s. During the winter some TriStars are transferred to Eastern Air Lines. Over 10,000,000 passengers are carried annually.

Aircraft carriers

Aircraft carriers are in effect floating airfields with hangar space beneath a flat-topped flying deck for up to 200 aircraft. A superstructure on the starboard side of the deck, called the island, houses the flying and navigational controls. The true aircraft carrier, as distinguished from the early seaplane carriers and present-day helicopter carriers, operates fixed-wing aircraft in attack, fleet defense, reconnaissance, and antisubmarine roles. The aircraft may take off under their own power, or with the aid of a catapult; when returning landing speed can be retarded by ARRESTER SYSTEMS. On modern carriers the deck is angled to port away from the island and a crash barrier can be quickly brought into action in an emergency.

Naval aviation dates from 1911, when a Curtiss biplane landed on a wooden platform on the U.S. cruiser *Pennsylvania* and then took off again. The first true carrier with an unobstructed flight deck joined the British Navy at the end of World War I, and during World War II the United States, Britain, and Japan were the world's carrier powers. At present the United States has overwhelming superiority in carriers with the nuclear-powered *Nimitz* and *Enterprise* (ships of over 85,000 tons and some 1,100 ft.—336 m—long) and 19 fleet carriers of the Forrestal, Midway, Hancock, and Essex classes. In comparison, Russia has only the two new Kuril Class carriers; France has two, and a nuclear-powered one building; Britain has the *Ark Royal* for fixed-wing aircraft; and four countries, Argentina, Australia, Brazil, and India, each have an ex-British carrier.

Airfoil

The special shape of an airplane wing section, which has a rounded upper sur-

Nuclear-powered aircraft carrier U.S.S. Enterprise

5

face and a relatively flat under surface. The front, or leading edge, is broad while the rear, or trailing edge, is sharply tapered. Airplane propellers and helicopter rotors also have an airfoil cross section (see AERODYNAMICS; WINGS).

Airframe

The body-shell of an aircraft, including its WINGS, FUSELAGE and TAIL, that is, the structure that carries the engine(s), fuel tanks, and payload.

In the early days of aviation airframes were made of wood or, eventually, metal girders, covered with fabric and sometimes wire-braced. This system was largely replaced by a metal stressed-skin riveted to a substructure of longerons (carrying end loads), stringers, bulkheads, and frames, with cantilever wings usually based upon a main spar, a rear spar, and internal ribs (to form the airfoil section).

The construction of the airframe and the choice of materials depend on the proposed function of the aircraft. The stresses and strains to which it will be subjected have to be calculated, cutting windows or other openings, which weakens the structure, must also be taken into account. METAL FATIGUE is an important factor to be considered, and milling skins and stabilizing members from solid billets can reduce the number of weakening joints.

Air France

France's national airline was formed in 1933 by uniting CGTA (Compagnie Générale de Transports Aériens, formerly Lignes Aériennes Farman), which had begun operations in Europe in 1919, CIDNA (Compagnie Internationale de Navigation Aérienne), which originally operated services to Istanbul and Prague, Air Union (consisting of Messageries Aériennes and Grands Express Aériens), Air Orient (serving the Far East), and Aéropostale (formerly Lignes Aériennes Latécoère), which operated in West Africa.

Air France began with some 23,600 mi. (37,980 km) of route and 259 aircraft. Older types of aircraft were replaced (notably by Dewoitine 338s and Bloch 220s), and the network was extended throughout Africa, as well as to Hanoi and Hong Kong, and to Stockholm. North Atlantic survey flights using Latécoère 521 flying boats were halted by World War II, which resulted in the eventual withdrawal of what was left of Air France to North Africa.

European services resumed in 1946, and DC-4s were introduced on Paris–

The last of Air France's flying boats, the Latécoère 631, used briefly on flights to the Caribbean in 1947

New York flights in the same year. As the company's lines were reestablished and extended, Constellations came into use, followed by Viscounts (1953) and Caravelles (1959). Boeing 707s arrived in 1960, and by 1972 Air France had more than 100 jet airliners. A.300 Airbus services started in May 1974, and the first commercial Air France Concorde flight took place early in 1976.

The airline's routes cover 291,135 unduplicated miles (489,753 km); over 8,000,000 passengers are carried annually together with nearly 700,000 tons of freight.

Airfreight

Airfreight is usually defined as including all types of cargo other than mail and baggage. The development of large, long-range transport aircraft during World War II made possible the establishment of the modern air cargo industry in the years following the war; prior to this time aircraft had lacked the size and performance to carry really bulky freight. A striking example of the new freight-carrying capabilities of aircraft was the BERLIN AIRLIFT of 1948–49, during which over 2 million tons of freight were flown into the beleaguered German capital. Airfreight on a commercial basis nonetheless was responsible for only a small proportion of the business of the growing commercial aviation industry: between 1927 and 1941 Pan American's air cargo business never represented more than about 5 percent of the airline's total revenue, and in 1947 the figure was still only just over 10 percent. Today it is some 15 percent (over $226 million) of a far larger total.

The advantage of speed in airfreighting perishable goods, medical supplies, livestock, or urgently needed tools was evident, but in the immediate postwar years there was never more than a limited market for specially adapted

cargo aircraft such as the Douglas DC-6 Liftmaster, Boeing Stratofreighter, or modified DC-4s, Constellations, or DC-3s. Several commercial operators were, however, specialist freight carriers, notably Seaboard World and the Flying Tiger Line.

The big Boeing and Douglas jetliners, which began operating in the late 1950s, offered substantial below-deck freight accommodation, and when the even larger Boeing 747s and DC-10s appeared a decade later they were equipped with more below-deck room than most routes required: over the North Atlantic they operated with cargo load factors as low as 35 percent. Boeing 747s with main-deck cargo capability began to appear early in the 1970s, and Lufthansa's nose-loading 747F started to operate in 1972. By 1975 air cargo (which includes mail and baggage) was contributing 17 percent of British Airways' total revenue and 25 percent of Lufthansa's.

Wide-bodied freighters take standard 8 ft. by 8 ft. (2.4 m × 2.4 m) containers in lengths of 10, 20, and 40 ft. (3.05 m, 6.1 m, and 12.2 m), together with various pallets of similar width and height, or "igloos" of 88 in. by 125 in. (223 cm × 317 cm) or 88 in. by 108 in. (223 cm × 274 cm). As airfreight traffic increases, assembly of loads away from the airport, probably using intermodal containers that can be transferred directly to and from trucks, will become increasingly common, although the use of pallets and "igloos" is likely to be favored by some operators.

Despite mechanization at cargo ports and computerized documentation, handling costs reached $120 per metric ton in 1975 (double the 1968 rate), and the average international shipment spent six days in transit, with only eight percent of this time being actually spent in the air.

Large freight aircraft can carry bulky engineering equipment to remote locali-

Loading the giant Boeing 747 freighter

ties where oil or mineral prospecting is taking place—for example, the Russian An-22s that airlift oil drilling machinery to Siberia. Boeing 747Fs on the New York–London route have carried complete loads weighing nearly 245,000 lb. (111,130 kg), and items as large as a pipe-laying machine measuring 22 ft. by 9 ft. 9 in. by 10 ft. (6.7 m × 3.0 m × 3.05 m).

The world's largest airfreight operator is the giant Russian airline Aeroflot, which carries over 2 million tons of cargo annually. Pan American is the second-biggest, followed by the Flying Tiger Line and United Air Lines.

Air-India

The Indian international airline was established as Air-India International after the country's independence in 1948. With the nationalization of all Indian airline operations in 1953, it had a fleet that included 74 DC-3s, 12 Vikings, and 3 DC-4s; by 1955 it was operating routes through Europe to London and eastward to Tokyo. Viscounts came into use on domestic services in 1957, and by 1960 routes had been established to New York via London, to Moscow via Tashkent, and to Sydney via Djakarta.

The airline's name became Air-India in 1962, and by the mid-1970s it was operating a fleet of 5 Boeing 747s and 9 Boeing 707s, with 10,390 employees. It carries some 600,000 passengers annually.

Airlines

Airlines operate commercial scheduled or charter services carrying passengers or freight or both. While an air carrier may call itself an airline, the term is generally restricted to operators of aircraft seating 30 or more passengers. Smaller carriers are generally known as feeder lines or air-taxi services.

German airships had operated regular passenger services as early as 1910–12, but the first regular service carrying passengers by airplane was flown by a Benoist XIV flying boat operating between St. Petersburg and Tampa, Fla., from January to April 1914. During World War I converted bombers were used to fly passengers across the English Channel, and these operations later continued in private hands after the Armistice of 1918, becoming the first airline services in a form recognizable today. Aircraft Transport and Travel Ltd., which opened on August 25, 1919, with London–Paris services is generally considered the first international airline. Airmail deliveries were established by Austria (Vienna–Kiev in March 1918, the first scheduled international airmail service), France, Italy, and the United States, providing a basis for pioneer airline services. London, Paris, and Berlin became centers of early European operations, but the many small firms soon began to amalgamate into larger, more economically viable units as services were extended.

The three-engined Fokker F.VII became one of the most widely used airliners of the interwar years. In Germany the predominantly Junkers-equipped Deutsche Luft Hansa was formed in 1926, and by 1931 it had become one of the world's most important airlines. France was the second-largest European carrier, with lines now extending to Africa, followed by Italy and by Britain (whose Imperial Airways was establishing services to India, Australia, and Africa).

In the United States a transcontinental airmail route was pioneered soon after World War I, using single-engined DH 4s, and there had been some passenger flying in the Caribbean between 1919 and 1924. The Pratt & Whitney-powered Boeing Type 40 was introduced on the Chicago–San Francisco service in 1927, and by 1930 amalgamations of small operators had produced United Air Lines, American Airways, and TWA (originally Transcontinental and Western). U.S. services were reorganized in 1934 when the large aviation holding companies were broken up, and domestic routes were soon being flown by DC-3s, BOEING 247s, and Lockheed ELECTRAS and Lodestars. Pan American had expanded through the Caribbean to South America by 1930 and across the Pacific to Hong Kong and Auckland by 1940. German sponsorship helped to pioneer air services in South America, and other European nations encouraged their airlines to establish routes to overseas colonies.

The 1930s were the heyday of the large flying boats, but the development of new military transports during World War II provided postwar airlines with a new generation of long-range land-planes, principally of American design, and including the STRATOCRUISER, DC-4 and CONSTELLATION. Regular passenger services were now inaugurated over intercontinental routes; Pan Am, which enjoyed a virtual monopoly of American international routes until the early 1940s, now had to compete with other U.S. operators, and other nations quickly found that a "national" airline had a distinct prestige value, even if it was often American-equipped.

By the mid-1950s international airlines were flying stretched DC-7Cs and Super Constellations carrying up to 100 passengers on 4,000-mi. (6,400 km) stages (including routes over the Arctic pioneered by Scandinavian Airlines System). Turboprops enjoyed considerable popularity on domestic routes with the advent of the four-engined Vickers VISCOUNT in BEA service (1953), but after a pause following BOAC's unsuccessful introduction of the COMET in 1952, pure jets eventually began to supplant propeller-driven aircraft beginning in 1958. Passenger capacities of BOEING 707 and Douglas DC-8 jets rose to over 200 during the 1960s; in 1970 the first BOEING 747 ushered in the era of 400-passenger loads; and the DC-10, TRISTAR, and European AIRBUS soon followed.

At present there are some 650 airlines

operating at all levels, of which over 200 fly international services. The world's largest airline is AEROFLOT, the Russian international and domestic carrier responsible for almost all Soviet civil aviation and carrying nearly 100 million passengers annually. After Aeroflot, the airline with the largest fleet and greatest number of passengers is United, one of the "Big Four" American carriers. International fares and standards of service are established by IATA (International Air Transport Association), subject to the necessary government approval.

Airmail

In the early days of commercial flying, the carriage of letters by air usually preceded the regular transport of passengers over any given route. The first experimental airmail flight to be attempted was at Blackpool, England, in 1910, but it is generally accepted that the first authentic airmail (with properly franked envelopes) was the flight made in India by the Frenchman Henri Piquet, who on February 18, 1911, flew a small packet of mail from Allahabad to Naini (about 5 mi., 8 km) in a Humber biplane as a feature of the United Provinces Exhibition.

The coronation of King George V in England in 1911 was the occasion of an airmail service of three weeks' duration between Windsor and London, operated by the Grahame-White Aviation Co. Gustav Hamel flew either a Farman or a Blériot, and between September 9 and 26 carried about 25,000 letters and 90,000 postcards. The same year an Italian airmail service was operated from Venice to Bologna and Rimini, and in the United States a Blériot flown by Earle Ovington carried mail from Nassau Boulevard to Mineola, Long Island, between September 23 and October 2.

During and immediately after World War I, military aircraft carried mail and dispatches in Italy (including 480 lb., 217 kg, of mail flown from Turin to Rome on May 22, 1917, which bore specially surcharged stamps), Germany (Berlin to Cologne via Hannover), France (Paris to Saint-Nazaire via Le Mans), and across the English Channel (RAF aircraft).

The first scheduled international airmail service opened between Vienna and Kiev (later extended to Odessa) on March 11, 1918, and lasted until November of the same year. Regular U.S. airmail operations began between New York and Washington on May 15, 1918, using Curtiss Jennys. Italy remained in the forefront of pioneer air-

The DH 4A, pioneer of international airline services in 1919

Passenger cabin of the DC-3, the airliner that initiated a new era in air travel in the mid-1930s

BOAC's de Havilland Comet, first of the world's jet airliners

mail flying and in 1918–19 ran services from Venice to Fiume via Trieste and Pola (Posta Aerea Transadriatica), between Scutari and Corfu via Durazzo, between Padua and Vienna, and from Rome to Naples, Pisa, Milan, Turin, and Sardinia.

An appropriation of $100,000 had been made to the U.S. Post Office in 1917 for financing an experimental air service. Following the inaugural service between New York and Washington, the acquisition of new aircraft (including specially-built Standards and DH 4Ms) enabled the service to be extended to Chicago (1919) and Sacramento (1920, joining an existing service to San Francisco). In 1927 the Post Office transferred operation of its routes from U.S. Army flyers to contract carriers, which included the precursors of several present-day airlines. Reorganization of American domestic airmail services took place in 1934, and new mail contracts were awarded by the Post Office.

During the 1920s airmail services became established in Canada, Australia, and South America, often as a preliminary step in the initiation of passenger routes. Pioneer transoceanic airplane services were all begun with

airmail flights: Lufthansa across the South Atlantic in 1934 (Dornier Wal), and Pan American across the Pacific in 1935 (Martin 130) and the North Atlantic in 1939 (Boeing 314).

Airports

The structure, form, and facilities of modern international airports are dictated to a large extent by the operating characteristics of the largest jets, which may require runways up to 12,000 ft. (3,600 m) in length and carry 400 or more passengers. Airports usually serve large cities, and transportation for passengers to and from the adjacent metropolis is often a major problem. Road and rail systems are sometimes combined in efforts to overcome this difficulty, but the use of helicopters has not provided a solution because of their small capacity (very large helicopters cannot operate into city centers). With expansion in air travel many big cities have built a second or third airport to meet their requirements. These second-generation airports were designed to be modern first-class facilities from the start, instead of gradually growing up from primitive airfields like many first-generation terminals.

The airport's main runways are aligned to provide takeoffs and landings into the prevailing wind, the first two digits of their magnetic heading being marked on the threshold (04 corresponds to 040 degrees, 34 to 340 degrees). Often parallel runways are used to increase traffic capacity (04 left/22 right and 04 right/22 left at New York's Kennedy, for instance). Additional runways are frequently provided (e.g. 31/13 at Kennedy), but crosswind landings in powerful modern jets pose no special difficulty, and Frankfurt, for example, has only a single pair of long runways (07/25). Runway lengths must be increased at high-altitude airports because of the longer takeoff runs needed in the thin air: Nairobi, at an altitude of 5,000 ft. (1,525 m), has a runway of 13,507 ft. (4,115 m); Mexico City at 7,347 ft. (2,240 m), has a 10,824-ft. (3,300-m) runway.

Runway construction is usually of concrete some 2 ft. (0.7 m) thick to withstand the 775,000-lb. (350,000-kg) weight of a Boeing 747. Not all runways are absolutely level, and the surface may have transverse grooves to make braking easier in rainy conditions.

For night use there is a line of white lights along the edge of a runway, with a row of green lights across the near threshold and red lights across the far end. Airports certified for instrument landings with a cloudbase down to 100 ft. (30 m) and quarter-mile (0.4-km) visibility must have runway center-line lights, and for operations with still lower levels of visibility there must be lights demarcating a narrow touchdown zone on the runway. Calvert system approach lights are in three sections, the outer section being of triple lighting units, the middle section of double units, and the final section of single lights; six illuminated crossbars of decreasing length provide an artificial horizon, the widest being the farthest out and the narrowest the nearest to the runway. Sequenced flashing lights are also used on the approach, and Visual Approach Slope Indicators (VASI) are based on the use of a color code to keep the pilot on the correct glide slope.

Taxiways are carefully laid out so that aircraft can clear runways as quickly as possible after landing and can reach the apron and terminal facilities without having to encroach on any other runways. Terminal buildings may be located between a pair of parallel runways (which can be either staggered or exactly aligned), in the angle between two convergent but non-intersecting runways, or (especially at very large airports) in the center of a complex of parallel pairs of runways that intersect. Arrival and departure facilities are usually on different levels to speed passenger processing, and all the buildings need to be well grouped to facilitate transfers of passengers from one airline to another.

Each airline has administrative accommodation, and there may also be facilities for government services (customs and immigration for international airports) as well as shops and restaurants. Piers (fingers) incorporating lounges and boarding gates now often extend out onto the apron, and access for passengers is generally by covered walkways leading right out to the aircraft doors. A circular pattern terminal has been adopted by some operators (including Pan American at Kennedy), with aircraft pulling into positions around the perimeter. At some airports magnetic baggage tags are used in conjunction with automatic sorting systems, but generally baggage passes to a central sorting post where it is reassigned to the appropriate loading bay.

The focal point of the terminal area is the control tower, but with modern radar installations many air traffic control centers have been moved away from the airport.

Servicing and emergency facilities must also be readily available. Fuel is dispensed from 12,000-gallon (54,540-liter) tankers or a built-in hydrant system, and hangars big enough to accommodate the largest jets must be built for undertaking major servicing and overhauls.

The planning of modern airports must take into consideration the need for future expansion and the nature of the surrounding terrain. Approach and takeoff routes at many existing airports leave much to be desired; on Hong Kong's landward approach, for example, the landing lights follow a curve instead of a straight line because of the surrounding mountains, with airliners coming in low over Kowloon just before touchdown. At most major airports, however, there are restrictions which regulate throttle settings and rates of climb to minimize aircraft noise nuisance. Night operations are often limited.

Airports opened in the 1970s included Charles de Gaulle near Paris and the Dallas–Forth Worth Regional Airport. The Texas terminus measures 9 mi. (14.4 km) north to south and 8 mi. (12.8 km) east to west. There are two pairs of main north–south runways; the two outer ones will be lengthened if necessary from an initial 13,400 ft. (4,080 m) to 20,000 ft (6,000 m). Crosswind runways are located at either end of the row of 13 terminal buildings, and separate additional runways will be provided for executive and STOL (Short Take-Off and Landing) aircraft. The control tower is 200 ft. (60 m) high. This complex is able to operate 266 aircraft movements an hour in visual meteorological conditions, and by the end of the century will probably handle 40 million passengers and $3\frac{1}{2}$ million tons of cargo a year, with 234 gates capable of accepting jumbo jets.

Airline terminals at New York's John F. Kennedy International Airport

Britain's ill-fated R-101, *shortly before leaving on the flight that ended in disaster*

ing "able to be steered." The first fully successful airships were built at about the turn of the century and came into widespread use in the 1920s and 1930s. They pioneered long-distance air travel when the airplane was in its infancy, and they were also the first effective bombers. However, by the late 1930s a number of airships had suffered spectacular accidents in various countries, and interest in them waned. By then, also, long-distance airplane travel had become more reliable. Today only a few airships are flying, including blimps built by Goodyear. One of their principal uses is to provide a stable platform for aerial television cameras. It has been suggested that airships would make excellent large-capacity cargo carriers or be valuable for specialist uses such as transporting industrial equipment or rockets. Several designs for massive airships (one of which resembles a huge flying saucer) have been created in recent years, but none has yet been built to full scale.

Semirigid airships had a metal keel along the length of the envelope to help keep the airship in shape. The flying shape of the craft was also maintained by the gas pressure. Nonrigid airships, or blimps, the only types still flying today, have no supporting framework whatsoever. The blimp has an envelope of rubberized fabric to retain the lifting gas. Its front end is stiffened by battens to withstand wind pressure, and in the front and rear are air bags, or ballonets, used to control lift. They are vented to lighten the craft and make it rise, and filled to make the craft heavier and cause it to descend.

Airships are powered by engines that drive propellers, usually rearward facing pushers. Directional control is provided by rudder and elevators, much as it is in an airplane. Beneath the bag is slung a car, or gondola, for cargo and passengers, and engines to provide propulsion.

When a bag is filled with a gas that is lighter than air, it experiences a net upthrust, or lift, that causes it to rise. This follows from the principle of buoyancy which applies to fluids in general—gases as well as liquids. The best gas for airships is the lightest of all, hydrogen, which has a weight of only 0.09 g/liter (air weighs 1.2 g/liter). The great disadvantage of hydrogen is that it is highly flammable, which led to disasters caused by the hydrogen gas exploding. The alternative to hydrogen is the next-lightest gas, helium, which has a weight of 0.18 g/liter. Helium provides less lift, and it is expensive to produce on a large scale, but it has the great advantage of being nonflammable.

As the range of airliners has increased, a number of airports that were once important as refueling stops have declined in significance. On the North Atlantic route, Goose Bay in Labrador, Gander in Newfoundland, Søndrestrømfjord (Bluie West 8) in Greenland, and Prestwick in Scotland have all lost business, while across the Pacific the airports on Guam, Wake Island, and Midway are no longer obligatory refueling stops. In the Mediterranean, long-distance services now fly over Malta and Cyprus, and the expensive runway built on the Cocos Islands for airliners on the South Africa–Australia route is largely derelict.

In addition to the big international airports serving major cities, there are numerous smaller fields, some (in the United States and Canada, for example) well equipped, others (like many in South America and the Australian outback) with only unpaved dirt runways. Heliports have been built in a number of cities such as New York, Brussels, and Melbourne, but with the limited scale of helicopter operations they are of only small size. Special STOL runways are being incorporated in some airport layouts (such as New York's La Guardia), so that STOL aircraft need not use the long runways required by conventional airliners.

Airship

A steerable lighter-than-air aircraft with its own engines. An alternative name for the airship is *dirigible*, mean-

There are, or rather were, three types of airships—rigid, semirigid, and nonrigid. The largest airships built—the famous Zeppelins—were of rigid construction. The main body or hull of the ship consisted of a rigid framework of light metal (usually aluminum alloys) made up of circular crosswise members linked by longitudinal struts and cables. Fabric was stretched over the framework, forming the envelope, or skin. The lifting gas was contained in a number of gas bags inside the frame. Because of their structure, rigid airships kept their shape regardless of whether they were filled with gas. Control of lift was provided by venting the gas (to go down) or releasing water ballast (to go up).

The first successful airship was built by the Frenchman Henri Giffard in 1852. A 144-ft. (44-m) long hydrogen filled dirigible powered by a 3-hp steam engine, it was cigar-shaped and traveled a distance of 17 mi. (27 km) at an average speed of 5 mph (8 km/h).

This was followed by the German engineer Paul Haenlein's airship of 1872, with an internal combustion engine that was fueled by hydrogen from the gas bag—and by those of Albert and Gaston Tissandier (1883) and Charles Renard and A. C. Krebs (1884, *La France*), both of which were powered by electric motors. The Tissandiers' ship was hopelessly underpowered, but *La France*—with lighter batteries, a more efficient motor, and greater lifting power—was the first airship to make a controlled flight in a closed circuit.

The last and greatest exponent of the early nonrigid airships was the Brazilian-born pioneer of flight, Alberto SANTOS-DUMONT. Apart from the enormous public interest he aroused, his contributions to airship development included the adoption of the gasoline engine, the use of steel wire on which to suspend the car (reducing drag considerably), and the invention of an improved method of tilting the airship. On earlier ships the aeronaut shifted sandbags up and down the car to vary the angle of flight for ascent or descent. Santos-Dumont simply pulled a cord, which swung a heavy guide rope fore or aft to alter the center of gravity. In his *Airship No. 6*, powered by a lightweight 16-hp gasoline engine, he won a prize for the first return flight from Saint-Cloud to the Eiffel Tower.

During the first decade of the 20th century a number of successful semirigid airships were built in France and Germany (and supplied to Russia and Britain). These, and less successful airships constructed by the American journalist, Walter Wellman, and his colleague, Melville Vaniman, were all some 200–225 ft. (61–69 m) long. But by 1910, Count Ferdinand von ZEPPELIN, the greatest airship designer of all, had perfected the rigid airship.

As early as the mid-1870s, Zeppelin had realized the possibilities of huge rigid airships, and in 1900 he launched his first—*LZ-1*. It was 420 ft. (128 m) long and had a maximum diameter of 38 ft. (11.6 m). Despite many early problems, Zeppelin persevered, and in the period 1910–14 five Zeppelins logged a total of over 3,000 flying hours, carrying 34,228 passengers without a single mishap. During World War I, Zeppelins were built in large numbers for the German army and navy. Their size and capabilities were vastly improved and they undertook both strategic bombing and naval reconnaissance missions.

Most of the airships built after World War I were based on wartime Zeppelin designs, including Britain's R-34 (1919, wrecked in a gale while moored in 1921), America's SHENANDOAH (built 1923, broke up in a storm 1925), *Los Angeles* (built 1924, retired 1932), *Akron* (built 1931, crashed 1933) and *Macon* (launched 1933, sank 1935). The *Macon* was the last of America's rigid airships. The highly successful British *R-100* made a return flight across the Atlantic in 1930, but its rival, the *R-101*, crashed in 1930, ending British interest in passenger-carrying airships. Germany's GRAF ZEPPELIN had nine years of continuous successful service before it was retired in 1937, but the much larger HINDENBURG burst into flames as it was about to moor at Lakehurst, N.J., on May 6, 1937. The 36 fatalities were the first in the entire history of commercial airship operation, but with the destruction of the *Hindenburg*, the era of passenger airship flights was over.

Air shows
The large international air shows of western Europe have become major commercial events where aerospace manufacturers display and demonstrate their products to potential buyers.

The Paris Air Show is held in alternate summers (1977, 1979, etc.) and is organized by an executive committee appointed by GIFAS (Groupement des Industries Français Aéronautiques et Spatiales). The West German show at Hannover is also staged biennially (summer 1976, 1978, etc.) under the auspices of the Deutsche Messe und Ausstellungs AG; the British international event at Farnborough, run by the Society of British Aerospace Companies, likewise takes place in 1976, 1978, etc., but in September.

Light aircraft shows are designed to attract prospective private purchasers, and large crowds attend the annual National Maintenance and Operations Meeting at Reading, Pa.

Many more modest air shows consist almost entirely of flying displays to attract visitors, who may find stands selling aviation books, models, and souvenirs. Shows of this nature feature spectacular aerobatics, parachute drops, and the presence of exciting high performance airplanes in order to draw spectators. They may in some cases commemorate events in aviation history (Battle of Britain Day in England), or be designed to raise money for charity (an air force fund, for example, or a preservation appeal for historic aircraft).

Air-traffic control
The volume of modern air traffic makes it essential to operate an orderly system of regulating flights. This function is exercised by air-traffic control, which directs the movement of airplanes within the congested areas of the sky near busy airports and along the airways linking important places (controlled air space). Air-traffic control also provides a general information service for pilots. Virtually all transmissions are in English, the generally recognized language of international air-traffic control.

Airways provide controlled routes for flights between specific places. They extend for 5 nautical miles (30,380 ft. or 9,260 m) either side of a straight line joining the points of origin and destination. Their upper and lower limits are also specified, so that other traffic may cross above or below. Identification is by means of a color combined with a number (e.g. Amber 2).

Where the airways converge on an airport they enter a terminal control area (with an area control center), which also has upper and lower limits, although it is irregularly shaped according to requirements. Immediately surrounding an airport is a control zone, extending upward from ground level to a specified height and supervised by a zone control unit. Airport traffic zones begin about 9,000 ft. (2,750 m) from the boundary of the airport and normally extend up to 2,000 ft. (610 m); they have an approach control for aircraft coming in to land or taking-off, together with an airport control unit.

Circuits are normally left-handed, since the first pilot sits in the aircraft's left-hand seat and thus has the better view of the airport. If large numbers of airplanes are waiting to land, they may

11

be subject to stacking above a demarcated holding point until they are cleared by the approach controller to come in over the radio beacon (which is usually located about 4 mi., 6.4 km, downwind of the runway). If visibility permits, VFR (Visual Flight Rules) operate. Outside controlled airspace a pilot flying at or below 3,000 ft. (915 m) may use VFR providing he remains clear of cloud and in visual contact with the ground; above this height he must be able to see at least 5 nautical miles (30,380 ft. or 9,260 m) and he must keep 1 nautical mile (6,076 ft. or 1,852 m) horizontally and 1,000 ft. (305 m) vertically away from cloud. Within controlled airspace (an area where pilots are under the jurisdiction of air-traffic control), the visibility criteria applicable above 3,000 ft. in uncontrolled airspace apply at any height.

If visibility deteriorates to less than these values by day, IFR (Instrument Flight Rules) apply, and at night they are always mandatory. Under this system, pilots in controlled airspace follow the instructions of the air-traffic controller. Outside controlled airspace aircraft are required to fly according to a set of rules stipulating that below 25,000 ft. (7,620 m) pilots on a magnetic track of less than 90° cruise at a level corresponding to an odd number of thousands of feet, between 90° and 180° at odd thousands plus 500 ft. (152 m), between 180° and 270° at an even number of thousands of feet, and between 270° and 360° at even thousands plus 500 ft. For greater heights, magnetic tracks up to 180° are allocated a different series of cruising levels from tracks between 180° and 360°.

ALTIMETERS are calibrated before flight by setting the atmospheric pressure (in millibars) in a special window contained within the dial. The pilot may request from air-traffic control either a QNH setting (which will give him altitude above sea level), or QFE (height above a reference datum—usually the airfield). QFE is normally used only for takeoffs, landings, and flying in the circuit. Once above 3,000 ft. (the transition altitude), the international standard atmosphere of 1013.2 mb can be set, which will enable the pilot to use flight levels (flight level 30 is 3,000 ft., flight level 200 is 20,000 ft., etc.). All pilots proposing to make a flight must notify the airport air-traffic controller of their intention ("booking out") and request permission to taxi. Filing a flight plan is recommended for even the smallest plane if the pilot will be flying more than 10 mi. (16 km) from the coast or over sparsely populated areas:

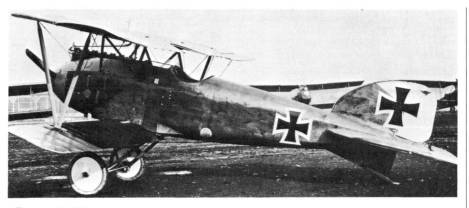
Albatros D.III

in the event of a forced landing, the rescue services will know where to look for the plane as soon as it is reported overdue at its destination. For flights in controlled airspace, flight plans are obligatory. Flight over certain areas is expressly prohibited or subject to special restrictions; their nature and location are notified to pilots in the information bulletins of the national aviation authorities.

Modern airliners can operate even in very poor weather conditions, using GROUND-CONTROLLED APPROACH techniques and INSTRUMENT LANDING SYSTEMS, together with RADAR aids.

Albatros

German aircraft manufacturer which played an important role in aviation during World War I. Established in 1910, Albatros was already building 150 aircraft annually three years later.

During the early months of 1917 the most formidable scout on the western front was undoubtedly the Albatros D.III "vee strutter," powered by a 175-hp Mercedes engine and flown by RICHTHOFEN and other German aces. The previous Albatros D.I and D.II scouts of 1916 had seen extensive service, although pilots were advised not to indulge in prolonged dives since the wings were prone to fail (the spar was positioned too far aft in the lower wing, causing vibration). By the summer of 1917 new Allied aircraft (the SE5 and Sopwith CAMEL) had proved their superiority over the D.III, and the D.V and D.Va were produced to redress the balance. Although improved aerodynamically and fitted with a 200-hp Mercedes engine, the D.V was inferior to the Fokker D.VII and remained subject to wing failure; it was, however, flown by front-line German squadrons until the end of the war.

Albatros also produced a number of single-engined two-seat reconnaissance planes, beginning with the B.II (which

was employed as a dual-control trainer until the end of the war), and including the C.I (widely used from 1915 onward), the C.III (1916, produced in considerable numbers), the C.V (with an unreliable Mercedes D.IV engine), the C.VII (350 of which were in service by February 1917), the C.X (replacing the C.VII at the front in the fall of 1917), and the C.XII (1918, the last of the series). Albatros also produced the J.I ground-attack plane of 1917, with twin downward-firing machine guns, and the W4 floatplane of 1916, based on the D.I and used for protecting naval air stations.

A new firm bearing the name Albatros was established in 1923. It built mostly trainers until its amalgamation with Focke-Wulf in 1931.

Alcock and Brown

The first nonstop flight across the Atlantic, from Newfoundland to Ireland, was made by two British aviators, Alcock and Brown, in 1919. The aircraft was a converted twin-engined Vickers VIMY bomber; Capt. John William Alcock was pilot and Lt. Arthur Whitten Brown navigator.

They took off from St. John's, Newfoundland, on June 14, 1919, flying into a stiff gale. Their radio failed almost at once. When night came they ran into fog, and the aircraft went into a spin; Alcock recovered the plane a few feet above the water, upside-down. Brown had to crawl out on the wings six times during the flight to chip off ice. Alcock put the plane down in a bog at Clifden, in Galway. Total flying time was 16 hours 12 minutes, with a coast-to-coast time of 15 hours 57 minutes; total distance covered was 1,890 mi. (3,032 km). Both men were knighted for their exploit, and they shared a £10,000 prize offered by the London *Daily Mail*.

Alcock (1892–1919) had served in the Royal Naval Service in World War I, and was Vickers' chief test pilot. He died in a flying accident in December 1919. Brown (1886–1948), born in Glasgow of American parents, was an engineer who served in the British army and air force in World War I.

Alitalia

The Italian national airline, Alitalia, was created in September 1946. It began operations with Fiat G.12s, SM.95s, and Lancastrians, all on domestic routes. Services were extended to Buenos Aires in 1948, when DC-4s became available. Early in the 1950s routes were being operated across the North and South Atlantic and to Africa, using DC-6Bs, and Convair 340s were appearing on European flights.

Linee Aeree Italiane (LAI) merged with Alitalia in 1957, and on June 1, 1960, the airline introduced DC-8 jets on North Atlantic services. By 1969 no piston-engined aircraft remained in service; in 1970 Boeing 747s began operating to New York. Alitalia's fleet includes 5 Boeing 747s, 21 DC-8s, 8 DC-10s, 35 DC-9s, and 13 Caravelles. Some 5,800,000 passengers are carried annually.

Allison

American aircraft engine manufacturer. During World War II, Allison, a subsidiary of General Motors, was principally responsible for the design and manufacture of high-powered liquid-cooled aircraft engines. During the late 1930s it had developed the V-12 V-1710 engine that was to power such warplanes as the P-39 AIRACOBRA, P-40 WARHAWK, P-38 LIGHTNING, and early P-51 MUSTANGS. After the war, Allison specialized in jet engines, producing both centrifugals (the Model 400 for the F-80 SHOOTING STAR and the J.33 for the Scorpion) and axials (the J.35 for the F-84 THUNDERJET).

In the 1970s, Allison was producing the 4,910-hp T.56 turboprop (used in the C-130 Hercules, the Orion, and the Hawkeye) and the 650-hp Model 250 (T.63) turboshaft/turboprop for helicopters and light aircraft. The TF-41-A-1 (14,500 lb., 6,568 kg, thrust) is a license-built version of the Rolls-Royce Spey, while another project with Rolls-Royce was the J.99 VTOL lift jet.

Alpha Jet

Jointly designed by DASSAULT-BRE-GUET and DORNIER, the Alpha Jet two-seat trainer and close-support aircraft made its maiden flight on July 26, 1973. By the end of the following year four prototypes had flown, and the aircraft

Historic moment in aviation: Alcock and Brown's Vimy at takeoff from St. John's, Newfoundland

was scheduled for service with the air forces of France, West Germany, and Belgium. Its wingspan is 29 ft. 10¾ in. (9.15 m) and length 40 ft. 3¾ in. (13.25 m). Powered by two Larzac 04 turbofans of 2,965 lb. (1,343 kg) thrust, the Alpha Jet has a maximum speed approaching Mach 1.

Altimeter

An airplane's flying height is registered by the altimeter. The instrument case is connected through the STATIC TUBE to the air outside, and contains a capsule (or series of capsules) from which the air has been evacuated. As the aircraft climbs, the air pressure in the instrument case falls, and the capsule expands and moves a needle on the dial. There may be up to three needles (hundreds, thousands, and tens of thousands of feet), and a digital display. A small panel set into the dial enables the pilot to calibrate the instrument by setting the ambient air pressure in milli-

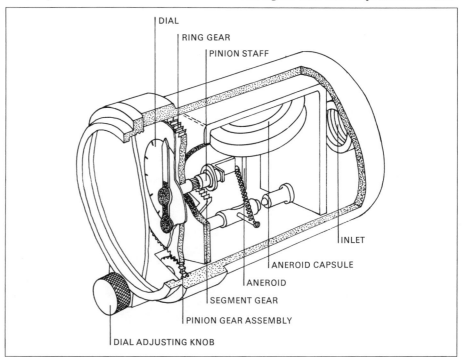

Changes in atmospheric pressure, causing expansion or contraction of an aneroid capsule, are converted into movement of one or more needles on the dial face of the altimeter

bars: QNH is the code for height above sea level, QFE for height above the airfield (see AIR-TRAFFIC CONTROL).

Radio altimeters measure the time radio pulses take to bounce back from the ground below. They are very accurate at low altitudes but have an upper limit of about 25,000 ft. (7,620 m). Their principal use is in automatic landing systems.

American Airlines

American Airways, as it was then known, was founded in 1930 as the operating subsidiary of the Aviation Corporation, formed the previous year and incorporating a large number of small pioneer American airlines. In 1934 American Airlines was created as an independent airline, with an extensive route network and a fleet that included the Orion, Condor Stinson A, Vultee V1, and Pilgrim 100. In 1934 the DC-2 was introduced, followed in 1935 by the DST (DC-3). By 1942 the fleet consisted entirely of DC-3s. The route network was extended to Toronto and Mexico City, and a scheduled transcontinental air cargo service was opened on October 15, 1944. DC-4s came into use on March 7, 1946, and DC-6s the following year; by the late 1950s the fleet consisted of 86 DC-6s, 58 DC-7s, and 58 twin-engined Convairliners.

The first U.S. transcontinental jet service was started with Boeing 707s on January 25, 1959, with Convair CV-990s coming into use during 1962. AA was the first airline to fly DC-10s (Los Angeles–Chicago, August 5, 1971), and its present equipment includes 9 Boeing 747s (one a freighter), 25 DC-10s, 90 Boeing 707s, and more than 100 Boeing 727s. One of the world's largest airlines, AA's network extends across the United States, with additional routes to Toronto, Mexico, the Caribbean, and the Pacific. Passengers total over 20,000,000 annually.

Amphibian

Aircraft capable of operating from both land and water, amphibians are FLOATPLANES or FLYING BOATS equipped with landing gear, which is usually retractable. In World War I, the American Grover Loening produced a number of amphibians with single central floats, small stabilizer floats near the wing tips, and landing gear that folded alongside or into the central float. This configuration was further developed by GRUMMAN, culminating in the J2F Duck of World War II.

Many examples of the boat-hull amphibian with wheels retracting into the hull were also made by Grumman (Goose, Widgeon, Mallard, Albatross), but the largest of all World War II amphibians was the PBY-5A version of the CATALINA flying boat. Present-day amphibians include the Japanese SHIN MEIWA SS-2A and Soviet BERIEV Be-12.

Antiaircraft defenses

The first nation to be subjected to sustained air attack was Britain, bombed by stray aircraft in 1914, by Zeppelin and Schütte-Lanz airships from 1915, and by bomber airplanes from 1916. Defense at first was negligible, but a day-and-night system was gradually set up with ground observers (using blind people with good judgement of distance and direction), messengers, searchlights, and antiaircraft guns—the defended area being divided into lettered and numbered squares.

This scheme was refined between the world wars, and great improvements were made in range finders, aiming mechanisms, muzzle velocities, and so on. But the basic problem of pinpointing hostile aircraft remained unsolved—the gun crew only knew the identity of the square the raiders were thought to be flying over. The vital factor that enabled the British to win the Battle of Britain was RADAR, which provided exact information no matter how bad the visibility.

By late 1940 the RAF had night fighters equipped with airborne radar, and similar aircraft were used by the United States and the Luftwaffe later in World War II. Typical antiaircraft guns of this war were 40-mm rapid firing automatics capable of firing 120 rounds a minute to a height of 2 mi. (3.2 km). Larger crew-loaded guns were also used, including the American 120-mm "stratosphere" gun—which fired 50 lb. (22.7 kg) projectiles to a height of 50,000 ft. (15,200 m). The proximity fuze (VT fuze), a radar carried in the cone of a shell which exploded it as it approached the airplane, was not used by the Americans in Europe until late 1944.

By 1945 the Germans were developing antiaircraft guided MISSILES, but the first of this new generation of weapons to enter service (in the mid-1950s) were the U.S. Army's Nike Ajax and the U.S. Navy's Terrier. An antiaircraft gun of the same period, the U.S. Army Skysweeper, was aimed and fired automatically by a radar-computer system.

Today virtually automatic radar and computer-controlled antiaircraft de-

Grumman SA-16A Albatross amphibian

fense systems firing surface-to-air or air-to-air guided or homing missiles have rendered obsolete the searchlights, the barrages of flak from fast firing guns, and the guesswork of the past. With modern electronics every single missile is likely to hit—unless the aircraft can "confuse" or interfere electronically with the missile's guiding or homing system.

Antonov, Oleg Konstantinovich (1906–)

Russian aircraft designer, head of a postwar design bureau that has specialized in transports and utility aircraft. Antonov himself began building gliders in 1923. The first, *Golub*, was followed by some 60 training and high-performance gliders over the period between 1923 and 1960. Antonov became chief designer at the Central Glider Design Bureau in Moscow in 1930. The bureau closed in 1938, and Antonov joined the YAKOVLEV team, his first task being to prepare the Fieseler Fi 156 Storch German military lightplane for production in the Soviet Union. In 1940 he designed the A-7 transport glider, and during World War II was involved in building Yakovlev fighters.

The world's last mass-produced biplane, the An-2 utility aircraft (Nato code name "Colt"), was the product of a new design bureau set up under Antonov's direction in 1946. An estimated 5,000 to 10,000 An-2s were built in the Soviet Union and Poland between 1948 and 1975. Other important designs produced by the Antonov bureau include the An-12 "Cub" four-turboprop transport, the twin-turboprop An-24 "Coke" light transport, and the An-22 "Cock." Powered by four 15,000-shp turboprops, the "Cock" was the largest aircraft in the world when it appeared in 1967, and is today surpassed in size only by the Lockheed C-5 GALAXY. With a wingspan of 211 ft. 3½ in. (64.40 m) and length of 188 ft. (57.30 m), the aircraft has a payload of some 175,000 lb. (80,000 kg) and maximum takeoff weight of 550,000 lb. (250,000 kg).

Arado

The German aircraft manufacturer Arado was the successor to Werft Warnemünde of 1917. In the 1930s it produced the Ar 68 biplane, one of the Luftwaffe's first fighters, and the Ar 66 biplane trainer. It also built floatplanes, the most important of which was the cannon-armed Ar 196, catapulted from large German warships and flown from shore bases. The staple Arado product was the Ar 96B, the Luftwaffe's standard

Antonov An-2 "Colt"

Antonov An-22 "Cock"

Arado Ar 234B-2

wartime trainer, of which 11,546 were delivered. A development of the Ar 96B, the Ar 396, became a widely used French postwar trainer, the SIPA S.10.

The Ar 240 twin-engined long-range fighter was built in small numbers, but saw service on the Russian front, and the Ar 232 assault transport, which had soft-field landing gear and rear doors, was also produced in limited quantities. Arado's outstanding technical achievement of the war was the Ar 234, generally recognized as the world's first jet bomber, and first flown on August 25, 1943. After much redesign the aircraft went into Luftwaffe service in

July 1944 as the Ar 234B Blitz. It was a high-wing single-seater, powered by two Jumo 004 turbojets of 1,980 lb. (900 kg) thrust with optional rocket-assisted takeoff, and served in several versions as a high-altitude reconnaissance bomber. Capable of speeds up to 460 mph (740 km/h), and with a ceiling of 33,000 ft. (10,000 m), it proved almost immune to interception unless caught at low altitude.

Area rule

A method of designing the contour of an aircraft fuselage to achieve the lowest

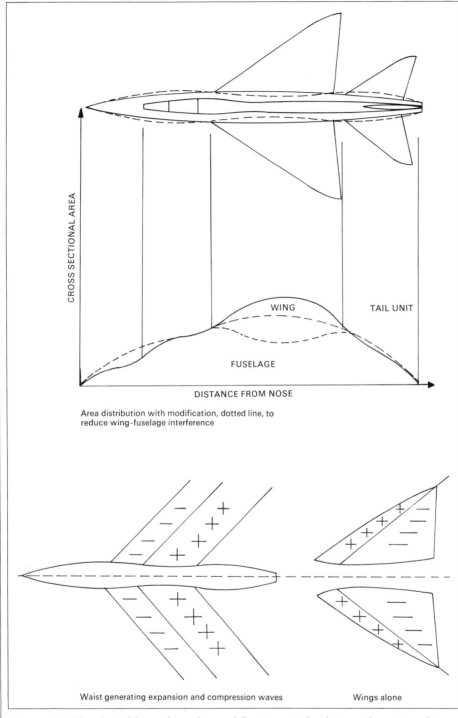

CROSS SECTIONAL AREA

WING

TAIL UNIT

FUSELAGE

DISTANCE FROM NOSE

Area distribution with modification, dotted line, to reduce wing-fuselage interference

Waist generating expansion and compression waves

Wings alone

Area rule: The dotted lines show the modifications in the design of an aircraft's fuselage needed to create a smooth cross-sectional contour

possible transonic wave drag. The increase in cross-sectional area where the wings and tail join the fuselage is compensated for by waisting the fuselage. The expansion and compression shock waves from the waisted section of the fuselage are canceled out by compressions and expansions generated by the wings. Area rule was first applied when difficulty was experienced in breaking the sound barrier in the 1950s.

(See SUPERSONIC FLIGHT.)

Armament

At the beginning of World War I aircraft were regarded by military experts purely as reconnaissance machines. Pilots and observers soon, however, began carrying pistols or rifles to fire at their opposite numbers, until eventually the British started using a drum-fed

.303 Lewis gun, mounted in the front (observer's) cockpit of pushers, or above the upper wing in tractor biplanes.

The French made experiments with such aircraft as the Morane-Saulnier, in which they armor-plated the blades of the propellers so that a machine gun could be fired between them. In Germany, Anthony FOKKER devised an INTERRUPTER GEAR that prevented bullets from striking the propeller blades at all, and for the rest of World War I the standard weapons for single-seat scouts were machine guns (Spandau on German aircraft, Vickers on British) along the top of the engine cowling, sometimes with subsidiary armament on the upper-wing center section. Two-seaters had a swiveling gun (on German machines a Parabellum) in the back cockpit, and bombers employed several defensive machine guns in various positions. This basic aircraft armament remained unchanged until the mid-1930s, when monoplane fighters appeared with multiple wing-mounted machine guns (eight in the SPITFIRE and HURRICANE) and cannon guns both in the propeller hub (BF 109, P-39 AIRACOBRA) and in the wings.

By 1941 most fighters were being fitted with cannon guns, since small-caliber machine guns alone were proving insufficient to bring down well-armored fighters or bombers. American fighters such as the P-47 THUNDERBOLT and P-51 MUSTANG used heavy-caliber .50 machine guns (eight in the case of the Thunderbolt). Twin-engined fighters employed a combination of nose-mounted cannon and machine guns (Lockheed P-38 LIGHTNING, ZERSTÖRER, Beaufighter).

Bombers continued to rely on machine guns for defense, often in power-operated turrets (LANCASTER), sometimes installed in remote-controlled barbettes (B-29 SUPERFORTRESS). The use of very heavy caliber weapons in ground-attack aircraft was not particularly successful (75-mm cannon in B-25 MITCHELLS). After World War II, jet bombers tended to rely on speed for safety, although the B-52 STRATOFORTRESS had provision for four machine guns or a 20-mm cannon in the tail, and Russian long-range aircraft such as the "Bear" and "Badger" 23-mm cannon in barbettes.

Nose-mounted cannon guns remained the standard jet fighter armament in the early postwar years (30-mm in the HUNTER and MiG 19), the retention of six .50 machine guns in the F-86 SABRE giving rise to adverse criticism. Guided missiles were expected to replace guns entirely at one time, and the BAC LIGHTNING, the U.S. F-104 PHANTOM, and the Russian "Fiddler" and "Flagon" relied principally on air-to-air missiles.

Combat experience in Vietnam and the Middle East led to a reintroduction of cannon guns, including rotating multibarrelled weapons (F-15 EAGLE, F-14 TOMCAT, F-111).

Armstrong Whitworth

The British firm of Sir W. G. Armstrong, Whitworth & Co. became involved in the field of aviation on the eve of World War I. It built aircraft engines (1912), propellers (1913), and airplanes under license (BE2s). During the first part of the war, Frederick Koolhoven, as Armstrong Whitworth's aircraft designer, originated the FK8 reconnaissance biplane (1916), which served with British forces in France and later helped to establish Qantas Airways in Australia. Koolhoven's wartime triplanes (FK5, FK6) and quadruplanes (FK9, FK10) were unsuccessful, but the company became involved in airship manufacture, producing nonrigid Submarine Scout (SS) ships for the Royal Navy, followed by three rigid ships: the *R-25* of 1917, the *R-29* of 1918, and the *R-33* of 1919 (which completed 735 hours' flying before being dismantled in 1928).

In 1919 Armstrong Whitworth took over Siddeley Deasy, which had designed the SR2 Siskin fighter. The Siskin served in the RAF (1924–32) and Royal Canadian Air Force; between 1927 and 1933 a total of 478 Atlas trainer military observation aircraft were built for the RAF and RCAF. Armstrong Whitworth's next major design was the trimotor Argosy airliner, which flew with Imperial Airways from 1926 to 1934. The four-engined Atalanta monoplane first appeared in 1932, and after serving with Imperial Airways (later BOAC) until 1941, it was pressed into service with the RAF. The big, four-engined AW 27 Ensign airliner (first flown in 1938) had a wingspan of 123 ft. (37.78 m) and seated up to 40 passengers, but it was found to be underpowered. After 1941, Cyclone engines were fitted to improve performance, and the aircraft continued in service until 1946.

The unsuccessful AW 29 single-seat day bomber and AW 23 twin-engined bomber transport of 1935 were followed by the AW 38 Whitley bomber, of which 1,814 were produced between 1936 and 1943. The Whitley later became a maritime patrol aircraft, transport, and glider tug. Postwar experiments with flying wings (AW 52G glider, AW 52 twin-jet) were unsuccessful, and the

Armstrong Whitworth Whitley V bomber

AW 55 Apollo four-turboprop airliner proved to be too small (maximum 31 passengers), unstable, and without reliable power units.

Factory resources in the early 1950s were largely devoted to manufacture of Gloster-designed Meteor night fighters (NF 11, NF 14), and the last true Armstrong Whitworth airplane was the AW 650 Argosy four-turboprop freighter (first flown 1959), which saw service with the RAF from 1962 to 1971, as well as with BEA (1961–1970) and on Logair services between U.S. Air Force bases in America.

Armstrong Whitworth had become part of Hawker Siddeley Aircraft in 1935; in 1963 Armstrong Whitworth was completely absorbed within HAWKER SIDDELEY Aviation.

Arrester systems

Owing to the short landing area of a flight deck, the use of some type of arrester system has been standard practice on aircraft carriers since Eugene Ely landed a Curtiss biplane on the converted deck of the U.S. cruiser *Pennsylvania* in 1911. All types of arresting systems incorporate energy absorption units. The simplest operate on the linear friction principle (an obsolete method involving an anchor chain dragging along the ground); others employ rotary friction (disk brakes on the same shaft as the drum that stores the connecting tape or cable); some use a rotary hydraulic mechanism (a vaned rotor or impeller in a fluid-filled housing), or work by a linear hydraulic method (a piston operating in a fluid-filled tube). The arresting gear itself generally consists of a cable $\frac{7}{8}$ to $1\frac{1}{4}$ in. (2.2–3.2 cm) in diameter, which is engaged by a special hook on the aircraft. Sometimes there may also be a cable in the overrun area of the runway, triggered by an aircraft's nose wheel and springing up to catch the main undercarriage legs.

Barriers that act as a backstop in-corporate a net (often of nylon) that will usually stop an aircraft traveling at up to 120 knots; above this speed the net normally bursts or breaks free of its anchorage.

Artificial horizon

The attitude director or artificial horizon indicates whether the aircraft is rolling, climbing, or diving. It is based on a gravity-orientated gyroscope maintained in position by torque motors governed by mercury-level switches. The flight director consists of two moving lines at right angles to each other superimposed on the face of the artificial horizon. They are controlled by a computer receiving information from the other flight instruments; the pilot maintains his course by aligning the aircraft symbol on the instrument's glass with the intersection of the moving lines.

Attack helicopters

In the early years of their development, the low speed and power of helicopters confined them to light transport tasks. With the general introduction of high-power turbine engines in the mid-1950s, however, helicopters could at last have the potential for battlefield assault and fire-suppression duties. Korea, and later Vietnam, showed that while troop-carrying helicopters could salvage a desperate tactical situation, they were vulnerable to antiaircraft fire because of their low operating altitude and speed.

The first helicopter gunships were no more than heavily armed versions of conventional rotorcraft. The first attack helicopter conceived as an actual weapons platform was the outcome of the U.S. Defense Department's 1965 order from BELL Helicopters for a stripped-down version of its UH-1 "Huey" platoon-transport helicopter to be known as the AH-1G HueyCobra. It was built in six months using the engine, rotors, and transmission of the UH-1,

Mil Mi-24 "Hind" attack helicopter

Bell AH-1G HueyCobra, carrying the new ALLD (Airborne Laser Locator Designator) system beneath its stub wing

but with a new, ultra-slim (only 4 ft., 1.2 m, wide) two-seat fuselage. The gunner controlled a chin turret mounting two rapid-firing six-barrel guns or two grenade launchers, and racks under the tiny stub wings carried rocket packs, guns, or missiles. Maximum speed was 219 mph (352 km/h).

The HueyCobra gunship was always regarded as an interim solution to the need for an advanced attack helicopter. The definitive gunship was to have been the very advanced Lockheed AH-56A Cheyenne, a 250-mph (402-km/h) heli-

copter begun at the same time as the HueyCobra. Technical difficulties, however, forced its cancellation in 1969. The U.S. Army eventually ordered instead the extremely angular AAH (Armed Attack Helicopter) design known as the YAH-64A from Hughes in 1977. The primary mission of the YAH-64A is tank-destroying, and it carries armor protection for the crew and for the engines and rotors. In addition to a nose-mounted 30-mm cannon it is armed with the Army's principal antitank missile, the Hellfire,

guided to its target by a laser beam.

While this new U.S. helicopter will not become operational until the early 1980s, the Soviet Union has already put into service the Mi-24 gunship, known in the West under its Nato designation "Hind," and produced by the MIL helicopter design bureau. It appears to be powered by two 1,500-hp engines (similar in size and rating to those of the YAH-64A), and to be virtually identical in size to the U.S. machine.

Autogiro

An autogiro, or gyroplane, is a wingless aircraft supported in the air by freely revolving rotors that are turned by the slipstream (the principle of autorotation), not powered by direct engine drive as in the case of the HELICOPTER. An autogiro has a conventional propeller, fuselage, and undercarriage, but lacks wings. The rotor (which can be

Avro Cierva C30A Rota, RAF autogiro of World War II

thought of as a set of rotating wings) is mounted above the fuselage, and in the most advanced models could be clutched into the engine before takeoff and brought up to a speed of 180 rpm. The wheel brakes would then be released, the rotors disengaged, and the engine revved up and linked to the propeller. The autogiro would taxi a few yards and then lift as passage through the air and propeller slipstream drove the rotors still faster. The autogiro's horizontal and vertical stabilizers were like those of an airplane, but some autogiros lacked a rudder or elevators since control could be effected by alterations to the axis of the rotor shaft.

The first successful autogiro flight was made at Gatafe, near Madrid, on January 9, 1923, by Juan de la CIERVA, whose designs were subsequently manufactured in Britain, France, Germany and Japan. The first military autogiros, Avro Cierva C30s, entered RAF service in 1934. The U.S. Army Air Corps ordered a few Kellett autogiros in 1937, and other types, which included the Japanese Navy's Kayaba Ka 1, were used in World War II.

The autogiro's speed ranged from about 30 to 125 mph (50–190 km/h). It could not hover motionlessly like the helicopter, which has for all practical purposes supplanted it. There has, however, been a recent renewal of interest in the autogiro, but only as a one- or two-seat aircraft for sport use, sometimes marketed in kit form. Unpowered, kite-like models have even been produced for towing behind a car or motorboat.

Automatic pilot

Some of the fatigue and monotony of long-distance flights can be relieved by the automatic pilot. The instrument consists basically of three gyroscopes orientated to detect variations in pitch, roll, and yaw. If the aircraft deviates from its flight path in any of these planes, the relevant gyroscope will move and electrically activate the control surfaces to rectify the error, either by means of electric motors or by operating selector valves on hydraulic jacks.

Autopilots can also automatically trim an aircraft so that the flight attitude is maintained. Trim indicators in the cockpit tell the pilot whether the system is functioning correctly. The rate at which an aircraft is deviating from its required flight path can also be measured by autopilots, so that any necessary correction is applied before the aircraft has moved far from its normal course or altitude.

Avenger

American torpedo-bomber of World War II. When the United States entered the war, the Navy had only 100 carrier-borne torpedo aircraft (Douglas Devastators), and these were obsolescent. Grumman produced a replacement, the TBF-1 Avenger, which was hastened into production before prototype tests, but proved to be a sound design. With a wingspan of 54 ft. 2 in. (16.51 m) and length of 40 ft. (13.14 m), the Avenger was the largest Allied aircraft used for regular operations from carriers during World War II. It was powered by a 1,900-hp Wright radial engine, with a maximum speed of 270 mph (430 km/h). The crew of three included a gunner controlling a power-operated turret (the first on an operational Navy aircraft). Unlike its predecessor, the Avenger carried its torpedo within a closed bay.

Avengers entered active service at the Battle of MIDWAY (June 1942), and they

Avenger torpedo-bombers of the U.S. Navy over the Golden Gate

became the principal Allied shipboard torpedo-bombers of the war. A total of 9,836 were built (more than three-fourths under license by General Motors), with about 1,000 going to Britain's Royal Navy.

Aviation medicine

The effects of flight on human physiology are the subjects of aviation medicine. The most immediate medical problem likely to face a flyer is the progressive rarefication of the atmosphere and the consequent lack of oxygen. Above 4,000 ft. (1,200 m) night vision becomes impaired; at 10,000 ft. (3,000 m) anoxia begins to appear if heavy work is undertaken, and from 14,000–18,000 ft. (4,200–5,400 m) efficiency is seriously affected. Between 18,000 and 34,000 ft. (5,400–9,300 m) oxygen equipment will maintain sea-level capability, but above 34,000 ft. (9,300 m) even 100 percent oxygen will not prevent some degree of anoxia. To combat this, even fighter cockpits are pressurized to an altitude corresponding to 25,000 ft. (7,500 m).

Air pressure variation can cause injury to the ear structure and air passages; flying with a cold in an atmospheric pressure less than that corresponding to about 8,000 ft. (2,400 m) can cause sinus barotrauma.

Accelerative and decelerative forces are measured in terms of G (1 G is equal to the force of gravity at the earth's surface). Substantial positive G (which drains blood away from the head) is experienced in high speed pullouts and tight turns (and also during lift-off in space missions). Individual tolerances to G forces vary considerably, but at $+4G$ most subjects suffer impaired vision (grayout) due to reduced retinal blood supply, followed by blackout at $+5G$ and unconsciousness at $+6G$. Blackout thresholds as low as $+2.5G$ or as high as $+9G$ do occur, however. Unconsciousness due to negative G (which forces blood into the head) occurs at -4 or $-5G$ if this value is maintained for about 5 seconds. Research has also been undertaken into the very high accelerative forces experienced over short periods during the use of ejection seats. These can be as high as $+25G$ for a fraction of a second; the pilot's tolerance is dictated by the strength of his spine.

Aviation medicine is also concerned with more fundamental physiological and psychological aspects of flying, such as the cause of motion sickness, the selection of trainees for aircrew training, the effect of noise and vibration on a pilot's capability, spacial disorientation, decompression sickness, and the layout of controls and instruments (*ergonomics*).

Aviation museums

Recognition that obsolete airplanes are of historical importance has led to the widespread establishment and expansion of aviation museums.

In the United States, one of the largest and most valuable aviation collections

The Wright Brothers' Flyer *and Lindbergh's* Spirit of St. Louis *displayed in the Milestones of Flight Gallery of the National Air and Space Museum*

in the world is housed in the National Air and Space Museum in Washington, D.C. Exhibits include the Bell x-1, Lindbergh's SPIRIT OF SAINT LOUIS, a DC-3, a Ford Trimotor, a Boeing 247, Wiley Post's *Winnie Mae*, and the 1903 Wright FLYER.

The USAF museum at Wright-Patterson AFB, Dayton, Ohio, includes a wealth of rare warplanes, among them a MiG-15, a Martin B-10, the Superfortress *Bockscar*, an Fw 190, and a Zero.

The U.S. Navy has its own museum at Pensacola, Fla., where many warbirds of the Pacific conflict are on display, together with prewar biplanes and the early Navy jets of the 1950s.

The Confederate Air Force at Harlingen, Texas, maintains a fleet of World War II combat planes in airworthy condition.

Aviation museums in Europe include the Science Museum in London, which contains ALCOCK AND BROWN's Vimy, the first British jet plane (the Gloster E.28/39), and Amy JOHNSON's de Havilland Moth. The Imperial War Museum is devoted to military aircraft, foreign as well as British, and has an establishment in London and further displays at Duxford in Cambridgeshire. The RAF has its own museum just outside London at Hendon, with additional exhibits (including World War II German rockets) at Cosford in Staffordshire, and regular flying displays of World War I aircraft are given by the Shuttleworth Collection at Old Warden, in Bedfordshire. There is a Royal Navy aviation museum at Yeovilton.

On the continent of Europe, the Belgian Air Force maintain a museum in Brussels, there is a Musée de l'Air in Paris (with BLÉRIOT's cross-Channel

monoplane, a Demoiselle, and a POLIKARPOV "Chicka"), a good exhibit at the Deutsches Museum in Munich, and a museum (Museo del Volo) in Turin which houses the MACCHI-Castoldi MC 72 and CAPRONI-Campini CC2.

Elsewhere in the world there are good collections at the South African National War Museum, Johannesburg, and at the Australian War Memorial in Canberra. The Fokker F.VII *Southern Cross* is at Brisbane, while in Canada there is a National Aeronautical Collection in Ottawa.

Aviation world records

Seven records are recognized by the FAI as absolute world records. The holders are as follows:

Distance in a closed circuit
11,337 mi. (18,245.05 km) by Captain W. M. Stevenson in a Boeing B-52H on June 6–7, 1962 (Bermuda–Sondestrom, Greenland–Anchorage, Alaska–Key West).

Altitude
118,898 ft. (36,240 m) by A. Fedotov in the E-266 (MiG-25) on July 25, 1973.

Altitude in sustained horizontal flight
86,000 ft. (26,213 m) by Captain R. C. Helt and Major L. A. Elliott in an SR-71A on July 27, 1976.

Altitude after launch from a "mother ship"
314,750 ft. (95,935.99 m) by Major R. White in the X-15A-3 on July 17, 1962.

Speed in a straight line
2,189 mph (3,523 km/h) by Captain E. W. Joersz and Major G. T. Morgan

in an SR-71A on July 27, 1976.

Speed in a closed circuit
2,086 mph (3,357 km/h) over a 1,000-km circuit by Major A. H. Bledsoe and Major J. T. Fuller in an SR-71A on July 27, 1976.

Distance in a straight line
12,532 mi. (20,168.78 km) by Major C. P. Evely in a Boeing B-52H on January 10–11, 1962 (Okinawa–Madrid).

Avionics

A term derived from aviation and electronics, to describe the use of electronic systems in aircraft and the associated field of applied research. Until the 1940s, the systems involved in operating aircraft were purely mechanical, electric, or magnetic, with radio apparatus the most sophisticated instrumentation.

The advent of radar and the great advance made in airborne detection during World War II led to the general adoption of electronic distance measuring and navigational aids. In military aircraft such devices improve weapon delivery accuracy and in commercial aircraft provide greater safety in operation.

A basic aircraft instrument, such as the ALTIMETER, provides a typical example of the application of avionics. Earlier altimeters were, in effect, aneroid barometers graduated to read off on a height scale by atmospheric pressure, giving a guide to height above sea level. The radio altimeter, by sending and measuring a radio pulse from the aircraft to the ground and back, gives a true reading of the aircraft's actual height over the terrain below.

Avionics play an important part in NAVIGATION, in which previously magnetism—in the form of the ordinary compass—was used to indicate direction. The radio compass can identify the direction of a ground station and home an aircraft accordingly. Certain avionic equipment is now mandatory on civil aircraft to comply with international safety regulations. Automatic landing equipment, used on some civil airliners (British Airway's Trident, for example), is a logical outgrowth of avionic capability. Automatic takeoff is also a possibility.

Nevertheless, it is in the sphere of military flying that avionics have reached their most sophisticated level. A military aircraft, such as the Grumman A-6 Intruder, fitted with an electronic device like Norden multi-mode AN/APQ-148 radar, can simultaneously perform ground-mapping, identification, tracking, and ranging of moving targets. Advanced bombers, such as the B-1, not only have the capacity to navigate automatically to their targets in extremely low-level flight, but are also equipped with devices to confuse enemy radar and missiles.

Many avionic devices are based on television transmission and reception principles with similar cathode-ray tube displays; alternatively there may be a digital display, or linking to bring other systems, including weapon delivery, into action.

Avro

Founded by A. V. ROE in 1909, the British aircraft manufacturer Avro first achieved fame with the versatile Avro 504 biplane, which remained in production from July 1913 until mid-1933; 8,970 were constructed in Britain and an additional number were built in the Soviet Union. The 504 was used as both a bomber and a fighter in the early stages of World War I, and soon became the standard trainer of many nations' air forces. Other important early Avro products included the large single-engined Aldershot bomber and Andover ambulance aircraft, the Tutor and Prefect biplane trainers, the 636 multi-role biplane, the Rota (CIERVA autogiro), the Avian lightplane and Cadet trainer, and a number of airliners such as the Ten and 642.

In 1934 two Avro 652 monoplane airliners were supplied to Imperial Airways. Powered by twin Armstrong Siddeley Cheetah radials, they were highly advanced, featuring a low-wing layout and retractable landing gear. From this design stemmed the Anson, which was first flown in March 1935 and ordered for the RAF as a coastal-reconnaissance aircraft. It was the RAF's first monoplane and its first aircraft with a retractable under-carriage. Eventually nearly 7,000 Anson I aircraft were built in Britain, most of which were employed as crew trainers or utility transports. Shipped to Canada and equipped with American engines they became Anson IIIs and IVs; the Anson II was of all-Canadian construction. By May 1952 22 Anson variants had been produced; total output was 11,020. Later models of the Anson had stressed-skin wings of revised shape, deeper fuselages, and more powerful Cheetah engines with constant-speed propellers.

Probably Avro's greatest contribution to World War II was the four-engined LANCASTER heavy bomber, from which were derived the Lancastrian and

York transports. The Lincoln, originally designated the Lancaster IV, was the RAF's standard heavy bomber of the immediate postwar years. It had an extended wingspan, high-altitude Merlin engines, better defensive armament, and greater range. From the Lincoln was developed the Shackleton maritime reconnaissance aircraft, first flown in March 1949. Powered by 2,450-hp Rolls-Royce Griffon engines driving contra-rotating propellers, the Shackleton initially had radar under its nose and twin 20-mm guns in a dorsal turret. The MR2 variant moved the guns to the nose of the lengthened fuselage, with retractable radar placed at the rear. Further improvements on the MR3 included nosewheel landing gear, wing-tip tanks, and viper turbojets for boosting takeoff. Mark I Shackletons were later converted to serve as T4 trainers, and Mark 2s became AEW2 airborne early-warning aircraft.

The last Avro combat aircraft to see production was the VULCAN delta-wing jet bomber; the Avro 730 supersonic bomber program was canceled in 1956. In 1960 the Avro 748 (now Hawker Siddeley 748) marked the start of an extremely successful series of twin-turboprop aircraft, which has gone on to include Andover military transports and Coastguarder reconnaissance aircraft. At the beginning of the 1960s Avro lost its separate identity when it was merged into Hawker Siddeley.

Avro activities in Canada began during World War II, when Victory Aircraft of Toronto was involved in

production of Lancasters; in 1946 the firm became Avro Canada Ltd., and later designed the first all-Canadian combat aircraft, the CF-100 all-weather fighter, which was first flown in 1950. A total of 692 CF-100s were built in five chief versions, most of them being semiautomatic interceptors with Hughes radar fire-control and missiles. The extremely advanced CF-105 Arrow, a 1,500-mph delta interceptor, was canceled in 1959 after five had flown very successfully.

B

B-1

A variable-geometry (SWING-WING) U.S. Air Force bomber for supersonic low-level nuclear delivery. The B-1 has been under development by Rockwell for several years, but its great cost, together with the ongoing debate about the value of piloted strategic weapons, has made its future uncertain. At present it is likely that from the early 1980s a significant portion of America's nuclear deterrent capability may be vested in a fleet of 214 B-1 bombers. They will complement the land-based and submarine-launched arsenal of long-range nuclear missiles to form a system known as Triad.

Until the mid-1950s the United States, like Britain and Russia, had relied for its long-range, airborne strike capability on high-altitude subsonic bombers such

Avro Shackleton AEW2

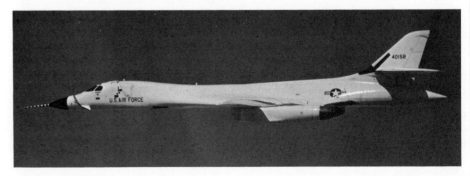

U.S. Air Force B-1

as the Boeing B-47 STRATOJET and, more recently, the B-52 STRATOFORTRESS. But SAMs (surface-to-air missiles) are now able to bring down such aircraft. To stand any chance of reaching their targets, the bombers have had to be assigned to low-level roles, so that they can fly beneath the detecting radar beams. The B-1 has been designed from the outset for very high-speed penetration of enemy territory at treetop level while carrying as great a warload as the larger B-52. Its variable-geometry wings, with a span of 137 ft. (41.76 m), are designed to be swept fully back to give the least drag when contour-following at the speed of sound, but will be rotated forward to provide the most efficient shape during the high-altitude cruise back to base. The aircraft's length is 143 ft. (43.59 m).

The B-1 carries no defensive armament, relying on its speed and highly sophisticated electronic countermeasures to evade both fighters and missiles; consequently it is flown by a crew of only four. Low-level flight to the target, navigation, and the bombing run itself are all carried out automatically, with the crew only monitoring the equipment and the progress of the flight.

B-17. See FLYING FORTRESS

B-24. See LIBERATOR

B-25. See MITCHELL

B-26. See INVADER; MARAUDER

B-29. See SUPERFORTRESS

B-47. See STRATOJET

B-52. See STRATOFORTRESS

B-57. See CANBERRA

BAC 1-11

A highly successful short/medium range airliner with twin rear-mounted engines, the British Aircraft Corporation's BAC 1-11 made its first flight in 1963. The series 200, with Spey engines rated at 10,330 lb. (4,686 kg) thrust, seated 89 passengers and entered airline service with British United and Braniff in April 1965. Its wingspan was 88 ft.

BAC 1-11: a series 200 airliner of Braniff International Airways

6 in. (26.97 m) and length 93 ft. 6 in. (28.50 m). Subsequent models of the aircraft were the series 300 (11,400-lb., 5,165-kg, Speys), with a longer range and higher payload; the series 400 (1965), essentially a "300" modified for American operators; and the series 500 (1967) with a 107-ft. (32.6-m) stretched fuselage accommodating 119 passengers, wing-tip extensions, and engines of 12,550 lb. (5,670 kg) thrust. The series 475 (1971) combines a "400" fuselage with "500" wings and modified landing gear for low-strength runways with poor surfaces. Maximum speed in all series is 548 mph (882 km/h).

Backfire

The latest and most advanced of the heavy strategic Soviet bombers built by TUPOLEV has been given the code name "Backfire" by Nato. In the late 1960s evidence began to accumulate that Russia was building a new bomber to replace its subsonic bombers, which were by then becoming obsolescent. About a year later American spy satellites were tracking the plane on its test flights, sending back information on its shape, performance, and capabilities.

It was soon confirmed that the new plane, which had variable-geometry wings (see SWING-WING DESIGN) and a top speed of about Mach 2.5, was smaller but quite similar to the new U.S. swing-wing bomber then being planned, known initially as AMSA (the Advanced Manned Strategic Aircraft) and later as the B-1. This similarity was to be expected because both countries had similar requirements—very long range (some or most of it flown subsonically) with a supersonic dash at low altitude to the target.

About a dozen prototypes were built

and flown, but difficulties must have been encountered because an improved version, known as "Backfire B," was introduced. This appears to be the definitive version, and some 80 aircraft had been established in squadron service by the beginning of 1977; by comparison, its U.S. counterpart will not be operational until the early 1980s.

"Backfire" weighs about 275,000 lb. (125,000 kg) and is equipped to carry the standard armory of Soviet weapons, notably the "Kitchen" air-to-surface missile. From bases inside the Arctic Circle, its 5,500-mi. (8,850-km) range would enable it to reach any target in the United States. For this reason "Backfire" is an important bargaining point in the continuing American-Soviet attempts at mutual reduction of armaments.

Bader, Douglas Robert Stewart (1910–   )

British fighter ace of World War II. In 1931 he crashed a Bulldog fighter while performing low-level aerobatics

RAF ace Douglas Bader

and injured his legs so seriously that they both had to be amputated. He rejoined the RAF at the end of 1939 and returned to flying fighters, despite the fact that he had artificial legs. By the summer of 1941 he had destroyed 20 enemy aircraft, but on August 9 he was shot down over Nazi-occupied France and taken prisoner, claiming two further victories in his last engagement.

Badger. See TU 16

Balbo, Italo (1896–1940)

A prominent Italian Fascist leader, Italo Balbo achieved world fame by forming and commanding large formations of SAVOIA-MARCHETTI S.55X twin-hull flying boats in the early 1930s. (As a result, any large formation of aircraft is still often called a "Balbo.") He took a leading interest in building up the REGIA AERONAUTICA (Italian air force) and became air minister in 1929.

Balbo personally led 12 flying boats on a flight from Italy to Rio de Janeiro (1930–31), and in 1933 commanded a formation flight of 25 S.55Xs, which flew on a 12,430-mi. (20,000-km) journey across the North Atlantic to New York and Chicago and home via the Azores and Lisbon.

In 1933 Balbo was appointed governor-general of Libya. He died when the aircraft in which he was flying was shot down by Italian gunfire over Tobruk.

Ball, Albert (1896–1917)

Britain's first great fighter ace of World War I, credited with 44 victories, Ball achieved enormous fame with the public for his blind courage in taking on enemy formations single-handedly. Ball joined the army in September 1914, was commissioned in October, and learned to fly the following year. He transferred to the Royal Flying Corps in January 1916, and was sent to serve in France initially with 13 Squadron (flying BE2Cs), soon transferring to 11 Squadron, with which he quickly established his reputation as a scout pilot, scoring his first victories flying his favorite NIEUPORTS. After a brief spell with 8 Squadron he joined 60 Squadron, still flying Nieuports, and his toll of German planes mounted rapidly. His favorite tactic was to fly beneath enemy airplanes and then fire his Lewis gun directly upward.

From October 1916 to February 1917 Ball was in England as an instructor, but returned to France with 56 Squadron on April 7, 1917, flying the new

Captain Albert Ball in the cockpit of his SE5

SE5. After scoring the unit's first two victories, Ball added several more successes to his tally, but was killed in a crash behind German lines on May 7, 1917, the cause of which has never been satisfactorily established.

Balloon

Lighter-than-air aircraft consisting of a large bag or "envelope" filled with a gas which is lighter than the surrounding atmosphere. Hydrogen, the lightest of all gases, provides the greatest lift. But it is highly flammable and also very penetrating, so the bag must be relatively thick and heavy. Helium, the next lightest gas, provides about 90% of the lift of hydrogen but is nonflammable. It is, however, very expensive. The third, and for amateur balloonists the most practical, alternative is hot air, though it provides only about 25% the lift of hydrogen.

The gas balloon consists of a spherical skin or envelope made of fabric impregnated with rubber, over which is slung a net. The net is attached to a load ring from which a light wicker basket for the crew is suspended. The net distributes the weight of the payload—the basket and passengers—evenly over the envelope.

At the bottom of the envelope is a narrow open-ended tube called the appendix, which acts as a safety valve. As the balloon rises, the surrounding air pressure falls and the gas in the balloon expands. If the envelope were sealed, pressure would build up inside and would eventually cause it to rupture, but the appendix allows gas to escape. A balloon will continue to rise until the decreasing density of the surrounding atmosphere equals the overall density of the balloon. At this point the balloon is in equilibrium. To rise higher, the balloon pilot must throw the ballast, usually sand, overboard. When all the ballast has been used up, the balloon can go no higher. To descend, the pilot releases gas from the top of the envelope through a valve. When the pilot has reached the ground, he rips open a ripping panel near the top of the envelope. This lets the gas inside escape and allows the envelope to deflate quickly before it is carried away by the wind.

The original Montgolfier hot-air balloon consisted of an envelope of fabric and paper with a wide opening at the bottom. Below the opening was a furnace burning straw. Since all the materials were highly combustible, it was not surprising that many of the early hot-air balloons caught fire. The modern hot-air balloon has an open lower end into which hot air is directed from a burner (fueled by bottled propane gas) slung beneath it.

The envelope is usually made of light nylon, treated to make it more airtight. Unlike gas balloons, hot-air balloons require no ballast and have no need of a release valve to make them go up or down. To make the hot-air balloon ascend, the pilot switches on the burner to heat up the air inside. To make the balloon descend, the pilot allows the air inside to cool. Like the gas balloon, the hot-air balloon is fitted with a

Hot-air balloon

ripping panel for quick deflation on landing.

The hot-air balloon can lift less than the gas balloon and stays in the air for a much shorter period, but it can be more readily inflated, it is simpler and cheaper to run, and inherently safer.

The altitude record for a hot-air balloon is held by two British pilots who in 1974 ascended to 45,838 ft. (8.68 mi.; 13.97 km) over India.

Although there are records of an indoor hot-air balloon flight in Portugal as early as 1709, the history of balloons is normally considered to date from 1783. On June 5 of that year the MONTGOLFIER BROTHERS launched the world's first large scale balloon in France. Some 30 ft. (9 m) in diameter, the "Montgolfière" was held over a fire until it filled with hot air, then it rose high into the air.

On August 27, 1783, another Frenchman, Jacques A. C. Charles, launched the first hydrogen balloon—or "Charlière," as it was called. This made a far more spectacular flight, traveling some 15 mi. (24 km). Three weeks later a Montgolfière carried a sheep, rooster, and duck aloft—to see whether they would survive—and the first manned flight followed on November 21. It lasted 23 minutes, reached an altitude of some 3,000 ft. (914 m), and covered almost 10 mi. (16 km).

On December 1 a Charlière carried its inventor and one other passenger a distance of 27 mi. (43 km) in a flight lasting two hours—convincingly demonstrating the supremacy of hydrogen (which was not supplanted by helium until the late 1930s, after the HINDENBURG disaster).

During the next century and a half balloons were used for military observation, for upper atmosphere research (starting in 1804 with a series of ascents sponsored by the French Academy of Sciences), and—during World War II—to provide protective "barrages" against low flying bomber aircraft.

Today, balloons continue to provide a valuable aid to scientific research and meteorology—lifting a variety of instruments high into the upper atmosphere to monitor conditions. Many carry radiosondes which collect data and radio it back to ground stations automatically. In addition ballooning is becoming an increasingly popular sport. International meetings are held regularly, and a significant feature of recent years has been the resurgence of the hot-air balloon.

Manned atmospheric research flights during the 1930s reached altitudes of over 13 mi. (21 km). The official record has stood since 1961 at 113,740 ft. (21.54 mi.; 34.68 km), although an unofficial 123,800 ft. (23.45 mi.; 37.7 km) was attained in 1966 by Nicholas Piantanida of Bricktown, N.J. In 1972 an unmanned balloon launched at Chico, California, reached 170,000 ft. (32.19 mi.; 51.8 km).

Baracca, Francesco (1888–1918)

Italy's leading fighter ace of World War I, credited with 34 victories, Baracca was a professional soldier who had trained as a pilot in 1912. Baracca scored his first combat victory on April 7, 1916, while flying a Nieuport, and by November had added four more to his tally and painted his personal emblem, a black prancing horse, on the first of his aircraft. In June 1917 he was appointed commander of 91ª *Squadriglia*, flying Spad scouts, and brought his score to 30 by Christmas 1917. He was then withdrawn from operational flying to test-fly for the Italian manufacturer Ansaldo, but in the spring of 1918 returned to the front and reopened his scoring on May 3. Baracca's two final victories were achieved on June 15. On the evening of June 19, 1918, he flew a ground-attack mission near Montello, and failed to return. After the Austrian retreat, Baracca's burned-out Spad was located, with his body nearby; he had apparently been killed by a bullet in the head fired from the trenches.

Barnstorming

When World War I came to an end, many young military pilots returning to civilian life sought to earn a living at the only trade for which they had any training—flying. In many parts of the United States an airplane was still a novelty, and it was not difficult to make a reasonable wage simply by flying an old Curtiss JENNY or Standard from town to town selling joyrides to the local inhabitants at a dollar a time. As the number of itinerant pilots increased, however, competition began to develop, while at the same time an airplane became an increasingly commonplace phenomenon.

As a result, pilots started staging spectacular air shows to attract prospective joyriders, employing risky low flying, daring aerobatics, and parachute jumps. Some operators formed groups known as flying circuses and evolved such hazardous activities as standing or walking on the wings, or climbing from one plane to another in midair.

Professional stunt pilots had existed before World War I. They included Lincoln Beachey (killed at San Francisco on March 14, 1915), Eugene Ely (on January 18, 1911, the first man to land on a ship, killed 10 months later), and Arch Hoxsey (killed at Los Angeles on December 31, 1910). In the 1920s, however, barnstormers became a regular feature of the American scene, their number including such personalities as Charles LINDBERGH, Roscoe Turner, and Frank Hawks.

Ivan R. Gates ran a 13-plane flying circus that was said to have introduced over a million people to flying before it went out of business. In 1928, one of Gates' pilots (Bill Brooks) made 490 flights in one day at Steubenville, Ohio, carrying two passengers on each trip. The Doug Davis Flying Circus joined forces with the rival Mabel Cody Flying Circus to produce one of the biggest display teams on the air show circuit (Davis was killed racing in 1934).

Later barnstormers included Skeet Sliter (who took up flying in 1927), Milo Burcham (later killed testing a P-80 Shooting Star jet), and crash specialist "Bowser" Frakes.

Many stunt flyers found lucrative employment in Hollywood, staging crashes, landing on moving trains, taking off from skyscraper roofs, and making hazardous parachute jumps. Their number included Frank Clarke (killed in 1948), Bobby Rose, Paul Mantz (killed in 1965), Hank Coffin, and Dick Grace. In the mid-1920s there was even a team of movie stunt flyers calling themselves the 13 Black Cats.

Batten, Jean (1909–   )

New Zealand-born woman pilot famed for setting long-distance records in the 1930s. In 1934 she flew from England to Australia in 14 days 22 hours 30 minutes, returning the following year (in 17 days 15 hours). Later in 1935 she set a new record for an England–Brazil solo (2 days 13 hours 15 minutes) in a PERCIVAL Gull, becoming the first woman to fly solo across the South Atlantic.

During 1936 she set new solo records: England–New Zealand (11 days 1 hour 25 minutes) and England–Australia (5 days 21 hours 19 minutes), both on the same flight. In 1937 she cut the Australia–England time to 5 days 18 hours 15 minutes.

Beagle. See IL-28

Bear. See TU-20

Bearcat

Produced by Grumman as a replacement for its F6F HELLCAT, the F8F Bearcat shipboard fighter was designed to achieve a performance that would surpass any anticipated improvements in Japanese fighters. The airframe was smaller and more than 2,000 lb. (900 kg) lighter than the Hellcat's, while fixed armament was reduced to four .50 machine guns; a Pratt & Whitney R-2800-34W radial engine gave it a top speed of 421 mph (675 km/h). The Bearcat had a wingspan of 35 ft 10 in. (10.92 m) and length of 28 ft. 3 in. (8.61 m).

The first deliveries from Grumman were made in February 1945—too late for the fighter to see action with the U.S. Navy in World War II. Additional production by General Motors was canceled, but a total of 1,266 Bearcats were eventually built (some postwar). The Bearcat was used by the French as a ground-attack aircraft in Indochina in the early 1950s.

Beechcraft

The marketing name of Beech Aircraft Corporation, one of the three principal American builders of light aircraft. Compared with its rivals Cessna and Piper, Beech builds in smaller numbers, and with greater concentration on business users.

The company was founded by Walter H. Beech (1891–1950) and his wife Olive Ann Beech in 1932. By 1941 it had already become well known for its back-staggered ("Staggerwing") cabin

F8F-2N Bearcat

Beechcraft Bonanza

biplanes, mostly Type 17s, and the "Twin Beech" twin-engined transport (Type 18). During World War II military versions of several Beechcraft planes were produced in large numbers. From the Type 18 were derived the C-45 Expediter light transport, and the T-7 and T-11 Kansan navigational trainers.

In 1945 Beech flew the first Bonanza cabin monoplane, with its distinctive V-shaped "butterfly" tail which is still built at the Wichita, Kansas, plant, with deliveries now having exceeded 10,000. From the Twin Bonanza of 1949 stemmed a popular series of twins (including the Travel Air, Baron, Type 99 Airliner, Queen Air, King Air 100 and T-tail 200, and Duke), several being powered by turboprops and with pressurized cabins. The T-34 Mentor military trainer of 1948, a descendant of the Bonanza, is still in production

(with a turboprop engine), and the T-42 instrument trainer, a twin-engined aircraft, is derived from the Baron. Beech Aircraft also carries out extensive subcontracting work, producing major parts of the F-4 Phantom and the airframe of the Jet Ranger helicopter.

Bell

Founded in 1935 by Lawrence Bell (1894–1956), a former vice-president of Consolidated Aircraft, what is today Bell Aerospace is one of the world's largest producers of helicopters. Bell's first aircraft designs included the twin-engined YFM-1 Airacuda pusher-propeller fighter (only nine were built, the project being abandoned due to unserviceability and poor performance), the P-39 AIRACOBRA, the P-63 Kingcobra, the XP-77 all-wood lightweight fighter (no production), and the P-59

AIRACOMET, the first U.S. jet aircraft.

After the war Bell produced a series of high-speed rocket-powered research aircraft. The first of these was the Bell X-1, which in 1947 became the first aircraft to break the sound barrier. It was followed by the X-2 (1953) and swing-wing X-5 (1951).

In the years since World War II Bell has largely specialized in the production of helicopters. The BELL 47 was the world's first helicopter to receive a commercial license and is still produced by foreign manufacturers over 30 years after its first appearance. The Model 204/205 Iroquois, a general-purpose helicopter introduced at the end of the 1950s, saw considerable success in U.S. Army service as the UH-1, and the six-seat Model 206 Jet Ranger, first flown in 1966, found a considerable civil market as well as being used by the Army as the OH-58 Kiowa observation helicopter. Later Bell helicopters have included the Model 209 AH-1G Huey Cobra ATTACK HELICOPTER, Model 212 (twin-rotor Iroquois), Model 214, and Model 309 (King Cobra). Bell also built the experimental XV-3 and X-22A CONVERTIPLANES.

Bell 47

The little Bell Model 47, with a length of only 31 ft. 7 in. (9.63 m) and empty weight of less than a ton, is the world's most widely used helicopter. The prototype made its first flight four months after the end of World War II, and in March 1946 the Bell 47G became the first helicopter to be licensed for civil operations by the FAA. Continuously improved, the helicopter remained in production in the United States for the next 28 years, a record for any type of aircraft. Over 5,000 have been built, and though the assembly line at Fort Worth came to a halt in 1974, the aircraft still emerge from factories in Italy (Agusta) and Japan (Kawasaki), where licenses to build the design were taken out many years ago.

The Bell 47G was a three-seat helicopter initially powered by a single 178-hp Franklin piston engine. Over the years more powerful Lycoming engines were fitted, and the seating capacity increased to four, but the basic shape, with its distinctive "goldfish bowl" cabin, remained unaltered, and the patented two-blade rotor was retained.

The success of the Model 47 is due to its versatility as a general-purpose helicopter. It can be found on traffic-control, transmission-line repair, crop-spraying, military and civil observation, rescue, light transport, sightseeing, and numerous other duties. The British Army operates some 250 (built by the English firm Westland under sublicense from Agusta) for observation and target spotting under the name of Sioux, and Agusta itself builds the 47B, with a five-seat cabin and covered tail boom.

Bellanca

In December 1927 Bellanca Aircraft Corporation, founded by G. M. Bellanca, began building a series of small, high-performance aircraft for civilian customers. Even the Bellanca lightplanes were exceptionally fast, and from a racer that entered the 1934 MacRobertson England-Australia race was derived the Model 28 series, initially planned for setting transatlantic records. Several of these aircraft were eventually sold to Spain and China as high-speed bombers, able to carry 1,600 lb. (725 kg) of bombs and an armament of five machine guns and still capable of flying at 280 mph (450 km/h).

The Airbus (and related models) were bigger twin-engined machines with a high wing which led to the Model 77-140 bomber, which could carry 2,300 lb. (1,040 kg) of bombs.

During World War II Bellanca built the Fairchild AT-21 trainer and undertook major subcontracting work. After the war, it continued manufacturing light aircraft, notably the Model 14-19 Cruisemaster and the retractable-gear Model 14-13 Cruisair, but they were not outstanding successes, and the company ceased operation in its original form at the end of the 1950s. A new company, Bellanca Aircraft Engineering, developed several new light aircraft including the Model 19-25 Skyrocket built largely of fiberglass and first flown in 1975.

Bennett, Floyd (1890–1928)

Pioneer American Arctic flyer. As a young Navy pilot in 1925, Bennett was a member of the Arctic expedition mounted by Richard Byrd with Loening amphibians. The following year, Bennett was Byrd's copilot on the first flight ever made over the North Pole, but in 1927 he suffered a broken right leg and a punctured lung when the Fokker trimotor *America* crashed on a test flight from Teterboro, N.J., with Anthony Fokker at the controls. In the spring of 1928 Bennett caught influenza while testing a Ford trimotor for Byrd's South Polar expedition, but left his sick bed to fly the Ford to Greenly Island to pick up the crew of the Junkers *Bremen*, which had landed there after making the first east-west crossing of the North Atlantic by airplane. Bennett was taken seriously ill in Quebec, and died of pneumonia. Byrd's Ford was named *Floyd Bennett* for the first South Polar flight, and New York's new municipal airport became Floyd Bennett Field (it was eventually employed as a military training airfield).

Beriev, Georgi Mikhailovich (1903-)

Soviet aircraft designer, head of a design bureau which has specialized in seaplanes. Beriev joined Paul Aimé Richard's seaplane design bureau in 1928, and two years later became chief designer of the seaplane department at the Central Design Bureau. The first design under his direction was the MBR-2 single-engined biplane of the 1930s. This was the first successful Soviet flying boat, and more than 1,300 were built, giving good service in World War II. The later two-seat KOR-1 single-float reconnaissance biplane was catapult-launched from Soviet warships.

Beriev Be-12 "Mail"

Postwar aircraft included the Be-6 of 1947 (Nato code name "Madge"), a long-range maritime patrol boat used extensively until its replacement by the Be-12 turboprop amphibian, introduced in 1961. During the postwar years Beriev's design bureau also produced at least two jet flying boats, the R-1 of 1952, and the Be-10 "Mallow" which, in 1961, established a speed record for flying boats over 15–25 km (9.3–15.5 mi.) of 566.69 mph (912 km/h). One landplane, the Be-30 "Cuff" feederliner, was also the product of the Beriev design bureau.

Berlin Airlift

Remarkable chapter early in the Cold War in which combined U.S. and British air fleets supplied the city of West Berlin with all its essential needs over a period of several months during a Soviet blockade of all other transportation links. In June 1948 the Russians employed various political subterfuges to sever all road, rail, and water-borne communications between West Berlin and West Germany. The Western allies' right to use the three air corridors into West Berlin had, however, been established in a quadripartite agreement negotiated by the Allied Control Council that could only be abrogated by what would amount to an act of war. In the face of mounting Russian intransigence, the Americans had been flying in supplies for their own forces and their dependents since April 1, 1948, and a similar British airlift began on June 28 after four days of total Russian blockade.

The Russian objective was to obtain control of West Berlin, and to frustrate this intention the RAF and the U.S. Air Force proposed to fly in all the commodities needed by West Berlin's civilian population: food, clothes, coal, medical supplies, household goods, tobacco, liquid fuel, and candles (electricity and gas supplies in West Berlin were severely restricted).

Initially both air forces used C-47 Dakotas, but it was quickly apparent that larger aircraft were needed. The U.S. Air Force (and from November 9 the U.S. Navy) assigned C-54 Skymasters to the operation (code-named Plainfare), while the RAF sent in its Avro Yorks.

The transports flew into both Gatow and Tempelhof from eight airfields in western Germany: Wunstorf, Wiesbaden, Rhein/Main (Frankfurt), Fassberg, Lübeck, Fuhlsbüttel, Celle, and Schleswigland (which accommodated the first of the RAF's new Hastings transports). In addition, British Sunderland and Hythe flying boats based on the Elbe River at Finkenwerder (near Hamburg) were flying onto the Havel.

As the tempo of the airlift increased, civil operators were contracted to carry in supplies (especially liquid fuel), and some of the latest service transports were dispatched to the Berlin airlift as soon as they became available (e.g. C-74 Globemasters, C-97 Stratofreighters).

In December 1948 a third airport was opened in West Berlin at Tegel. Inevitably a combination of winter weather, fatigued crews, and aircraft beginning to suffer from mechanical wear and tear exacerbated by a shortage of spares led to accidents: there were 68 fatalities to airlift personnel (18 RAF, 10 British civilians, 31 Americans, 9 Germans), U.S. accidents totaling 126 (50 of them minor) and RAF accidents 130.

In May 1949 the Russians admitted defeat and lifted the blockade. Supplies continued to be airlifted into Berlin on a diminishing scale throughout the summer, and when operations finally ceased a total of 277,804 sorties (65,857 RAF, 21,984 British civil, 189,963 U.S.) had flown in a total of 2,325,809 tons of supplies.

Betty

Although the MITSUBISHI G4M Type 1 bomber of the Japanese navy had been flying since 1939, and was used over China in May 1941, its appearance in the Pacific far beyond the presumed range of known Japanese bombers took American Forces by surprise in December 1941.

Dubbed "Betty" by the Allies, the G4M had been endowed with a range of over 2,000 mi. by filling its wings with fuel tanks and omitting almost all armor protection for the crew. With two 14-cylinder two-row radial engines of over 1,500 hp, the "Betty" had a top speed of nearly 270 mph (435 km/h) and a service ceiling of 30,250 ft. (9,220 m). The G4M acquired a formidable reputation as the aircraft that bombed Clark Field in the Philippines from Formosa and subsequently sank the British capital ships *Prince of Wales* and *Repulse* in December 1941, but despite these initial successes the G4M soon revealed a crippling weakness. Total absence of protection for the fuel tanks caused it to burst into flames so easily when hit that it became known to its crews as the "Type 1 Lighter" or the Hamaki ("flying cigar").

The G4M2 of late 1942 featured methanol/water-injection engines, laminar-flow wings, and another 330 gallons (1,250 liters) of fuel capacity in a fuselage tank. Some attempt to leak-proof the tanks was made, and defensive armament included a power-operated dorsal turret with a 20-mm cannon. Losses of G4M2s were nonetheless high, and the G4M2a (with the more economical Kasei 25 engines) was produced. It had a wingspan of 81 ft. 8 in. (24.89 m) and length of 64 ft. 4¾ in. (19.62 m). This was followed by the G4M3, in which range was subordinated to protection for the crew and fuel tanks. Total production was 2,479 aircraft of all variants, making the "Betty" the most widely produced medium bomber of the Japanese navy.

Beurling, George Frederick
(1922–1948)

Canadian-born RAF fighter ace of World War II, who was credited with 32 victories. Beurling left school in Montreal at the first opportunity so that he could earn enough money to take flying lessons. When war broke out he had done 150 hours' solo flying, but was rejected by the Royal Canadian Air Force on educational grounds. Beurling then worked his passage to England on a Swedish ship in 1940 and joined the RAF. He volunteered for overseas service and was sent to Malta, where he shot down 27 enemy aircraft before being wounded and returned to Britain. He later flew with the RCAF. Beurling died when an aircraft he was ferrying to Israel for use in the conflict with the Arabs crashed on takeoff from Rome.

Bf 109

Manufactured in greater numbers than any other airplane during World War II, the MESSERSCHMITT 109 fighter was first flown in September 1935. Initial production machines (Bf 109Bs) began to reach the Luftwaffe in 1937, and some aircraft served in the SPANISH CIVIL WAR with the Condor Legion. The Bf 109C also fought in Spain, and by the outbreak of World War II in 1939, the D variant had been largely supplanted by the 1,100-hp Bf 109E—the standard German single-seat fighter during the invasions of Poland and France and in the Battle of Britain.

The Bf 109F was an aerodynamically improved variant with a 1,300-hp Daimler-Benz engine that began to reach fighter squadrons in January 1941. It proved an excellent machine, but its successor, the Bf 109G of 1942, weighed 7,438 lb. (2,370 kg) loaded due to demands for heavier firepower and additional equipment, and its handling

Bf 109E

characteristics consequently deteriorated despite the installation of a 1,475-hp DB 605A engine. As on most 109 types, armament usually consisted of a cannon firing through the propeller hub and two machine guns in the engine cowling, but this was varied on special ground-attack or reconnaissance versions. The Bf 109H was an unsuccessful high-altitude fighter with extended-span wings that developed flutter in a dive. The final production version was the Bf 109K (a slightly modified Bf 109G).

The Messerschmitt 109 also served with the air forces of Spain (Hispano-Suiza-engined HA 1109 and two-seat HA 1110, Merlin-engined HA 1112), Switzerland, Finland, Czechoslovakia (Junkers Jumo-engined Avia C.210 and two-seat C.110), and Israel (Avia C.210).

Bf 110. See ZERSTÖRER

Biplane

Airplane with two sets of mainplanes (wings) placed one above the other. The first successful airplane, the FLYER, built by the Wright Brothers, was a biplane based on a box kite, which gave this configuration initial popularity with pioneer manufacturers. The biplane also provided considerable lift, a distinct advantage in the early days of feeble engines. The two wings of a biplane, however, with their interconnecting struts, experience more drag than a single wing of equivalent area. The few biplane types still built are mainly for sport flying and aerobatic displays, in which control and handling qualities are more important than speed.

Bishop, William Avery (1894–1956)

Second most successful RAF fighter pilot of World War I, credited with 72 victories, Bishop was a Canadian who arrived in England with the Canadian Mounted Rifles in 1915. He transferred to the Royal Flying Corps in July 1915 and flew as an observer with 21 Squadron in France. After pilot training he joined 60 Squadron (equipped with Nieuport 17s) in March 1917, scored his first victory on March 25, and quickly ran up a prodigious score.

On June 2, 1917, he flew a lone dawn sortie to attack a German airfield, and shot down three aircraft, damaging others on the ground. For this action he was awarded the Victoria Cross. In August 1917, with his tally at 47, Bishop was sent home to Canada to assist in recruiting, but returned to England early in 1918. He served briefly as an instructor, and was then given command of 85 Squadron (flying SE5As), which went to France on May 22. By June 21 Bishop had added a final 25 victories to his score, and he was then recalled to England to help in forming the Royal Canadian Air Force and saw no further combat.

Black box

Almost any intricate apparatus, such as a bombsight, automatic pilot, control box, or electronic device, which can be removed as a complete package, has at one time or another been known as a black box. In recent years the term has come to refer specifically to the flight recorder. These instruments were first used to take down the verbal reports of test pilots while trying out new designs, and their use was extended to monitor instruments and record readings. Test pilots previously had to scribble their findings on knee pads at a time when their utmost concentration was needed. Since recordings of crew conversation during flight might throw light upon accident causes and so contribute to future flight safety, airliners are now universally fitted with these recorders, which also monitor certain instrument readings. Made in specially strengthened cases to withstand impact, they are anything but black, in order to be conspicuous in the wreckage of a crash.

Blackburn

The British aircraft designer Robert Blackburn (1885–1955) began his career by constructing a series of successful monoplanes in 1911–12. He followed these with the Type L biplane seaplane (1913), the twin-fuselage TB (1914), a triplane (1916), and the GP (General Purpose) seaplane (1916). The GP was followed 18 months later by its landplane development, the Kangaroo, of which 20 were built for the RAF, some finding a civil use after World War I (with Grahame-White Aviation Co. and North Sea Aerial Navigation Co.).

The association of Blackburn with naval aircraft became firmly established after the war. Neither the Shirl (1918) nor the Blackbird (1918) achieved production status, but the Swift single-engined biplane torpedo-bomber (1920) led to the Dart (1921), and eventually to the Blackburn (1922), 33 of which served with the Royal Navy until 1931. The Velos (1925) was designed for the Greek navy, and the Iris flying boat of 1926 was the predecessor of the Sydney (1930) and the Perth (1933).

Nearly 100 Ripon torpedo/reconnaissance biplanes (1926) saw extensive service with the Royal Navy. The Baffin (1932) was a replacement for the Ripon and also flew with the New Zealand air force, while the Shark (1933) was built in large numbers for both the Royal Navy and the Canadian air force.

A classic biplane in production for 20 years: the Avro 504

Light aircraft were not neglected; they included the Bluebird of 1927–31 and its B-2 trainer development.

Experiments with the B-20 retractable-float seaplane fighter ended in a crash during April 1940, and the company's resources were devoted to the production of Skua fighter/dive-bombers and Roc turret fighters. The Botha (1938) was a twin-engined trainer/target tug orignally designed as a torpedo-bomber; the cannon-armed Firebrand torpedo-fighter arrived too late in the war to see operational service, although the Centaurus-engined TF5 version continued in use until 1953.

The B-54 (YA-5) was an unsuccessful turboprop antisubmarine aircraft, but the jet-engined Buccaneer low-level strike aircraft (first flown 1958) found extensive employment with the Royal Navy, the RAF, and the South African air force through the 1970s.

Blackburn also built the large Beverley four-engined military transport, used by the RAF from 1956 until 1967, but the company had become a part of HAWKER SIDDELEY in 1960 and the name Blackburn was dropped in 1965.

Blériot, Louis (1872–1936)

A pioneer French aviator, and designer of the first fully successful monoplanes, Blériot is probably best known for accomplishing the first flight across the English Channel (July 25, 1909). An engineer by training, he began to study aeronautics in 1896. Ten years later he established what is considered the first French airplane factory, at Issy-les-Moulineaux, where he concentrated on the development of monoplanes. By September 1907 his Blériot VI was able to fly 603 ft. (184 m); his type VII covered distances of up to 1,600 ft. (510 m) by December. The 50-hp Blériot VII is in fact generally considered the world's first successful monoplane. With its tractor engine and partly enclosed fuselage it set the basic pattern that aircraft designers, after several decades of producing biplanes, would return to.

Blériot's historic flight in a 25-hp Blériot XI across the English Channel from Les Baraques, near Calais, to Northfall Meadow, near Dover Castle (a distance of 23.5 mi., 37.8 km), took 36½ minutes and earned him the prize of £1,000 offered by the London *Daily Mail*. It was an event that placed aviation in the public imagination as never before, and made vivid the military potential of the airplane. At the end of 1909, Blériot also held the world speed record of 47.85 mph (76.99 km/h)

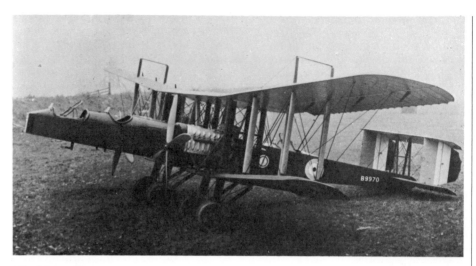

Blackburn Kangaroo

The Blériot XI monoplane

in a type XIII monoplane (established during the Reims aviation meeting of August 1909). Blériot went on to found several notable flying schools, and among later aircraft he produced were some of the first to be sold in quantity to military buyers.

Blind flying

During the first decades of powered flight, pilots had only very limited ability to navigate, or even to retain control of their aircraft, unless they could see the ground. During the 1920s, however, instruments were developed that began to confer a measure of "blind flying" capability (flight without the need to see outside the cockpit). One instrument was the ARTIFICIAL HORIZON; another was the turn/slip indicator, with a needle and either a heavy ball or a bubble in a transverse liquid tube, which showed whether the airplane was slipping into a turn or skidding out of it. By 1929, such pioneer pilots as Jimmy DOOLITTLE in the United States

and Pat Johnson in Britain were flying complete circuits "under the hood:" that is, they had fitted opaque covers over their cockpits, making their flying totally "blind."

During the 1930s Britain's RAF developed a standardized blind-flying panel containing six instruments: turn/slip indicator, artificial horizon, directional gyro, climb/descent indicator, airspeed indicator, and altimeter. With this group of instruments, a high standard of instrument flying was possible, and other nations soon adopted similar panels. At about the same time in the United States, Elmer Sperry, Paul Kollsman, and others were able to make instruments so accurate and reliable that they could be trusted to give correct indications in any conditions. After World War II, electronic devices made further major contributions to blind flying capability.

Blitz

Blitz is a shortened form of the

German *Blitzkrieg* ("lightning war"), the term used to describe the German method of attack at the outset of World War II. This called for the concentration of all available resources to eliminate a single specific limited objective before moving on to another target and repeating the process. The *Blitzkrieg* technique ensured complete local numerical superiority for a short time, but this could not be maintained if a campaign was unduly prolonged.

The so-called Blitz was the period of German night raids against Britain commencing in September 1940 and continuing until the spring of 1941. Faced with heavy losses inflicted on day bombers by RAF fighters, the LUFT-WAFFE began night attacks against London on September 7, 1940, with a force of about 700 twin-engined bombers.

For the first two months of the Blitz, an average of over 150 bombers a night attacked London, using *Knickebein*—a radio-beam bombing aid to which an aircraft's blind landing equipment could be tuned. In mid-November 1940 the objectives were extended to include industrial centers in the English Midlands, and on November 14 the first use was made of *X-Gerät* ("Ruffians")—a main radio-beam crossed by two oblique subsidiary beams, the time taken by a bomber to fly between the points of intersection giving an accurate ground speed so that the bombing point could be determined. This initial use of *X-Gerät* resulted in widespread destruction at Coventry, where 500 tons of high explosive and 900 incendiary canisters killed or seriously injured 1,400 people in a 10-hour raid.

Subsequently, heavy raids were again made on London (the central area of which was severely damaged by fire on the night of December 29), as well as on Liverpool, Birmingham, Plymouth, Bristol, Glasgow, Belfast, Sheffield, Cardiff, and Newcastle.

An inadequate number of antiaircraft guns and the availability only of makeshift night fighters (Blenheims, Defiants, Hurricanes) without accurate radio-location aids enabled the Luftwaffe to escape with very light battle casualties, although mechanical wear and tear led to an unserviceability rate of up to 50 percent by January 1941. Some success was, however, achieved by the British at jamming *Knickebein*.

The advent of shorter nights and the transfer of bombers to the Balkans and the newly opened eastern front in 1941 brought the Blitz to an end. The last (and heaviest) raid was on London on May 10, when 708 tons of high explosive and 86,700 incendiaries were dropped by 550 aircraft.

The German bombers available (Junkers JU 88s and Heinkel HE 111s) were not large enough nor sufficiently numerous to inflict really crippling damage to British industry on any scale, and public morale, for the most part, remained high.

Bloch

One of the most successful French aircraft manufacturers of the years between the wars. Marcel Bloch (born 1892) turned to aircraft production in the 1920s and quickly became a leading constructor of transports for Air France (the trimotor Bloch 120 and the Bloch 220 twin of 1937), and warplanes for the Armée de l'Air. The Bloch 151 and 152 radial-engined monoplanes, armed with four machine guns or two machine guns and two cannon, equipped seven fighter squadrons of the French air force in May 1940. Many notable bombers were also built. The Bloch 210 was a heavy night bomber, the 131 a reconnaissance bomber, the 162 a four-engined bomber, the 134 a medium bomber capable of 350 mph (560 km/h), the 174 a multirole machine, and the 175 probably the best of all France's attack bombers.

After the fall of France, Bloch was eventually arrested by the Nazis and sent to Buchenwald. Released in 1945, he began to rebuild his business interests under the name Dassault (from his brother's Resistance cover code "Char d'Assault"), which he also adopted himself. The company is now known as DASSAULT-BREGUET, and is France's largest private aircraft manufacturer.

Blohm und Voss

In July 1933 the Hamburg shipyard of Blohm und Voss formed an aircraft construction subsidiary, Hamburger Flugzeugbau. During World War II most of its manpower was devoted to subcontracting work for Junkers, Messerschmitt, and other manufacturers, but it also built several machines of its own design. The only plane to achieve true quantity production (279 aircraft) was the three-engined Bv 138 maritime reconnaissance flying boat. First flown in 1937, and used extensively by the Luftwaffe during World War II, the Bv 138 had an unusually short central hull with cannon-armed turrets and twin tails carried on long booms. Another seaplane designed in the years immediately before the war was the four-engined Ha 139. This was a huge floatplane capable of being launched by catapult and used to carry transatlantic mail on Lufthansa routes. Biggest of all Blohm und Voss seaplanes was the giant six-engined Bv 222 Wiking, the largest flying boat of World War II, with a wingspan of 150 ft. 11 in. (46.00 m). Only 14 Wikings were ever built, and began transport and reconnaissance duties in 1941.

Blohm und Voss also built several conventional landplane transports. The Bv 141, first flown in February 1938, was, however, one of the most unorthodox airplanes ever built. Intended as an observation machine for the Luftwaffe, it had an asymmetric layout with a completely glassed-in nacelle on one side of the centerline and a single engine and tail boom on the other. The wings were unequal in span, and the tailplane (stabilizer) was almost entirely on the left side of the fin. Only small numbers of this remarkable plane were built. Its final power unit was the 1,580-hp BMW 801A radial engine; provision was made for machine guns, cannon, bombs, and an assortment of cameras.

In 1956 Hamburger Flugzeugbau returned to aviation by taking part in production of Nord Noratlas transports for the new Luftwaffe. It later designed and produced its own HFB 320 Hansa executive jet, and took part in other cooperative European efforts. HFB is now part of Messerschmitt-Bölkow-Blohm, the largest German aircraft manufacturer.

Bloch 175

BMW

Founded in 1916, Bayerische Moto-ren Werke AG has long been renowned for its automobiles and motorcycles. But early in its history it was already producing the 6-cylinder BMW III, a 185-hp engine which powered the Fok-ker D. VII in the final year of World War I. Production of aircraft engines was begun again during the 1930s, and in World War II BMW engines were used in the Ju 90, Ju 86, Ju 52, Hs 126, Hs 123, He 115, Ar 196, and Fw 200 KONDOR (all powered by the 9-cylinder type 132 supercharged air-cooled radial), and in the Ju 290 and 390, the JU 88 series, the FOCKE-WULF 190, and the Do 217 (powered by a 14-cylinder two-row type 801 supercharged radial with fuel injection). Neither the type 802 (an 18-cylinder two-row radial) nor the type 803 (a 28-cylinder liquid-cooled four-bank two-stage super-charged radial) was used for production aircraft.

After the war BMW acquired a license to manufacture and service Lycoming and General Electric J79 engines, and to overhaul Orendas. It ceased involvement in aircraft engine manufacturing when its interests in the field were taken over by Man Turbo GmbH in the mid-1960s.

Boeing

What is today probably the world's largest and most successful manu-facturer of civil airliners was founded by William E. Boeing (1881–1956), who became actively interested in aviation in his thirties, and collaborated with Commander G. Conrad Westervelt in building two B. & W. float biplanes at Lake Union, Seattle, in 1916. Westervelt was then sent to a new post by the Navy, and Boeing established his own Pacific Aero Products Company. Early in 1917 an order was received for 50 Type C two-seat trainers, followed by a contract to build 50 (later reduced to 25) Curtiss HS-2L flying boats under license. The firm became the Boeing Airplane Company on April 26, 1917, and occupied premises south of Seattle on the Duwamish River, landplanes being test flown at adjacent military airfields.

In 1921 an order came to build 200 Thomas-Morse MB-3A fighters. The firm was also engaged at the same time in construction and conversion of de Havilland DH 4s. William Boeing was, however, primarily interested in build-ing aircraft designed by his firm, and in 1923 the Model 15 fighter biplane made its maiden flight. It was manufactured for the Army as the PW-9, and for the

Boeing 247D

Navy as the FB. The Model 21 trainer also saw service with the Navy, some remaining in use as late as 1929 for anti-mosquito spraying in Puerto Rico.

The Model 40 was designed as a replacement for the DH 4 to fly mail services. When the U.S. Post Office handed over its airmail routes to private operators in the summer of 1927, the Model 40 was redesigned with a two-seat passenger cabin and secured the Chicago–San Francisco route. The Model 40 continued in production until 1932; later examples carried four passengers and served on many early airline routes, including those of United Air Lines and Western Airlines.

Successful fighter biplanes continued to be built by Boeing in the late 1920s, among them the Model 69 (F2B) and its developments, the Models 74 and 77 (F3B). These were followed by the outstanding F4B/P-12 series, which saw extensive service with the Army as well as the Navy. Nearly 600 were built before the series was supplanted in 1933 by the P-26 fighter, the U.S. Army's first all-metal monoplane, nick-named the PEASHOOTER by its pilots.

Boeing's future interest in large air-craft was foreshadowed by the three-engined 12-passenger Model 80 of 1928. The single-engined Monomail mono-plane of 1930 was designed for high-speed mail services, but only two machines were built. The B-9 mono-plane bomber of 1931 also failed to attain production status. In the com-mercial field, however, the BOEING 247 and the 307 STRATOLINER were successes, and a new series of very large four-engined long-range bombers was initi-ated with the XB-15, of which only a prototype was constructed, since the performance proved disappointing be-cause of the airplane's great size and weight. The Model 299 of 1935 later became the famous B-17 FLYING FOR-TRESS, which was followed in turn by the B-29 SUPERFORTRESS.

Boeing's answer to demands in the late 1930s for long-range passenger flying boats had been the BOEING 314 (1938), but postwar Boeing production concentrated on landplanes—the 377 STRATOCRUISER and eventually the jet-propelled BOEING 707, both of which were developed from military transport projects (the C-97 and the C-135 re-spectively).

Of Boeing's military designs, neither the experimental Sea Ranger flying boat of 1942, nor the XF8B carrier-borne single-engined long-range attack aircraft attracted orders. But the B-47 STRATOJET of 1947 became the Strategic Air Command's principal bomber until replaced by the B-52 STRATOFORTRESS.

Outside the military field, airliners designed for medium and short hauls were developed in the 1960s to supple-ment the successful 707 (BOEING 727, BOEING 737), followed by the BOEING 747 jumbo jet in 1969.

Involvement in helicopters through the Vertol Division (originally Piasecki Helicopter Corporation) led to the twin-rotor Model 107 of 1958, which found both civil (carrying 25 passengers) and military applications, and to the big CH-47 CHINOOK "battlefield mo-bility" helicopter, which has been adop-ted by various air forces as well as by the U.S. Army.

Boeing 247

Often considered the first "modern" airliner, the Boeing 247 heralded the great revolution in air transport that ended the age of the biplanes and the ungainly high-wing trimotors. The 247 was a low-wing all-metal aircraft carry-ing 10 passengers and powered by two Pratt & Whitney Wasp radial engines of 550 hp, which gave a cruising speed of 155 mph (249 km/h) and range of 485 mi. (780 km). Wingspan was 74 ft. (22.56 m) and length 51 ft. 4 in. (15.64 m). The 247 was among the world's first civil aircraft to employ flaps and was Boeing's first airliner to feature a retractable undercarriage. Faster, more comfortable, and more reliable than the earlier generation of passenger aircraft, the first of 60 Model 247s entered service with United Air Lines only a few months after making its maiden flight in 1933.

Problems of performance at high altitude and in hot weather led to development of a new version, the Model 247D, in 1934. Thirteen 247Ds were produced (also for United); they had increased fuel capacity, variable-pitch propellers, and other modifications. These innovations boosted cruising speed to 189 mph (304 km/h) and range to 745 mi. (1,200 km), and were retrospectively incorporated in the earlier aircraft.

The advance marked by the Boeing 247 was quickly appreciated. Unable itself to acquire the 247, TWA approached Douglas with specifications for a comparable aircraft: the result was the 14-passenger DC-2 of 1934. To meet this new competition American Airlines (which had been flying Curtiss CONDOR biplanes) ordered the 21-passenger DC-3 the following year.

Boeing 307. See STRATOLINER

Boeing 314

The four-engined Boeing 314 was the last large American transoceanic flying boat. Six of these 74-passenger aircraft were ordered by Pan American in July 1936. Each had a wingspan of 152 ft. (46.33 m) and length of 106 ft. (31.31 m). Sponsons on the side of the hull gave stability afloat, contained extra fuel, and served as loading/boarding platforms.

First flown on June 7, 1938, the Model 314 initially had a single fin and rudder, but it was found to require twin fins, to which an additional central fin was later added. The first six 314s were delivered in the early months of 1939, and they inaugurated a North Atlantic mail service on May 20, followed by the first regular passenger service (June 28). They also flew a transpacific route (San Francisco–Hong Kong). On March 20, 1941, the first of a second batch of six 314s (Model 314A) made its maiden flight. The 314As had more powerful engines, larger-diameter propellers, extra fuel capacity, and seating for 77 passengers. Five of the first machines were brought up to 314A standard, and in May 1943 three of the second batch of boats were sold to BOAC. All of these aircraft saw only very brief commercial service after the end of World War II.

Boeing 377. See STRATOCRUISER

Boeing 707

The first American commercial jet transport, the Boeing 707 has been one of the most successful airliners in the history of aviation. The prototype, the Boeing 367-80, first flew in 1954, and from it were developed the C-135 Stratolifter military transport and KC-135 Stratotanker (which both entered service in 1957), as well as the 707.

The first flight of the 707-120 airliner took place in December 1957, and though it was intended for domestic operations it inaugurated American-operated North Atlantic jet services with Pan American in October 1958. Within a few months American Airlines put it into service on its transcontinental routes. The 707-320 Intercontinental appeared in 1959 and became the standard long-range version of the aircraft. The 707-320B of 1962, seating 189, had improved takeoff performance and better payload range characteristics. The 707-320C, with four wing-mounted 18,000 lb. (8,165 kg) Pratt & Whitney engines, was fitted with side-loading freight doors and could be used for air cargo; with a gross weight of 333,600 lb. (151,300 kg) it cruises at 550 mph (885 km/h) for 4,300 mi. (7,000 km) and carries as many as 219 passengers. Wingspan is 145 ft. 9 in. (44.43 m) and length 152 ft. 11 in. (46.61 m). The 720 is a smaller variant of the 707 which carries up to 167 passengers; the 720B has uprated engines that improve its performance.

Boeing 727

The Boeing 727 short/medium-range trijet airliner is powered by three Pratt & Whitney tail-mounted turbofans. First flown on February 9, 1963, the series 100, which seated 131 passengers, entered service with Eastern Air Lines in 1964. The 727C and 727QC versions have side-loading freight doors, the QC ("quick change") being convertible to a passenger aircraft by the installation of palletized seating.

The fuselage of the 727-200 was stretched by 20 ft. (6.09 m); its length is 153 ft. 2 in. (46.68 m) and wingspan 108 ft. (32.91 m). Cruising speed is 568 mph (914 km/h). The aircraft seats up to 189 passengers and made its first flight on July 27, 1967. Over 1,000 727s have been sold to world airlines.

Boeing 314A operated by BOAC, taking-off from Baltimore

Boeing 707

Boeing 737

The Boeing 737 short-haul airliner, with two Pratt & Whitney turbofans mounted under the wings, made its maiden flight on April 9, 1967. It is the smallest in the Boeing family of jet airliners. The first production batch (series 100) seated 115 passengers; the series 200 (first flown on August 18, 1967) had its fuselage stretched by 6 ft. 4 in. (1.84 m) to seat up to 130 passengers. The wingspan of the 737-200 is 93 ft. (28.35 m) and length 100 ft. (30.48 m). Cruising speed is 568 mph (914 km/h).

Variants include 737C passenger/cargo convertibles, 737QC "quick-change" aircraft with palletized seats, and business jets. The advanced 737-200 incorporates aerodynamic improvements that permit operation from runways as short as 4,000 ft. (1,200 m). Over 700 Boeing 737s have been sold.

Boeing 747

The first wide-bodied jet airliner ("jumbo jet") to enter service, the Boeing 747 carried over 7,000,000 passengers in its first 12 months of service and has revolutionized civil aviation. The 747 is powered by four Pratt & Whitney turbofans and made its maiden flight on February 9, 1969. With a wingspan of 195 ft. 8 in. (59.64 m) and a length of 231 ft. 4 in. (70.51 m), the 747 was designed to carry up to 490 passengers in a double-deck fuselage that can accommodate 10 seats across its 20-ft. (6.1-m) width; the 57 first class passengers had a lounge on the upper deck.

Equipped with engines of 43,500 lb. (19,700 kg) thrust, the 747 entered service with Pan American across the North Atlantic on January 22, 1970. The 747B is powered by uprated engines of 45,500 lb. (20,600 kg) thrust; with a gross weight of 775,000 lb. (351,000 kg), it has a range of 4,330 mi. (6,965 km),

Boeing 727

Boeing 737

Boeing 747

and maximum speed is over 600 mph (965 km/h) at 30,000 ft. (9,000 m).

A convertible freight/passenger variant is known as the 747C; both this aircraft and the 747F freighter have loading doors at the nose. A short-range model, the 747SR, can carry up to 498 passengers and has been specially stressed for a high frequency of takeoffs and landings; the 747SP is a short-fuselaged variant for long-haul, low density routes.

Because of its great size the 747 needs 18 undercarriage wheels (four four-wheeled main assemblies and a twin nose-wheel).

Boelcke, Oswald (1891–1916)

One of the first German fighter aces, credited with 40 victories and an important role in developing combat tactics for the earliest effective fighters, Boelcke was a professional soldier, awarded his pilot's badge on August 15, 1914. He first flew two-seat reconnaissance aircraft with *Flieger-Abteilung* 13, often with his brother as observer. He joined *Fl-Abt* 62 on April 25, 1915, and with Max IMMELMANN became one of the first pilots to fly the new Fokker EINDECKER in action. Boelcke scored his first victory on July 4, 1915, and by May 21, 1916, his tally had risen to 18; he was promoted and withdrawn from operational flying.

In July 1916 Boelcke was recalled to assist in the formation of the first *Jagdstaffeln* (fighting units), and to prepare a set of guidelines for aerial combat. On August 11 he was appointed to command *Jagdstaffel* 2, with RICHT-HOFEN as one of his pilots. Flying Fokker D.IIIs and ALBATROS D.Is and D.IIs, Boelcke continued his string of victories, shooting down his 40th victim on October 26, 1916. Two days later he died after a collision with one of his own pilots, Erwin Boehme, during combat with British DH 2s.

Oswald Boelcke

Cluster bombs beneath the wing of an RAF Phantom: each 600-lb. (270-kg) bomb ejects over 100 "bomblets" that blanket targets such as tank formations and parked aircraft.

Bomb

The first aerial bombs were constructed locally by Italian forces engaged in the war against Turkey in Libya in 1911–12. The aircraft were mainly Taubes, and the pilot merely pulled out the pin and dropped the grenade or bomb over the side. In 1913 the Turks were attacked with bombs properly designed as such when Bulgaria used bomber aircraft in the Balkan War. Early in World War I the British Royal Naval Air Service possessed a handful of small Sopwith Tabloid biplanes able to drop specially designed 20-lb. (9 kg) streamlined bombs, which had stabilizing fins and a contact fuze in the nose. On October 8, 1914, two Tabloids flew to Friedrichshafen and destroyed a Zeppelin airship in its hanger as well as dropping a bomb on Cologne railroad station. On November 21, 1914, four Avro 504 biplanes each released four bombs of this early type on airship hangars and a gasworks. Gradually the British increased the size of their bombs until a weight of 1,650 lb. (750 kg) was reached in 1918 (this was the bomb intended to be carried to Berlin by the four-engined Handley Page V/1500). Most of the bombs dropped in World War I were of course much smaller, weighing from 50 to 112 lb. (23–50 kg). Strategic bombing was pioneered by the Italians and Russians, and by the spring of 1917 heavy attacks were being made on Britain by German GOTHAS carrying bombs of up to 1,100 lb. (500 kg),

while the R-PLANES sometimes carried monster bombs of 2,205 lb. (1,000 kg). Special bombs were developed for piercing the armored decks of ships, and for sinking submarines.

Major increases in the size and destructive capability of bombs took place early in World War II, made possible at least partly by the development of heavy strategic bombers by the United States and Britain. The RAF made massive attacks on Germany, generally using a mixture of incendiary (fire) bombs and light-case blast bombs of 4,000 lb. (1,800 kg). In some attacks 4,000-pounders were bolted together to give "blockbusters" of 8,000 or 12,000 lb. By 1945 the biggest RAF bomb was the so-called "Grand Slam" or "Earthquake" of 22,000 lb. (10,000 kg), which dropped at supersonic speed, penetrated more than 50 ft. (15 m) into the ground, and demolished difficult targets (such as viaducts) by actually shaking the earth. A profound change accompanied the development by the United States of nuclear weapons. The only atomic bombs used in war were the five-ton "Little Boy" dropped on Hiroshima (by the B-29 SUPERFORTRESS *Enola Gay*) and "Fat Man," which destroyed Nagasaki (carried by another Superfortress, *Bockscar*). Since 1945 nuclear weapons have decreased considerably in size, so that nominal 20-kiloton weapons (with an explosive power roughly equal to 20,000 tons of TNT) have shrunk from 10 ft. (3 m) in diameter and 25 ft. (7.5 m) in length to today's 1 ft. (0.3 m) in diameter and 6 ft. (1.8 m) in length, with a special streamlined shape for carriage under supersonic aircraft. The far more devastating hydrogen weapons have yields measured in megatons (millions of tons of TNT). They are of course larger, but the B-1 bomber can still carry 28 missiles, each with a megaton-class warhead. Other developments include the retarded bomb, whose descent is slowed to help the bomber escape its effects, and cluster bombs, which open in flight to release hundreds of "bomblets."

The laser-guided "smart bomb" has already proved its superior accuracy in Vietnam and the Middle East. Less sophisticated, but still very effective in close-support operations against troops and armor, is the napalm-type bomb. Even more terrifying is the new "concussion" bomb, which sprays out and then ignites an explosive aerosol mixture, killing everything for dozens of yards around—even troops sheltering well underground.

Bombers

The earliest bombers were simply aircraft whose pilots could throw grenade-like bombs out of the cockpit. Probably the first was a German Taube used by Italy against the Turks in 1911 (see BOMB). Although many early warplanes were rigged up to carry light bombs, the first real bombers were the French Voisins, a series of pusher biplanes with all-steel structure. These proved so effective as bombers that the Aviation Militaire had begun to establish special bomber squadrons by September 1914.

Czarist Russia began to form its own "Squadron of Flying Ships" using ILYA MOUROMETZ four-engined bombers. From February 1915 onward, considerable numbers of these big Russian aircraft were in action on the eastern front. In Italy, the Corpo Aeronautica Militare built up a large force of CAPRONI bombers, which undertook long missions across the Alps to bomb Austro-Hungarian targets. The best of the Capronis was the Ca 5 of early 1918, with three 300-hp Fiat engines, which could carry 1,500 lb. (680 kg) of bombs at up to 95 mph (152 km/h). Some earlier Capronis accommodated loads of up to 3,910 lb. (1,770 kg). The defensive gunners had to stand in the freezing propeller slipstream for up to seven hours on a raid. Other notable World War I bombers were the AEG, GOTHA, and Friedrichshafen aircraft of Germany, the huge German R-PLANES, and the British HANDLEY PAGE O/400

Caproni Ca 5, an Italian bomber of 1918

and V/1500.

Between the wars bombers grew larger and faster, and monoplanes replaced biplanes. A notable Soviet machine was the Tupolev TB-3, a four-engined monoplane built in large numbers between 1929 and 1940. In the United States the all-metal Martin B-10 appeared in 1932, and the first four-engined Boeing B-17 FLYING FORTRESS heralded the age of strategic bombing on a large scale. The Germans, however, concentrated on relatively small twin-engined bombers to support their blitzkrieg tactics, but the RAF soon acquired four-engined heavy bombers (STIRLING, HALIFAX, LANCASTER). Over Europe—and later, Japan— the U.S. Army Air Force favored daylight precision bombing, using B-17s, B-24 LIBERATORS, and (from 1944) B-29 SUPERFORTRESSES. Toward the end of World War II the Luftwaffe used ME 262 and ARADO 234 jet bombers, and in the early postwar years piston-engined bombers were rapidly replaced by jets. The U.S. Strategic Air Command employed the mixed piston/jet CONVAIR B-36, the B-47 STRATOJET and B-52 STRATOFORTRESS; Britain built the CANBERRA, Valiant, VULCAN, and Victor; France used the supersonic Mirage IVA. The Mach-2 B-58 Hustler had only a short U.S. Air Force career, and the aging B-52s were supplemented by small numbers of swing-wing F-111s. By the 1970s the Soviet Union had built up a powerful bomber force that included the Tu-20 Bear, the Mya-4 Bison and Tu-16 Badger jets, the supersonic Tu-22 Blinder, and (in 1974) the Tupolev BACKFIRE. The most advanced strategic bomber under development in the West in the mid-1970s was the Rockwell-International B-1.

Bombing

Aerial bombing during the early stages of World War I involved small hand-held bombs dropped over the sides of planes onto enemy gun-emplacements, troop concentrations, or airplanes. Later in the war Germany's Zeppelins were bombing London in the first long-range massed strategic attacks, while Britain's Royal Flying Corps had developed a number of quite effective tactical bombers such as the DH 9 and 9A, which could carry up to 450 lb. (200 kg) under the wings. As a reprisal for the Zeppelin attacks, the big HANDLEY PAGE O/400 bombers were introduced; these could carry 1,800 lb. (800 kg) of bombs, and as the Armistice was being signed British aircraft were being prepared to bomb Berlin. Germany's GOTHAS had a similar performance.

By the war's end it was clear that targets could be separated into two categories; tactical and strategic. In the first group were those which were associated with the day-to-day running of the war such as troop, gun, or tank dispositions, the destruction of which could result in a temporary and local advantage. In the second category came those targets which were essential to the long-term well-being of the enemy and his ability to prosecute the war indefinitely. These were represented by oil refineries, rail communications, harbors, and aircraft factories. The wearing down of civilian morale was also considered to be a strategically desirable objective. Strategic bombing was recognized as an essential prerequisite of large-scale wars.

In the 1920s and 1930s considerable advances were made in tactical bombing techniques, notably by Britain. Responsibilities in the Middle East and along India's Northwest Frontier, gave the RAF plenty of practice. While tactical bombing at least had made very little difference to the course of World War I, in the Middle East it was shown to be effective. It was there that the air force refined two of the principal methods in use today: ground attack and close support.

Ground attack is a rather loose term implying the use of bombs, guns or rockets on a tactical target. Very often it is carried out at the request of the local army commander to remove an enemy concentration which is blocking his advance or threatening him in some other way. This technique was brought to a high degree of sophistication during the Allied push through Europe in 1944: patrols of fighter-bombers, principally Typhoons, would be kept aloft ready to respond to calls from advancing troops. In this way reaction time was kept to a minimum and enemy forces were rarely able to gain the initiative.

About this time also a new technique known as interdiction came into general use. This was the cutting of enemy supply routes—railways, canals, roads—to a depth of about 100 mi. (160 km) behind hostile frontiers, so depriving him of supplies where they were most needed. Typhoons and MOSQUITOS from bases in Britain roamed at will over northern France bombing or strafing with rockets anything that moved and largely preventing the reinforcement of German efforts to thwart the Allied advance.

Meanwhile the strategic bombing of Germany gained momentum, the U.S. Eighth Air Force's FLYING FORTRESSES by day supplementing RAF Bomber Command's efforts by night. The RAF concentrated on the destruction of cities such as Hamburg, Berlin, and the industrial Ruhr, while the Americans went for the annihilation of pinpoint targets such as the oil refineries at Ploesti. At that time the standard weapons were the 500 lb. (225 kg), 1,000 lb. (450 kg), and 2,000 lb. (900 kg) bombs, and more specialized weapons such as the 4,000 lb. (1,800 kg) "Cookie," the 12,000 lb. (5,400 kg) "Blockbuster," and—biggest of them all—the 22,000 lb. (10,000 kg) "Grand Slam," the latter two carried only by the RAF's incomparable LANCASTER. The "Grand Slam" was an "earthquake" bomb for the destruction of special targets such as railroad viaducts. A rather similar weapon in its effect was the "bouncing bomb" (see DAMBUSTERS) used to rupture the Eder and Mohne dams at the head of the Ruhr valley in May 1943; these, again, were strategic targets.

With the introduction of nuclear devices in the late 1940s new methods of attack had to be developed to ensure the safety of the bombers from the destructive effects of both the blast and the radiation. In the loft or toss maneuver the attacking aircraft begins to pull up into a climb some miles short of the target. The bomb is released when the plane has reached about 45° and continues on toward the target with its initial speed. Meanwhile the attacker has made the first half of a loop and, rolling out, follows a reciprocal course back to base. Today intermediate-range and intercontinental ballistic missiles have taken over the task of delivering the largest warheads, and the main use of bombers in the nuclear role appears to be to deliver relatively small nuclear bombs. With quite small nuclear weapons, the bomber can be flown directly over the target in the so-called "lay-down" technique. Owing to the increasing effectiveness of surface-to-air missiles, the traditional method of delivering strategic attacks from high altitude has been abandoned in favor of the terrain-following techniques, in which the attacker makes his way to the target at the lowest possible height. This presents the defenses with a much more difficult problem, because the radar detection system has insufficient time in which to detect the target, lock onto it, and launch a missile. It presents the attacker with difficulties as well. Navigation becomes more critical, because at a typical height of 200 ft. (60 m) fewer landmarks can be distinguished. Since the target may be heavily defended the attacker can only afford to make one pass before losing the advantage of surprise, and so accurate navigation at treetop height at speeds of perhaps 12 miles a minute calls for a radar-controlled autopilot. The bombing attack itself may also be automatic, the point of release being computed from measurements of distance to target, wind, and the known ballistic characteristics of the weapons. See also BOMB; BOMBERS.

Bong, Richard Ira (1920–1945)

Highest-scoring American fighter ace of World War II, credited with 40 victories achieved in more than 200 missions. Bong was assigned to the 9th Fighter Squadron of the 49th Fighter Group, then in Australia, in September 1942. He claimed his first victory, a Japanese bomber, on December 27, 1942, over Dobodura in New Guinea. He continued operational flying in the Southwest Pacific until mid-April 1944, by which time, while flying a P-38 LIGHTNING, he had shot down 28 Japanese aircraft.

After leave and a gunnery course in the United States, Bong returned to the Pacific as a gunnery instructor. In October 1944 he volunteered to rejoin his former unit providing air support for the invasion of the Philippines. Between October 10 and November 15, Bong destroyed eight more aircraft, and for his conduct during this period was awarded the Congressional Medal of Honor. He then flew with the 475th Fighter Group, also equipped with P-38s, and during December destroyed another four aircraft to bring his total to 40, three-fourths of which were fighters. Bong was killed in a flying accident in California while a test pilot with Lockheed.

Boundary layer

When the wing of an aircraft in flight moves through the atmosphere, the particles of air are pushed aside to let it through. Those particles that are in contact with the surface of the wing retain their position relative to the wing's surface owing to skin friction and do not move. Particles in the next layer are restrained from moving by the air particles beneath them and have a very small relative movement. The next layer moves more quickly, and so on, until the full relative velocity of air is reached.

The region in which this velocity transition from zero to normal takes place is called the boundary layer. It may be only .01 in. (0.25 mm) deep, but if it can be kept free of TURBULENCE, the power-consuming drag generated as the aircraft moves through the air will be substantially reduced. To achieve this smooth flow, holes may be made in the wings to exert a suction effect, or air may be blown out from inside the wing to help the flow. At very low speeds the boundary layer becomes increasingly turbulent and eventually breaks away from the wing surface, causing a stall. Leading-edge slats opening at low speed increase the airflow and delay the onset of STALLING.

Boyington, Gregory (1912–   )

Highest-scoring U.S. Marine Corps fighter ace of World War II. In 1941 Boyington volunteered for service with the American Volunteer Group in China (the so-called FLYING TIGERS) and in 1942 shot down six Japanese aircraft over Burma, flying a P-40 WARHAWK. Later, with the Marines, he saw service in the Solomon Islands flying F4U CORSAIRS. In two months he shot down 14 enemy aircraft, including five in one day (September 16, 1943). During the last two weeks of 1943, Boyington led his squadron, VMF 214, in a series of fighter sweeps over the enemy base at Rabaul, and added another five victories to his score. On January 3, 1944, he brought his total (including victories with the Flying Tigers) to 28 by shooting down three more aircraft, but had to abandon his own fighter and parachute into the sea. Reported missing in action for many months, he had in fact been picked up by a Japanese submarine and made a prisoner of war.

Braniff International

Founded as Braniff Air Lines in 1928, and reestablished as an independent operator two years later, Braniff began services with a Tulsa-Wichita route, and expanded rapidly through the mid-1930s. It acquired the name Braniff International to mark the opening of a Houston-Havana-Lima route in 1948, and in 1967 absorbed Pan American-Grace Airways (PANAGRA), giving it a vastly expanded Latin American network. Braniff International now operates routes serving the United States, Mexico, and South America, with Pacific services inaugurated in 1969 to link five Southern cities with Honolulu.

In 1965, DC-6s, DC-7s, and Convairliners were replaced by a large fleet of BAC 1-11 and Boeing 707 jetliners for short and long hauls respectively. Braniff's domestic routes are now served almost exclusively by Boeing 727s, with DC-8s flying the South American and Hawaii services. The present fleet consists of 1 Boeing 747, 11 DC-8s, 79 Boeing 727s, and 1 BAC 1-11 (executive). Braniff routes total some 30,000 mi. (48,000 km), and 8,600,000 passengers are carried annually.

Bristol

One of the oldest British aircraft manufacturers, Bristol was founded in 1910, and in its earliest years produced FARMAN biplane designs. During World War I its outstanding product was the F2B Fighter, a two-seater that combined a robust structure, a powerful and reliable Rolls-Royce engine, and good front and rear armament. The F2B, known as the "Brisfit," remained in service with the RAF until 1932; over 4,000 were built.

In 1920 Bristol established an engine department, which soon became one of the world's leading producers of aircraft power plants. Its main product over its first 10 years was the Jupiter nine-cylinder radial, which turned out 386 hp

Bristol Blenheim IV

Bristol Brabazon

at its debut in 1921, but had been developed to produce 560 hp by 1930 and was used in 262 different types of aircraft. From the Jupiter was developed the Pegasus (which by World War II was giving 1,000 hp), the smaller Mercury (the power of which rose from 440 to 990 hp over eight years), and a radically new family of sleeve-valve engines that included the 905-hp Perseus, 1,200-hp Taurus, 1,400–2,000-hp Hercules, and 2,200–3,000-hp Centaurus.

Aircraft produced by Bristol between the wars tended to be for the military. In the late 1920s the Bristol Bulldog biplane began service as a standard fighter with the RAF; it also flew with the air forces of such nations as Denmark and Finland. Bristol also produced a family of very fast, twin-engined monoplanes that played a major role in World War II. A total of over 4,000 Blenheim light bombers, introduced in 1937, were eventually produced. From the same basic twin-engined design came the Beaufort torpedo-bomber and maritime reconnaissance plane, the Beaufighter escort and night fighter and ground-attack aircraft, and later airplanes including the Buckingham, Buckmaster, and Brigand.

One of Bristol's first postwar products was the twin-engined Freighter, intended as a military cargo-carrier but also flying as a civil freighter and car-ferry and as a passenger aircraft (the Wayfarer). Another post-1945 product was the ill-fated Brabazon, a giant airliner with eight engines driving four contra-rotating propellers and a wingspan of 230 ft. (70.10 m). The Brabazon made its maiden flight in 1949 and had an intended range of 5,500 mi. (8,850 km); it was too advanced and too large for its market, and the first and only prototype was scrapped in 1953 after 400 hours' test-flying. Bristol's most successful product was the four-turboprop Britannia medium/long-range transport (basis also of the Canadair CL-44 airliner and freighter, CC-106 Yukon military transport, and, in a modified form, CP-107 Argus patrol/antisubmarine machine).

After the war Bristol's engine department was reorganized as Bristol Aero-Engines; it produced the Theseus and Proteus turboprops; the Orpheus, Olympus, and Pegasus jets; and the Thor and Odin ramjets. It became Bristol Siddeley Engines in 1959, following a merger with Armstrong Siddeley Engines; in 1966 Bristol Siddeley became part of ROLLS-ROYCE. Bristol Aircraft was absorbed by the BRITISH AIRCRAFT CORPORATION (BAC); helicopter programs (including the Syca-

"Battle of Britain" memorial flight of Spitfires and Hurricanes

more and twin-rotor Belvedere) were taken over by WESTLAND.

Britain, Battle of

By the beginning of June 1940, Nazi forces had occupied the coast of the English Channel opposite England, and Hitler was planning the invasion of Britain. As an essential preliminary to the invasion, the LUFTWAFFE was given the task of securing air superiority by bringing the British Fighter Command to battle and destroying its aircraft and ground facilities.

Arrayed against Britain's 600–700 SPITFIRES and HURRICANES with 1,253 pilots, the German air force had 900 Messerschmitt BF 109 and ZERSTÖRER (Messerschmitt 110) fighters, 875 twin-engined bombers (HE 111s, Do 17 FLYING PENCILS, JU 88s), and 300 STUKAS. These Luftwaffe aircraft formed Luftflotte 2 (under Feldmarschall Kesselring, in Holland and Belgium) and Luftflotte 3 (under Feldmarschall Sperrle, in northern France); in addition, Luftflotte 5 (under General Stumpff, in Norway and Denmark) had a further 123 bombers and 34 Zerstörers at its disposal.

RAF Fighter Command under Air Chief Marshal Sir Hugh Dowding was divided into four geographical groups. London and the Southeast were part of 11 Group (Air Vice Marshal Park), with 12 Group (Air Vice Marshal Leigh Mallory) covering eastern England and the Midlands; 10 Group in the West Country and 13 Group (Scotland and the North) played a lesser part in the forthcoming battle.

Early in July, the Germans began raiding British shipping and ports. Radar stations around the coast of England gave early warning of enemy formations building up and approaching, but the importance of these installations was not fully realized by the German High Command. The commencement of the intensive pre-invasion German air attacks was scheduled for August 13, *Adlertag* ("Eagle Day"). Raids were launched principally at RAF airfields near the coast and against radar stations, 1,485 Luftwaffe sorties being flown on *Adlertag* itself and 1,786 on August 15. Heavy combat on these two days cost the RAF 33 pilots killed or wounded, and an adequate number of replacements could only be found by drawing personnel from bomber units. The Luftwaffe, however, had lost 106 aircraft on the same two days, including 13 Stukas and 34 Zerstörers, and Luftflotte 5 was so badly mauled that most of its remaining aircraft were transferred to Luftflotten 2 and 3.

German attacks now moved to RAF airfields further inland, and by the end of August the British fighter base at

Manston needed to be evacuated, and other important airfields were badly damaged. RAF Fighter Command had an adequate supply of aircraft, but the pilot shortage was becoming acute (300 casualties during August, but only 260 replacements leaving training schools).

At the end of August the RAF bombed Berlin by night, and partly in retaliation (partly also to draw all the remaining Spitfires and Hurricanes into battle) the Luftwaffe switched its attack to London on September 7. A typical German raid consisted of waves of from 20 to 40 bombers with a close fighter escort and a screen of high-altitude fighters above. British Spitfires engaged the high-level screen, while Hurricanes attacked the bomber escorts. Once the fighter resistance had been softened, further squadrons moved in against the bombers.

The distance of German bases from London restricted the time that the short-range Bf 109s could remain over England to escort the bombers. And while the Germans concentrated their attacks on London, RAF Fighter Command was able to repair its airfields. German losses consequently reached a catastrophic 60 machines on September 15, and Hitler postponed the invasion of England indefinitely. The Luftwaffe gradually shifted to night operations (the BLITZ) as the daylight hours shortened.

In the course of the Battle of Britain from July 10 to October 31, the Luftwaffe lost 1,733 aircraft; the RAF lost 1,003 Spitfires and Hurricanes—and 481 pilots (killed, missing, or captured). The quality of the combat aircraft and crews was fairly equally matched. The decisive factors of the battle were, first, Britain's use of radar early warning in conjunction with an effective system of fighter control (which deployed intercepting squadrons to the best possible advantage)—and second, the inferior quality of the German High Command. By mid-October the British situation had in fact become desperate—most fighter squadrons were manned by trainees without the experience to engage enemy fighters effectively. Had the Germans persevered a little longer, or had they pursued their policy of destroying RAF Fighter Command's airfields and fighters instead of turning their attention to London, the outcome might possibly have been reversed.

British Aircraft Corporation (BAC)

BAC was formed in the 1960s to take over the aviation interests of English Electric, BRISTOL, VICKERS, and Hunting

BAC 167 Strikemaster

Aircraft (see PERCIVAL). Manufacturing divisions deal with commercial aircraft, military aircraft, and guided weapons.

The new company continued manufacture of aircraft, such as the LIGHTNING and VC-10, which it inherited from its parent companies. It continued development of what became known as the BAC 1-11 short/medium-range airliner. BAC also produced the Strikemaster light attack aircraft (a development of the Percival P.84 Jet Provost trainer), powered by a Rolls-Royce Viper turbojet of 3,410 lb. (1,544 kg) thrust. The Strikemaster is specially designed for counterinsurgency duties and has two fuselage-mounted machine guns and provision for carrying rockets or bombs under the wings.

BAC is involved in the Panavia consortium (responsible for the TORNADO) and Sepecat (builders of the JAGUAR), and is a partner with Aérospatiale in building the CONCORDE. In 1977 it became part of the new government-owned aircraft manufacturer British Aerospace.

British Airways

The British national airline, British Airways, was formed in 1974 by the merger of BOAC (British Overseas Airways Corporation) and BEA (British European Airways).

British Airways owes its origins to the scheduled passenger services across the English Channel that had begun as early as August 1919. By February 1921, however, no further scheduled flights were being made by these pioneer civil operators. After only 18 months of cross-Channel flying, Aircraft Transport and Travel (later Daimler Airway), Handley Page Transport, Instone Air

Line, and British Marine Air Navigation required a government subsidy to reopen their services, and in 1923 they were merged to form Imperial Air Transport (Imperial Airways in 1924). Initially there were some 1,760 mi. (2,820 km) of routes and 18 aircraft, but by 1927 a new service to Cairo was opened with de Havilland Hercules aircraft, and two years later a route was inaugurated through Basra to Karachi from Paris, Basel, Genoa, and Alexandria (journey time seven days). In 1933 further route extensions reached Calcutta, Rangoon, and Singapore, with QANTAS initiating a connection between Singapore and Brisbane at the end of 1934.

Meanwhile services from London to Mwanza, in East Africa, had begun in February 1931, using Calcutta flying boats from Cairo southward, and in April 1932 there was a passenger service to Cape Town (10½ days' flying). In 1936 a route across Africa opened from Khartoum in the Sudan to Kano in Nigeria.

As a result of the empire airmail scheme of 1934, which dispensed with any airmail surcharge, Imperial Airways had by 1938 become the world's largest carrier of nondomestic mail. This enabled 28 Short C-class flying boats to be acquired for the main routes to Australia and South Africa. A special long-range C-class boat also inaugurated British North Atlantic airline services in 1937, with regular mail flights in 1939 that employed in-flight refueling. In November of that year Imperial Airways merged with British Airways (which had been expanding its routes across Europe since 1935) to form the British Overseas Airways Corporation.

During World War II many services were still maintained, including routes to Sweden (flown by Mosquitos), over the North Atlantic (Liberators returning ferry pilots to the United States) and to West Africa (Boeing 314 flying boats). After the war the England–Australia route was reorganized using Lancastrians; Yorks took over the South African service; and Haltons flew to West Africa. In 1946 Constellations were assigned to the New York route, and reequipment led to the introduction of Argonauts and Stratocruisers in 1949 and the Hermes in 1950. Comets appeared on the South African run in 1952 but were withdrawn in 1954 following the loss of several aircraft. Delays in the delivery of Britannia turboprops resulted in the purchase of DC-7Cs for the New York route, which was extended to San Francisco in March 1957. Redesigned Comets began the first North Atlantic jet service in October 1958, but from May 1960 onward they were gradually replaced by Boeing 707s.

In the mid-1960s VC-10s were introduced on both African and North Atlantic services, and 707s opened routes from London to Osaka via Anchorage and to Tokyo via Moscow. Boeing 747 flights began in 1971 (to New York), and in 1974 BEA merged with BOAC to form British Airways. Concordes began operating with the new airline in 1976. British Airways now carries some 13,800,000 passengers annually.

Brumowski, Godwin (1889–1937)

Leading Austrian fighter ace of World War I, credited with 40 victories by official records (35 according to other sources). A professional soldier, Brumowski served on the eastern front until 1915, when he was transferred to the air force as an observer. After qualifying as a pilot he was given command of a flying unit which served against the Italians, and which he tried to remodel on the lines of the German "flying circuses" he had seen when visiting the western front. Lack of official interest and shortage of airplanes largely frustrated his efforts, although at one time he led a formation of 18 aircraft, and Austro-German squadrons enjoyed a measure of air supremacy over the Italian front in the winter of 1917–18.

Brumowski achieved his early successes with a Brandenburg D.I, but used an ALBATROS D.III from mid-1917 onward.

Buffeting

Turbulent airflow impinging on any part of an aircraft structure causes the vibration known as buffeting. The causes are varied. With fixed-wing machines the approach to the stall is, or should be, marked by the onset of local turbulence, usually severe enough for the pilot to notice. Such buffeting can effect the ailerons or tail controls, so that the forces are felt directly on the pilot's controls if they are not power-assisted. With early jets it was not uncommon for the speed of sound actually to be reached at certain local points of low-pressure airflow (over the wing or canopy, or around the tail), and since designers had not foreseen such problems, buffeting resulted. With the X-1 buffeting was so severe that the pilot could barely retain control (see SUPERSONIC FLIGHT). Other causes of buffeting include gross changes in configuration or shape while in flight (opening rear freight doors, lowering landing gear, etc.), and the effect on an aircraft of high-frequency vibrations caused by its jet engines, which can lead to structural failure through metal fatigue.

Nearly all modern aircraft, especially jets, have suffered from buffeting problems at some time in their life. The flaps of B-52 Stratofortresses, for example, cracked and broke under the sonic buffeting from the engines at wet takeoff thrust. The VC-10's engine pods had to be completely reprofiled and moved further from the fuselage, and the Belfast military freighter needed to have its whole rear fuselage redesigned. The A-4 Skyhawk suffered severe tail buffeting problems that were cured by "temporary quick fixes," still in use more than 20 years later.

Business aviation

Although a few (generally very large) corporations owned their own aircraft during the 1930s, business aviation did not really become established until after World War II, when surplus military aircraft were rebuilt and internally restyled. Typical of such early business planes were the Douglas DC-3, Douglas A-26 Invader (rebuilt as the On Mark Marksman or Smith Tempo II), Lockheed Ventura (rebuilt as the Howard Super Ventura), and Twin Beech (in several popular rebuilds). The de Havilland Dove (1945) proved very successful, and the later Vickers Viscount found wide acceptance among business customers, even though it was a relatively large aircraft for the time, able to seat up to 80 passengers, and powered by four turboprops.

By 1960 Lockheed had sold the Electra turboprop, an even larger aircraft, to business customers, and built the first executive jet, the JetStar. This was slow to attract sales, but remained in production and was fitted with Garrett turbofan engines. Dassault in France built the Mystère 20; marketed as the Fanjet Falcon, it has sold well in many versions. Hawker Siddeley in Britain produced the popular HS 125; North American (now Rockwell) the Sabreliner; Lear (now Gates) the slim Learjet; Piaggio in Italy the PD-808; and Germany the HFB 320 Hansa. The American manufacturer Grumman built the big twin-turboprop Gulfstream, followed by the twin-jet Gulfstream II. Executive versions of conventional airliners are still popular (Fokker F-27, Hawker Siddeley 748, Caravelle, BAC 1-11, and DC-9), and with secondhand machines increasingly available, there are now many business-owned Boeing 707s and 727s, Convair 880s and 990s, and DC-8s.

The bulk of the market consists, however, of a large variety of light twins and twin-turboprobs such as the Swearingen Merlin, Mitsubishi Mu-2, Beechcraft Queen Air and King Air, and Cessna 441. One of the best-selling jets is the Cessna Citation 500, along with the larger long-range Citation 700 trijet.

Rotorcraft also play a major role in business aviation. A best-seller among executive helicopters is the Bell Jet Ranger series, but many other types are supplied by Bell, Hiller, Hughes, Aérospatiale, Enstrom, and other builders. To meet the need for all-weather reliability there are twin-engined helicopters such as the Bo 105, Westland 606, Agusta 109, and Vertol 179, together with twinned versions of earlier types.

Byrd, Richard Evelyn (1888–1957)

Pioneer American aviator, Byrd was the first man to fly over both the North and South Poles. On May 9, 1926, he and Floyd BENNETT accomplished the first flight to the North Pole (from Spitsbergen) in a Fokker F.VII-3m. The following year Byrd, with Bert Acosta, Bernt Balchen, and George O. Noville, flew the F.VII-3m *America* from Roosevelt Field, New York, to France (June 29–July 1, 42 hours' flying time), where fog and rain made a landing at Le Bourget impossible and forced them to ditch in the sea off Ver-sur-Mer in Normandy.

On November 28–29, 1929, Byrd, accompanied by A. C. McKinley, Balchen, and Harold June, flew the

Ford Trimotor *Floyd Bennett* to the South Pole from Little America, on the Bay of Whales. During a second expedition to the Antarctic, Byrd made five flights in 1934 from Little America over the Ross Ice Shelf and Marie Byrd Land.

C

C.5A. See GALAXY

CAA

The administration of British civil flying is the responsibility of the Civil Aviation Authority. This government body was established in 1971 and is responsible for issuing certificates of airworthiness and operating the National Air Traffic Services; it also undertakes research in connection with civil aviation operations, approves fares, licenses operators and advises on airport planning.

Camel

The first British fighter designed from the outset to carry twin synchronized machine guns, the Sopwith Camel biplane was also probably the most maneuverable aircraft to see combat in World War I. The first prototype was flown late in December 1916, and production machines began reaching France the following June. The 2F.1 variant—officially called the Ships' Camel—was carried by many Royal Navy vessels, and at least one Zeppelin (L.53) was destroyed by a 2F.1 taking off from a wood platform towed behind a naval destroyer. By the end of the war, Camel pilots had claimed nearly 3,000 victories, and in August 1918 almost half the RAF's front-line units were Camel-equipped squadrons.

With a 110-hp Rhône or 130-hp Clerget rotary engine, the Camel's top speed was 118 mph (189 km/h) and its ceiling 24,000 ft. (7,300 m). It had a wingspan of 28 ft. (8.53 m) and length of 18 ft. 8 in. (5.69 m).

Camm, Sir Sydney (1895–1966)

British aircraft designer, renowned for numerous highly successful fighter designs over a period of nearly half a century, Camm began his career by joining the Martynside Company during World War I. Between the wars he was responsible for Hawker (later HAWKER SIDDELEY) military aircraft, among them the Hart single-engined biplane bomber, the FURY biplane fighter, and the HURRICANE fighter of Battle of Britain fame. Camm's subsequent fighters in-

Sopwith Camel F.1

cluded the Typhoon, the TEMPEST, the Sea FURY naval fighter, the Sea Hawk jet fighter, and the HUNTER, one of the most successful of all European postwar combat aircraft. His last design was the P.1127 jet-lift V/STOL research airplane, flown in 1960. From this was developed the Kestrel, which in turn led to the remarkable HARRIER jump jet.

Canard

Term loosely describing a "tail-first" airplane, from the French word for "duck." More accurately, it refers to an aircraft in which the fin (vertical tail) is located at the back, but in which the stabilizer (tailplane, elevator, or slab "taileron") is at the front, where it is called a foreplane. The original Wright FLYER was a canard, as were many other pioneer flying machines. Gradually, however, the conventional rear-tailed layout became so universal that a canard, such as the Focke-Wulf Ente (Duck) of 1929, came to be regarded as a freak. During World War II a few machines in this class were built, including the Miles M.35 and M.39B Libellula and the Curtiss XP-55 Ascender fighter of 1944.

Modern examples of canards include the XB-70 Valkyrie Mach-3 bomber, the Saab-37 Viggen Mach-2 fighter, and the Rutan Vari-Viggen lightplane. The Soviet TU-144 supersonic transport is equipped with retractable canard surfaces, as is the Dassault Mirage-Milan

The retractable foreplanes of the Soviet Tu-144 supersonic transport, seen here extended on landing approach, provide a canard-type layout for low-speed flight

fighter. In such cases the foreplane is folded away at high speeds but at low speeds the additional lift it supplies holds up the nose, so that instead of the wing controls (ELEVONS) pushing the rear down they can be used to provide lift.

Canberra

The first British jet bomber to take to the air (May 13, 1949), the English Electric Canberra twin-jet medium bomber entered RAF service in 1951 as the B2, and was employed as a first-line combat aircraft until 1972. The B2 had a wingspan of 63 ft. 11½ in. (19.49 m) and length of 65 ft. 6 in. (19.96 m). Later Canberra variants included the B6 and the B(I)8 attack-bomber (with offset cockpit canopy and a ventral cannon gun pack). Photo-reconnaissance versions included the PR3, PR7, and PR9.

More than 400 Canberras were license-built in America as Martin B-57s. Some were attack-bombers (B-57A and B); others were specially modified high-altitude reconnaissance aircraft (RB-57D and RB-57F). The RB-57F had a wingspan of 122 ft. (37.18 m) and a reputed ceiling approaching 100,000 ft. (30,500 m).

Canberras manufactured in Britain numbered about 1,000, with a further 48 constructed in Australia. Canberras served with the air arms of more than a dozen nations.

Caproni

Count Giovanni Caproni (1886–1957) was founder of one of the earliest Italian aircraft manufacturers, and was himself a pioneering theorist of air power, friend of Giulio DOUHET and ad-

Caproni-Campini CC2

viser to the U.S. military. During World War I Caproni built a large number of strategic bombers used by many of the Allies. These aircraft all stemmed from the Ca 30 of 1913, a biplane with two tractor Gnôme engines and a third Gnôme arranged as a pusher at the rear of the nacelle, the tail being carried on long struts. Several hundred examples of Ca 30 developments were delivered. These included the Ca 33 and 36, with more powerful engines, a crew of three or four, several machine guns, and a bomb load of over 1,000 lb. (450 kg). Caproni's largest World War I bombers evolved from the Ca 40, with triplane wings spanning almost 100 ft. (30.5 m), a short nacelle with a pusher engine such as a 200-hp Isotta-Fraschini, and two more engines as tractors at the front of long tail booms. Other features were multiwheel landing gear and a bomb load of up to 3,910 lb. (1,770 kg) carried in a box on the bottom wing. Versions with more powerful engines became postwar passenger transports. Another group of wartime Caproni aircraft included the Ca 44 to 47 series of biplanes, many of which were made in the United States by Standard and Fisher Body.

Between the wars Caproni manufactured large bombers and passenger aircraft, together with a wide range of reconnaissance, ambulance, and utility machines, all of which served in considerable numbers in support of Mussolini's campaigns in Ethiopia and Albania. Caproni was also responsible for several outstanding technical achievements. The Ca 161 biplane, powered by an Italian-built Bristol Pegasus engine, was constructed specifically to set a new world altitude record and reached 56,046 ft. (17,083 m) in 1938—still a world record for piston-engined aircraft. In August 1940 the unusual Caproni-Campini CC2 made its first flight. This aircraft had a piston engine driving a compressor, which expelled a jet of air (in which extra fuel was burned) from a nozzle in the tail: it was an ingenious idea but not a true jet. Of many Caproni World War II models only the Ca 313 twin-engined reconnaissance bomber was made in large numbers. After the war the company flew a twin-pusher cabin machine the Ca 193 (1949), but went bankrupt in 1950.

Caravelle

A highly successful French-built medium-range airliner, the Caravelle was one of the first commercial aircraft to place its two jet engines at the tail. The Sud-Aviation prototype made its maiden flight on May 27, 1955, and the first of 19 80-passenger Caravelle Is entered service with Air France and SAS in May 1959. All these were rebuilt as Caravelle IIIs (with more powerful engines) and took their place alongside 78 newly-built Caravelle IIIs. Subsequent variants included 53 VINs (with Avon 531 engines and noise suppressors), 56 VIRs (with Avon 533s and thrust-reversers), and 22 104-passenger Super Bs of 1964 ("Super Caravelles," with longer fuselages, aerodynamic refinements, and turbofans). The Super B had a wingspan of 112 ft. 6 in. (34.29 m) and length of 108 ft. 4 in. (33.02 m). The

Canberra B(I)8

10R of 1965 was similar to the VIR but was powered by turbofans; the 11R of 1967 featured a stretched fuselage and forward freight door, and the 12—the final version—had a fuselage 118 ft. 10½ in. (36.23 m) long accommodating up to 140 passengers. Caravelles are in service with airlines throughout the world.

Catalina

An all-metal monoplane powered by two Pratt & Whitney R-1830 Wasp engines, the Consolidated Catalina was a large ocean patrol flying boat with retractable wing-tip floats. The prototype appeared in 1935, and prewar production for U.S. Navy patrol squadrons ran to 209 aircraft known under the designation PBY. The Catalina had a maximum speed of 196 mph (315 km/h); its wingspan was 104 ft. (31.70 m) and length 63 ft. 10 in. (19.20 m).

Manufactured in greater numbers than any other World War II flying boat, Catalinas served throughout the world. The PBY-5, produced between 1940 and 1944, was supplied to the Allies in substantial numbers. RAF Coastal Command, for example, employed Catalinas on convoy protection patrols from early 1941 onward. Catalinas were also built under license by the Soviet Union.

The PBY-5A was an amphibious version used extensively for air-sea rescue duties. Four additional factories (two in Canada) manufactured Catalinas under PB2B, PBN, PBV, PBY-6A, and 0A-10A designations, the last two of these being amphibians. Excluding Russian production, a total of 3,074 Catalinas were delivered, and a few were still in use with the air arms of some smaller nations in the 1970s.

Catapult

From the earliest days of aviation catapults have been used to assist takeoff. LANGLEY's ill-fated *Aerodrome* of 1903 fouled its launching mechanism on the roof of a houseboat on the Potomac; more successfully the WRIGHT BROTHERS frequently used a falling weight attached to a cable to speed their early biplanes along the ground. Much later, between the wars, gliders were often sent off by rubber catapults to give maximum acceleration.

Today most catapults are used to launch planes from AIRCRAFT CARRIERS, but until after World War II it was still common for aircraft, usually seaplanes, to be carried aboard ordinary surface ships and launched by catapult. Ocean-liners carried fast mailplanes; some

Catalina amphibian in RAF service

Do 18 Lufthansa mailplane catapulted from seaplane tender in mid-Atlantic

British merchant ships were equipped with defending fighters; large warships launched observation and scouting machines which were sometimes armed with bombs or a torpedo. Most of these catapults could swivel to point into the wind, and launched the aircraft by machinery working from hydraulic or compressed-air sources of power.

Aircraft carriers are now all equipped with "cats," and today's largest carriers each have four or five. The first catapults on carriers were hydraulic, and were subject to explosions and fires. In 1949 the Royal Navy introduced catapults powered by steam from the ship's own power plant; steam cats are now universal. Before each shot they are preset to accelerate aircraft of any selected weight to the appropriate speed, and can sustain a high rate of launching as long as necessary.

Caudron

The Caudron Brothers Gaston (1882–1915) and René (1884–1959) were among

the pioneers of the French aviation industry. By 1912 they had produced the G.II (G = Gaston), which was sold to many early aviators. Although the tail was carried on a network of struts, the engine (an 80-hp Gnôme) was on the nose of the nacelle, driving a tractor propeller. From the G.II was developed the G.III of World War I. Though slow (72 mph, 115 km/h), it was strong and reliable, and was widely used by the Allies for reconnaissance and training. The twin-engined G.IV, with a 220-lb. (100-kg) bomb load and speed of 82 mph (131 km/h), was also used by many Allied air arms, and René followed this in June 1915 with the R.4 twin-engined three-seater bomber. The R.11, with two 220-hp Hispano engines and capable of 114 mph (183 km/h), was used to escort French bombers and led to the big C.23 night bomber.

Caudron later produced many transports and lightplanes, and after constructing racers for the Coupe Deutsch competition (which handicapped entrants on a basis of engine size) built

Caudron G.IV

the remarkably light C.714 fighter of 1936, which had a wingspan of barely 29 ft. (8.8 m) and laden weight of 3,925 lb. (1,778 kg), and could reach a speed of 302 mph (484 km/h) on only 450 hp from a six-cylinder air-cooled Renault RE12 engine. This was in service in France and Finland by 1940.

The company became linked with Renault as Avions Caudron-Renault, and its last prewar product was the Goeland light twin-engined transport.

Cayley, Sir George (1773–1857)

A British pioneer of the science of aeronautics, Cayley was inspired as a boy by the balloon flights of the MONTGOLFIER BROTHERS and devoted much of his life to the problems of aerial navigation, calculating the lift, thrust, and drag of many potential designs. He was also the first scientist to point out that controlled flight would be impossible until a sufficiently light-weight power source could be invented, and actually suggested the use of an internal combustion engine.

Discarding the then-popular idea of imitating bird flight by flapping wings, Cayley designed a theoretical flying machine which closely resembled to-day's airplane in its essentials. By 1804 he had designed a model glider, the first of several monoplanes with a fixed wing and a fuselage terminating in vertical and horizontal surfaces. Toward the end of his life, in 1852–53, he produced a glider large enough to carry a person. The first manned glider flight was made by a boy, probably the son of one of his servants. Cayley had a team of men run down a slope towing the glider until it became airborne. He repeated the experiment with his coachman, John Appleby.

Cayley also devoted time to the problems of lighter-than-air flight, and was a pioneer of airship theory. He is known to have carried out experiments by whirling a flat plate around on the end of an arm to find the drag and the lift it produced, and from this experience he evolved the idea of curved wing sections and the principle of the airfoil's operation.

Ceiling

The maximum height to which a given aircraft can climb is known as its ceiling. It is determined by the ability of the wings to provide lift in the rarefied upper air, by the ability of the engine to function at heights at which oxygen is considerably less abundant, and—in piston engines and turboprops —by the amount of "bite" a propeller can obtain from the thin atmosphere at high altitudes.

The limitations of the propeller do not apply to aircraft with jet engines or rocket motors. Jet aircraft can operate at up to 100,000 ft. (30,000 m) if they are designed for high-perform-ance flight. Piston-engined aircraft will reach about 50,000 ft. (15,000 m) with the use of a SUPERCHARGER. Unlike jets

and piston engines, rocket motors do not require atmospheric oxygen, and rocket-powered craft can operate in the near-vacuum of space.

What is known as the *service ceiling* is the greatest altitude at which a rate of climb of 100 ft. (30.5 m) per minute can be maintained.

Cessna

Founded by the pioneer American aviator Clyde V. Cessna, who had been building airplanes since 1911, Cessna Aircraft was established in September 1927. It is now one of the world's largest builders of light aircraft and by 1976 had constructed over 140,000 airplanes. Its aircraft production is centered in Wichita, Kansas, though a small pro-portion of Cessnas are built by a subsidiary, Reims Aviation of France.

In the years before World War II Cessna established its reputation with a family of light high-wing cabin mono-planes known as Airmasters. Most were four-seaters of extremely clean design, with no bracing struts, the usual engine being the 145-hp Warner Super Scarab. A short time before the outbreak of war, the T-50 appeared. A twin-engined transport seating five or six, it became Cessna's principal wartime product designated the AT-17 as a trainer, or the UC-78 Bobcat, and was manufactured in the thousands.

Postwar production concentrated initially on single-engined, high-wing designs. The major models were the two-seat 140, the four-seat 170, and the radial-engined four/five-seat 190. In 1953 the Model 310 was introduced and marked the beginning of Cessna's series of twin-engined aircraft. Over these same years the company's single-engined models acquired tricycle and retractable landing gear. Neither the Cessna 620 four-engined transport nor the CH-1 helicopter entered production, which re-mains concentrated on small fixed-wing

Cessna 150

airplanes. Major products in the 1950s included the L-19 (O-1) Bird Dog military light aircraft, T-37A twin-jet trainer (Cessna's first jet aircraft), and the single-engined 170, 172, 175, 180 and 182.

Many of these aircraft remained in production in the 1970s, together with the twin-boom push-pull tandem engined 336 Skymaster and 337 Super Skymaster (used, as the O-2, as a Forward Air Control airplane in Vietnam), the large twin 340 with pressurized cabin, 402, 414 and 421 Golden Eagle, and the twin-fanjet Citation 500. The Cessna aircraft built in the largest numbers over the years have been the 172/Skyhawk family (about 25,000) and the basic 150, the least expensive model in the range (also about 25,000).

Newer products include the retractable-gear Cardinal RG four-seater, the low-wing AGwagon and related agricultural models, the A-37 Dragonfly light attack twin-jet, the luxurious 441 twin-turboprop, and the Citation 700 long-range business trijet. Cessna's annual production is about 8,000 airplanes.

CH-47. See CHINOOK

Chanute, Octave (1832–1910)

Chanute was one of the pioneer builders of gliders and one of the most important influences on the Wright Brothers. A French-born American railroad engineer, he was attracted by the work of Otto LILIENTHAL on man-carrying gliders and began experiments of his own in the mid-1890s. He soon concluded that control in flight could best be achieved by wing-warping rather than by the aviator shifting his own weight (as in Lilienthal's gliders). In 1896 Chanute built a series of five gliders (at least one a biplane), and in 1896–97 he and his associates made some 2,000 flights on the sand dunes along Lake Michigan. Chanute's book *Progress in Flying Machines* (1894) was highly influential, and he was in constant correspondence with the Wright Brothers when they were making their first experiments with gliders, even making several trips himself to Kitty Hawk. Chanute's visit to Europe in 1903, when he lectured on the progress he and the Wrights had made in glider design, was of considerable importance in the renewal of European interest in heavier-than-air flight.

Charger. See TU-144

Charles, Jacques Alexandre César (1746–1823)

The builder of the first hydrogen-filled balloon, Jacques Charles had been inspired by the hot-air balloon experiments of the MONTGOLFIER BROTHERS in 1783. He joined with the Robert Brothers in performing similar experiments, but using hydrogen. They sent up the first free hydrogen balloon, the *Charlière*, on August 27, 1783. It drifted for 45 minutes and came down at Gonesse, 15 mi. (24 km) from Paris, where it was attacked by terrified peasants. On December 1, 1783, Charles and a companion made the first manned flight in one of their balloons. They ascended in a balloon 27 ft. 6 in. (8.38 m) in diameter and remarkably like today's balloons in appearance, and drifted some 27 mi. (43.4 km).

Charter flying

Flying in which aircraft and crew are hired out to a customer is almost as old as commercial aviation itself. Charter flights operate over routes not served by a scheduled operator, fly at special times (in conditions of urgency), carry loads that normal scheduled operators cannot handle, and generally offer significantly cheaper fares than scheduled services, subject to certain conditions. Aircraft themselves may be leased to other operators—a "dry lease" being the aircraft alone and a "wet lease" including the crew. There are also "part charters," in which a customer hires only part of the capacity of an aircraft (possibly on a scheduled flight); inclusive tour charters (ITCs), which are low-fare passenger flights carrying vacationers who pay a flat price for the entire holiday; and affinity-group charters, which carry passengers belonging to an organization or club with enough members to hire the entire airplane.

An important class of charter business deals only with freight. Cargoes range from perishables to electronic goods, live animals, and special items which for reasons of security, their hazardous nature, or any other cause are unsuitable for transport on scheduled flights. They also include spare parts for almost every kind of machine and vehicle, rushed out to any place in the world if the need is urgent enough.

Charter flying began in earnest in the years after World War I, when there were many surplus military aircraft and veteran pilots who were eager to find a living in civil aviation. By 1928 various "air taxi" operators were in business in many countries—their clients included governments, police, newspapers, prospectors, and political cam-

paign managers.

After World War II the increased popularity of air travel brought new business to charter and scheduled services alike. Small airlines expanded their charter activities by entering contracts to carry troops and military freight, and this added business enabled them to lower their passenger fares (particularly on tourist flights). Their success led the big scheduled operators to introduce cheaper tourist and economy fares—which in turn initiated a temporary setback for charter flying.

Several supplemental (nonscheduled) airlines collapsed, and many others combined to form larger and more viable units. In the United States they concentrated on developing the affinity-group charter business, while European charter airlines promoted inclusive-tour vacations. On both continents expansion was rapid, and by the mid-1960s charter passengers often traveled in aircraft as comfortable and up to date as those used on scheduled flights.

Despite expansion of the network of scheduled airlines, the role of charter flying has continued to grow. Scheduled airlines now frequently offer block-bookings on their own regular flights as an answer to low-cost charters, and also operate charter subsidiaries themselves. By 1971 32 percent of world passenger travel was on chartered flights; three years later the figure had risen to 64 percent.

Chinese People's Republic Air Force

Since the split between China and Russia early in the 1960s, the Chinese defense establishment has greatly accelerated its plans to achieve self-sufficiency in all armaments. But even today most of the military equipment used by the PRC is still of Soviet design, though modified, made, and supported entirely in China. The national aerospace industry is centered on an industrial complex at Shenyang (Mukden, in what was formerly Manchuria).

The People's Liberation Army Air Force and the PRC Naval Air Force are independent of each other; the Air Force is much the larger. Eventually, the chief combat type of both forces will be the F-9, a multirole aircraft, originally based on the MiG-19. It resembles the American F-4 Phantom, and is powered by two afterburning Chinese-built Rolls-Royce Spey turbofans. The F-9 can carry area-coverage reconnaissance pods and considerable external stores; it also has an internal gun. Multimode radar fills the large pointed nose.

Before delivery of the F-9, the most

U.S. Army CH-47C Chinook

important combat machine in the PRC was the F-6, also based on the MiG-19. This aircraft went into production in China in 1960, and advanced versions were still being manufactured in the mid-1970s. Production of the F-8, the Chinese-produced MiG-21, ceased by 1971.

The PRC manufactured hundreds of Il-28 light bomber/reconnaissance aircraft and Il-28U trainers, as well as producing a small number of Tu-16 Badgers. The most common helicopter is the Russian-designed Mi-4, but it is being supplemented by an improved, Chinese-built version. France has supplied China with Aérospatiale Super Frelon heavy helicopters, and Britain has licensed manufacture of the Spey engine. Military aircraft are also involved in such activities as geophysical prospecting, mapping, coastal patrol, and sea, air, and mountain rescue.

Chinook

The twin-rotor Boeing Vertol CH-47 Chinook is the U.S. Army's standard medium-lift assault helicopter. It can carry 33 to 44 fully equipped troops into battle, or lift up to 13,450 lb. (6,000 kg) of cargo. Together with the Bell UH-1 "Huey," the Chinook undertook most of the heavy transport helicopter operations in Vietnam; deliveries to the U.S. Army had begun in August 1962. The Chinook proved an oustandingly successful helicopter. With their great weight-lifting capabilities Chinooks were able to carry back to base for repair more than 12,000 aircraft (including other helicopters) that had been brought down in the difficult terrain of Vietnam, and which would otherwise have had to be abandoned. Loads such as these, too bulky

to be accommodated inside the fuselage, were suspended beneath the helicopter; vehicles and howitzers were carried in the same way.

The Chinook is likely to remain in service for many years, since plans to build a much larger HLH (Heavy Lift Helicopter) were shelved indefinitely in 1976. The HLH was also to have been built by Boeing's Vertol division.

Cierva, Juan de la (1895–1936)

The designer of the AUTOGIRO or gyroplane, Cierva was a Spanish inventor who began his work with a search for a wing system impossible to stall on takeoff or landing. Cierva's work resulted in the flight of the world's first successful autogiro (his own trade name) on January 9, 1923. Instead of using a pair of conventional wings, the autogiro achieved its lift from a rotor (functioning as a set of rotating wings), which spins freely in flight, turned by the aerodynamic action of the slipstream (the principle of autorotation).

The autogiro could take-off and land within extremely confined spaces at slow speeds, although it was incapable of true vertical flight or motionless hovering. Until the revival of interest in the helicopter it seemed the best way to achieve V/STOL flight, and Cierva set up companies or licensees in numerous countries. His designs were manufactured by his own (British) company, the English firms Westland and Avro, and Pitcairn and Kellett in the United States.

Clostermann, Pierre Henri (1922–   )

Highest-scoring fighter ace of French nationality in World War II, Closter-

mann served with the RAF during the war, and his 33 aerial victories make him third-ranking RAF ace. He joined the RAF 341 (Alsace) Squadron in January 1943 as a sergeant pilot. Flying SPITFIRE IXs with this unit, he shot down two FOCKE-WULF 190s on July 27, 1943, for his first victories. After two more successes he was commissioned and assigned to 602 Squadron, serving in the Orkneys during early 1944, and then bombing German rocket sites immediately before the Allied landing in Normandy in June 1944. Based near the invasion beach-head soon after D-Day, Clostermann shot down five enemy aircraft in June and three more in July, after which he was awarded a Distinguished Flying Cross and rested from operations. He returned to combat with 122 (Tempest) Wing at Volkel in January 1945, flying initially with 274 Squadron. Clostermann was in continuous action until May 1945, claiming 21 more successes and bringing his war tally to 33.

Clouds

The two basic categories of clouds are stratiform, or sheet cloud, and cumuliform, or heaped cloud (caused by moisture condensing out of warm air rising as a result of CONVECTION initiated by the Sun's heat warming the ground surface). The prefix *strato-* is applied to cloud up to 8,000 ft. (2,400 m), *alto-* to cloud between 8,000 and 20,000 ft. (2,400–6,000 m), and *cirro-* to cloud above 20,000 ft. (6,000 m).

Cumuliform cloud is common at low and medium altitude during summer and usually signifies only some local instability; massive cumulonimbus formations indicate thunderstorms and are dangerously turbulent (see TURBULENCE).

An oncoming warm front is indicated first by high cirrus (mares' tails), then cirrostratus (composed of ice crystals) and altostratus. Nimbostratus is thick, dark, low-level raincloud. Warm fronts can be dangerous to light aircraft, causing them to ice up and become uncontrollable, or making the pilot lose his way (see ICING). Cumulonimbus are powerhouses in the sky, and even the largest aircraft stay clear of them. Violent ones can tear planes apart, or tip them into uncontrollable attitudes, and even less energetic clouds may have enough energy to damage aircraft by strong gusts. The electrical energy in lightning flashes can destroy equipment, and sometimes aircraft themselves.

Combat maneuvers

Combat flying had its birth in World War I. Tactics differed from unit to unit, and the Allies never adopted the German method of robbing regular fighter squadrons of their best pilots in order to build up "circuses," elite units that would overwhelm all opposition. But the objective of all pilots was to get on the enemy's tail and shoot him down with guns that were fixed to fire directly ahead. A common result was an orbiting ring of aircraft, each trying to turn more tightly in order to shoot down the plane in front. Fighter pilots learned to use full throttle, to employ harsh and extreme control movements, and to keep a sharp lookout to avoid being "jumped"—especially by an enemy in line with the Sun. As aircraft themselves became more agile, more complicated maneuvers became possible. Max IMMELMANN, for example, gave his name to a 180° turn achieved by half a loop followed by half a roll, but such stylized maneuvers were seldom possible in combat. Occasionally the behavior of the aircraft itself helped shape possible tactics; the CAMEL, for example, turned sluggishly to the left but very quickly to the right, the result of its rotary engine.

It quickly became apparent in the course of World War II that large, tightly-knit fighter formations could no longer be satisfactorily used in combat, and a loose formation such as the "finger-four," with four aircraft disposed like the tips of the outstretched fingers of a hand, became almost universal. Over Korea, two-aircraft sections were the standard combat units.

Major changes came in the early 1950s, when the U.S. Air Force introduced computer-controlled "collision course" interception, in which rockets or missiles were fired from the beam. A tendency to regard the traditional "curve of pursuit" and attack with fixed guns as obsolete led to a generation of fighters without fitted guns, but combat over Vietnam showed this to have been a mistake. Most fighters now carry a cannon, and are designed to have the maneuverability necessary for close-range dogfights of the traditional kind.

Comet

The first pure-jet airliner to enter commercial service was the four-engined medium-range de Havilland Model 106 Comet. The prototype made its first flight on July 27, 1949, and a scheduled BOAC service using Comet 1s opened between London and Johannes-

Comet 4C

Curtiss Commando in its postwar role as a freighter

burg in May 1952. A Far Eastern route was inaugurated in August of the same year. The Comet was powered by four Ghost engines of 5,050 lb. (2,290 kg) thrust; wingspan was 115 ft. (35.05 m) and length 93 ft. 1 in. (28.38 m). Fewer than 50 passengers could be carried.

Following the loss of three Comets between May 1953 and April 1954, all the aircraft were grounded and evidence of metal fatigue causing catastrophic pressurization failure was found. A limited number of Comet 1s were modified and subsequently used by the RAF.

The Comet 2, with uprated Avon engines and a longer fuselage, never entered commercial service because of the Comet 1 accidents but was used by the British Royal Air Force.

Only a single Comet 3 prototype was built, but the Comet 4 series, with four Avons of 10,500 lb. (4,750 kg) thrust, saw widespread use, inaugurating jet passenger service between London and New York (October 1958). The basic Comet 4, with leading-edge pannier fuel tanks, had seating for up to 81 passengers; the 4B and 4C had fuselages stretched to seat 102. Short-range clipped wings without pannier fuel tanks were employed on the 4B. Maximum speed was about 550 mph (885 km/h) and range up to 4,000 mi. (6,436 km).

The de Havilland 88 twin-engined racing monoplane (1934) also bore the name Comet. Five examples were built. One of them (G-ACSS) won first place in the MacRobertson race from England to Australia (October 1934) and subsequently established several long-distance records.

Commando

The Curtiss Commando was the military version of the Curtiss-Wright CW-20, a twin-engined 36-passenger civilian transport first flown in 1940. Production of the C-46, as it was designated in Army Air Force Service, only got into stride in 1944. Most of the 3,180 eventually built came from the Buffalo, N.Y., plant of Curtiss-Wright, but two other sources of production delivered some 300 Commandos before the end of hostilities. The Commando had a top speed of 241 mph (388 km/h). Its wingspan was 108 ft. 1 in. (32.94 m) and length 76 ft. 4 in. (23.26 m). The aircraft could carry 50 fully equipped troops or 8 tons of cargo in its bulbous fuselage.

Troop carrier units in the Pacific received C-46s late in 1944, and over 800 were on hand by August 1945. In Europe C-46s saw limited combat service, but played a major role carrying thousands of paratroops for the March

24, 1945, crossing of the Rhine near Wesel, the largest airborne assault of the war. The U.S. Air Force used a small force of C-46s during the Korean War (some were even recommissioned for use in Vietnam), and the aircraft remained on the strength of the air forces of the Dominican Republic, Honduras, Peru, Japan, and Taiwan into the 1960s. Many Commandos also found employment with civil operators, particularly as freighters.

Compass

The traditional magnetic compass embodies a needle which is attracted toward the north magnetic pole. To counteract capricious motion, the instrument is filled with a damping fluid and contains attached filaments that limit unwanted movement. All magnetic compasses are affected by surrounding metallic objects, and the inherent error (compass deviation) must be determined and noted on a compass correction card kept in the aircraft's cockpit.

In addition, when an airplane banks the needle is attracted toward the ground, so that pilots must allow for this incurred error when making turns. Acceleration and deceleration when on easterly or westerly headings also cause the needle to mis-read temporarily.

A gyrocompass relies on the Earth's rotation and the properties of gyroscopes to determine true north. But it too is affected by rapid turns, and is generally too heavy for aeronautical use. However a gyromagnetic compass is available, which incorporates a small gyroscope monitored by a powerful magnetic needle, and effectively smooths out most irregularities. In a radio compass an Automatic Direction Finder (ADF) detects the bearing of a known signal source and displays the aircraft's

bearing relative to that source on a compass dial. Modern practice is to combine the relative bearing with a signal from the gyromagnetic compass to give a magnetic bearing on a radio magnetic indicator (RMI). See also NAVIGATION; RADIO.

Concorde

Supersonic jet transport produced jointly by Aérospatiale of France and the British Aircraft Corporation. The first prototype flew at Toulouse, France, on March 2, 1969, with the second flying from Filton, England, shortly afterward. Two further pre-production examples incorporated design improvements, lengthened fuselage, and a revised nose shape. The first production Concorde flew on December 6, 1973. With a crew of three or four, and carrying up to 144 passengers, Concorde is 204 ft. (62.2 m) long and has a wingspan of 84 ft. (25.6 m). Four Rolls-Royce SNECMA Olympus turbojets give a cruising speed of 1,350 mph (2,170 km/h), and with maximum payload Concorde has a range of 4,000 mi. (6,440 km) at Mach 2. Concorde is delta-winged and droop-nosed at low speeds.

Concorde is designed to operate at altitudes of up to about 11 mi. (18 km). Problems of radiation and ozone at high altitude have been overcome by using solar-flare metering to advise pilots on safe height levels, and the pressurized cabin air system caters for the heat decomposition of ozone, with filters for dealing with any residue.

British Airways and Air France began operating Concordes in 1976, initially with a twice weekly service London to Bahrain by British Airways and Paris to Rio de Janeiro by Air France. Later that year both airlines

received trial permission for Concorde services to Washington, D.C., despite earlier regulations by the U.S. Federal Aviation Administration prohibiting SST flights over the United States or its territorial waters. The service was to be for a period of 16 months, with no supersonic flight over land, and no night takeoffs. The profitability of this highly expensive Anglo-French venture depends largely on permission to fly into New York, though other routes, for example, subsonic cross-country flights to the West Coast, have been discussed. Concorde's future has now been raised to a high-level political issue between the United States, Britain, and France. The principal environmental problems associated with Concorde—and other SSTs—are noise (Concorde is roughly four times as loud as a DC-10 at takeoff), and the possibility of damage to Earth's protective layer of ozone (see SST; SUPERSONIC FLIGHT).

Condor

The original Curtiss Model 18 Condor biplane airliner was a development of the 1927 XB-2 Condor bomber. The airliner made its maiden flight powered by two 600-hp Conqueror V-12 engines. It carried 18 passengers (in three cabins) and a crew of three; wingspan was 91 ft. 8 in. (27.94 m) and length 57 ft. 1 in. (17.40 m). The first Condors were initially operated by Eastern Air Transport and Transcontinental Air Transport.

Subsequent redesigning produced the T-32, based on a projected XT-32. First flown in 1933, the second-generation Condor resembled the Model 18 only superficially and had a single tail fin and 750-hp Cyclone engines. Twenty-one were built, followed in 1934 by the AT-32, with superchargers and variable-

BAC/Aérospatiale Concorde

pitch propellers. The AT-32 had a wing-span of 82 ft. (24.99 m) and length of 48 ft. 7 in. (14.80 m).

Users of the later Condors included Eastern Air Transport, American Airlines, and the Army Air Corps (for which the BT-32 bomber was evolved), but the monoplanes DC-2 and DC-3 quickly ousted the Condor biplane from major airlines.

Constellation

The Lockheed Constellation was one of the most important aircraft in the postwar recovery of commercial aviation. It owes its origins to TWA's specification for a new long-range four-engine airliner in 1939. The Lockheed Model 49 (later 049) was designed to fulfill these requirements, but in the meantime the United States had entered World War II and the new aircraft was developed as the C-69 military transport. It made its first flight in this role on January 9, 1943.

No C-69s reached the Army Air Force before the end of the war, and production was then diverted to civilian customers. With four 2,200-hp Wright R-3350 Cyclone engines, the L-049 was faster than the rival DC-4, had a greater range and payload, and incurred much lower operating costs. By 1946 this 60-seat aircraft was flying intercontinental routes with Pan American, TWA, Air France, KLM, and BOAC, as well as being used by a number of smaller airlines.

The next version of the Constellation, the L-649 of 1947 (with 2,500-hp engines), was sold only to Eastern Air Lines, but the long-range L-749, which was developed in parallel and used the same power plants, found an extensive market. In the first Super Constellation (L-1049) of 1950, the fuselage was stretched by 18 ft. (5.5 m); range was up to about 5,000 mi. (8,000 km) for the L-1049A. Super Constellations were extensively used by both the military (as C-121 transports and early warning aircraft) and by civil operators (as 100-seat airliners).

Turbocompound engines of 3,400 hp were fitted to the L-1049C (1953) and the L-1049G Super G (1954). Wingspan of the Super G was 123 ft. (37.49 m) and length 113 ft. 7 in. (34.62 m). The L-1049H was a freight version of the Super G. The final version of the Constellation series was the Starliner of 1956 (L-1649A), which had a wing-span increased to 150 ft. (45.72 m), cruised at 323 mph (520 km/h), and could range over 5,400 mi. (8,650 km) with a full payload. A total of 43 were built; operators included TWA, Air

Convair B-36, world's largest warplane

France, and Lufthansa.

The last Constellation sold from the factory to an airline was delivered in 1959 (an L-1049A to Slick Airways). A total of 856 Constellations of all types were manufactured (623 of them Super Constellations and L-1649As).

Controls

An airplane is controlled by means of ELEVATORS to govern flight in the looping plane, AILERONS to produce roll and bank about the longitudinal axis, and RUDDER(s) to control yaw. ELEVONS are a feature of DELTA-wing aircraft.

In HELICOPTERS, the basic controls consist of a collective pitch lever to effect forward, backward, or sideways movement by altering the cyclic pitch, and foot pedals to govern the pitch of the tail rotor blades for turns.

Convair

The origins of Convair go back to May 1923, when Maj. Reuben H. Fleet formed Consolidated Aircraft. The product that did most to establish Consolidated was the Husky, used as the principal primary trainer of the U.S. Army and Navy from 1925 to 1937. The XPY-1 of 1929 was the company's first large flying boat, and the Commodore civil version equipped NYRBA (New York/Rio/Buenos Aires), then the world's longest airline, which Fleet later sold to Pan Am. Commercial landplanes included the speedy Fleetster, and the company weathered the depression with the P2Y series of Navy flying boats, one squadron of which made the first-ever flight to Hawaii, taking 25 hours.

In October 1935 Consolidated moved from Buffalo to San Diego and went into production with the P-30 pursuit

and the first model of the highly successful CATALINA. In 1937 flight testing began on the PB2Y Coronado flying boat, followed in 1939 by the first B-24 LIBERATOR. In December 1941 Fleet sold Consolidated to Avco, the name Consolidated-Vultee Aircraft being adopted in 1943 to reflect the merger with Avco's aviation subsidiary VULTEE Aircraft, which itself included the STINSON company. During World War II, Consolidated-Vultee produced 33,000 aircraft, and in 1944 outproduced every other aircraft company in the world. In 1953 Convair, as the company had become known, was brought under the control of GENERAL DYNAMICS, and its products diversified. The B-36 bomber, the largest combat aircraft of all time powered by six piston engines and four jets (designed to bomb Germany from the U.S.), was followed by the F-102 DELTA DAGGER, the F-106 Delta Dart, and the world's first supersonic bomber, the Mach-2 B-58 Hustler, which established 19 world speed, load, and altitude records. Convair civil aircraft included the outstandingly successful Model 240, 340 and 440 Convairliners (METROPOLITANS) and the Model 880 and 990 Coronado jetliners (which were technically excellent but lost Convair over $450 million).

The company is now the Convair Aerospace Division of General Dynamics.

Convection

Heating of the Earth's surface by the Sun results in secondary warming of the air in contact with the ground (virtually no heat is directly absorbed by the atmosphere from the Sun's rays as they pass through it).

Increasing the temperature of a mass of air causes it to expand, so that it becomes less dense and thus lighter (volume for volume) than the colder surrounding air. Consequently the warmer air will rise, to initiate convection.

If the ascending air does not contain any visible moisture in the form of clouds, fog, etc., its expansion will cause it to cool. Once the warm air has lost sufficient heat to equalize its temperature with that of the surrounding air, it will cease to rise. But if the air is very moist, some of the water vapor begins to condense out as cloud when the temperature has fallen to such an extent that the air becomes saturated. If the air is particularly wet, clouds form at low altitude, but if it is dry, then they will form higher up. Latent heat is released by condensation and this may maintain the temperature of the rising convection current above that of the surrounding air, thus permitting very high ascents for SAIL-PLANES and BALLOONS.

The atmosphere itself normally cools at an average rate of 2°C per 1,000 ft. until a figure of about −67°F (−56°C) is reached at the tropopause. This rate varies according to local conditions and may even become positive, leading to an inversion. If this situation occurs, convection cannot take place: instead of decreasing, the ambient air temperature increases with height and quickly exceeds the temperature of any air seeking to ascend from the ground. The result is very stable, almost windless conditions.

Certain types of terrain heat up particularly quickly under the influence of the Sun and generate strong convection currents. These include such surfaces as bare rock, dry soil, masonry, road surfaces, and ripe crops. Conversely, heat is only absorbed slowly by water (seas, lakes, rivers), marshes, pasture, and woodland. Warm air rising from a rocky promontory will cause cool air to rush in from a surrounding wooded area to take its place, with cold air descending in turn over the woodlands, thus establishing a circulating convection current.

The warm rising air in convection currents gives rise to bumpy flying conditions and downdrafts known as "air pockets."

Convertiplanes

Convertiplanes are experimental aircraft intended to combine the conventional airplane's speed in level flight with the vertical takeoff and landing capabilities of the helicopter. Unlike other VTOL aircraft such as the HARRIER, which uses vectored thrust

LTV-Hiller-Ryan XC-142A Convertiplane

U.S. Aircraft carrier Lexington *ablaze after Japanese air strikes during the Battle of the Coral Sea*

from its jet engine to take-off, or a number of "vertiplanes" of the 1950s, which took off vertically from their tails, convertiplanes actually operate in two distinct modes, as machines with helicopter capabilities using rotors, fans, or jets, and as conventional aircraft.

One of the earliest convertiplanes was the McDonnell XV-1, with a propeller driven by a piston engine and rotors powered by tip-mounted jets. The first vertical flight of the XV-1 took place in April 1955, but the performance of such compound aircraft suffered from the added weight created by the wings in takeoff, and from the additional drag imposed by the rotors in ordinary flight. The first flights of the Bell XV-3 in August 1955 represented a new devlopment. The XV-3 had wingtip-mounted rotors that could tilt to act as propellers in level flight. This design concept was pursued by several firms, notably Hiller and Vertol, during 1957–58. A more ambitious scheme sponsored by NASA led to the

LTV-Hiller-Ryan XC-142A turboprop transport, which had wings that tilted through 90°, and thus converted itself from an airplane to a helicopter by transforming its propellers into rotors. The Canadian government later sponsored the Canadair CL-84 tilt-wing transport (1970), but no convertiplane has so far advanced beyond the trial stage.

Coral Sea, Battle of the

The first major naval battle in which opposing fleets were not involved in surface action and the outcome was decided by carrier-based strike aircraft. In the early spring of 1942 the Japanese planned to secure the southern limits of their empire by driving the Allies from their stronghold at Port Moresby in eastern New Guinea and taking the Solomon Islands 800 mi. (1,300 km) to the east. They began this campaign in May by occupying Florida Island in the Solomons, while another force of

12 transports, four cruisers, a carrier, and other support vessels set off for Port Moresby.

Allied intelligence was alerted to the threat, and Task Force 17, with carriers *Lexington* and *Yorktown* as major elements, was brought up from the south. On May 4 *Yorktown*'s SBD DAUNTLESS and TBD Devastator aircraft dive-bombed and torpedoed support vessels at Florida Island, hitting several and sinking a destroyer. On May 7 U.S. reconnaissance aircraft spotted the Moresby force and the carrier *Shoho*. The U.S. carriers were near enough to send out their Devastators and Dauntlesses with an F4F WILDCAT escort and were successful in sinking the *Shoho* with most of her aircraft still on board. American losses were three Dauntlesses and three Wildcats.

In the meantime the main Japanese support fleet had come into the Coral Sea south of San Cristobal Island in the Solomons, to prevent U.S. naval interference with the invasion forces. On May 8 the opposing carriers were in air range, and strikes were launched by both sides. Heavy clouds made it difficult to locate the Japanese ships, and the attacks by the SBDs and TBDs only resulted in three bomb hits on the carrier *Shokaku*. The cost was six of the U.S. strike aircraft and five fighters, chiefly to Zeros. The Japanese attack was more successful, and five hits on the *Lexington* made it necessary for her to be abandoned. With *Shokaku* badly damaged, the Japanese decided to withdraw their fleet and abandon the intended invasion by sea of Port Moresby.

Aircraft losses during the battle were 66 Allied and approximately 112 Japanese. While neither side was victorious in the battle, the U.S. Navy gave a major check to Japanese expansion in the Southwest Pacific.

Corsair

Several aircraft built by Vought have borne the name Corsair. The first of these was the O2U, which first flew in 1926. These single-engined biplanes, equipped with either a conventional undercarriage or floats, served as reconnaissance aircraft with the U.S. Navy. A later aircraft known as the Corsair was the Vought F4U, one of the best shipboard fighters of World War II. Its distinctive nose was of such length that its assignment to U.S. carrier squadrons was delayed because of anticipated landing difficulties.

First flown in 1940, and issued to Navy and Marine Corps units from July 1942 onward, the F4U-1's 18-cylinder Double Wasp 2,250-hp engine

gave this formidable fighter a maximum speed of 425 mph (684 km/h) at 20,000 ft. (6,100 m). Later Corsairs (F4U-4s), with 2,450-hp engines, were capable of 446 mph (717 km/h) and could claim a performance superior to that of the MUSTANG above 12,000 ft. (3,660 m). Wingspan of the F4U-4 was 42 ft. (12.80 m) and length 33 ft. 8 in. (10.16 m). During the course of the war in the Pacific F4Us were credited with

destroying 2,140 enemy aircraft for a loss of only 189 machines. The Corsair was also flown by the British Fleet Air Arm and the Royal New Zealand Air Force, and continued in production after 1945 (F4U-4, F4U-5, and F4U-7) eventually serving in the Korean War. Normal armament was six .50 machine guns, although some aircraft had four 20-mm cannon; bombs (up to 4,000 lb., 1,800 kg) and rockets could also be

F4U-4 Corsairs, each armed with six ·50 machine guns, eight rockets, and a 500-lb. (225-kg) bomb

A-7E Corsair II landing on the U.S.S. America *underway off North Vietnam, July 1970*

carried, while night-fighter versions had a radar pod on the starboard wing. Corsairs were supplied to the French navy in 1952–53, and eventually equipped various small air forces such as those of Argentina, Honduras, and El Salvador.

The name Corsair II has been given to the Ling-Temco-Vought A-7, a carrier-based subsonic light-attack jet aircraft. This was developed from the 1,000-mph (1,610 km/h) F-8 Crusader (1955), the principal U.S. Navy and Marine Corps fighter of the late 1950s and early 1960s.

The A-7 first flew on September 27, 1965, and has since been supplied to the U.S. Navy and the American and Greek air forces. Variants include the A-7A, A-7B (without an afterburner), A-7C and A-7E (U.S. Navy), A-7D (U.S. Air Force tactical fighter), and A-7H (Greek air force). The A-7D has a wingspan of 38 ft. 9 in. (11.81 m) and length of 46 ft. 1½ in. (14.06 m).

Cougar

The first jet fighter produced by Grumman was the straight-winged F9F-2, -3, -4, and -5 Panther (first flown 1947), which entered service with the U.S. Navy in 1949, and saw extensive use in the Korean War. From the Panther was developed the swept-wing F9F-6 Cougar (1951), which began to reach Navy and Marine Corps squadrons in 1952, and also served in Korea. The F9F-6 (later F9F) had a J48 engine of 6,500 lb. (2,944 kg) thrust; the F9F-7 (F9G) used a J33 power plant; and the F9F-8 (F9H) was powered by a J48 turbojet of 7,250 lb. (3,289 kg) thrust and could attain 714 mph (1,149 km/h) at sea level. Wingspan of the F9F-8 was 34 ft. 6 in. (10.52 m) and length 41 ft. 7 in. (12.67 m).

The F9F-8T (TF-9J) two-seater became the Navy's standard advanced trainer in 1955; a number of Cougars were eventually converted to use as target drones.

Cub

Designed by C. Gilbert Taylor during 1928–30, this classic light plane became the foundation of the Taylor Aircraft Company established in 1931. A braced, high-wing monoplane, the Cub was constructed of welded steel tube with fabric covering. It had two seats, one behind the other, in an enclosed cabin, and was powered by any of several low-powered engines (the most common was the 35-hp Continental A-40, but later models had 65-hp engines).

In 1936, Taylor set up TAYLORCRAFT Aviation, and the original plant was bought by the company's former secretary and treasurer, William T. PIPER, who began manufacture of improved Cubs. The new J3 Cub went into production in 1938, and by the end of 1941 a total of 10,000 Cubs had been delivered. During the war production was switched to the L-4 Grasshopper (an artillery spotter of which over 5,500 were built), and it in turn succeeded by still further new models of the Cub in 1946. Total output of the Cub series was 30,086.

Curtiss

One of America's earliest aircraft manufacturers, founded by Glenn Curtiss (1878–1930), member of a select group who helped the Aerial Experiment Association in 1907, and a pioneer of EARLY AVIATION. In 1908 Curtiss produced his first design, the *June Bug*, whose 1,266-ft. (420-m) flight in June 1908 made him the second American to fly after the Wright Brothers. To market the aircraft he had designed, Curtiss joined Charles Herring to form Curtiss-Herring Company, selling aircraft to such famed barnstormers as Lincoln Beachey, Charles K. Hamilton, and Eugene Ely. In 1911 Curtiss was equipping his seaplanes with ailerons, which he devised for lateral control. This brought a bitter lawsuit with the Wrights, who alleged patent infringement, claiming that their awkward wing-warping method of control covered Curtiss' innovation as well. Curtiss was awarded the patent and found an increasing world market; his Curtiss Aeroplane and Motor Company soon outgrew its small facilities at Hammondsport, N.Y., and new plants were opened at Garden City and Buffalo.

In 1918 Curtiss was in fact the largest aircraft manufacturer in the world and the chief producer of flying boats, the H-4 America and H-12 Large America types being the principal ocean patrol machines of the Allies. The JENNY was easily the Allies' most numerous primary trainer, with about 5,000 in service. One of Curtiss' most famous flying boats was the *NC-4*, a four-engined machine that made the first transatlantic crossing (in several stages) in May 1919 (see PIONEER FLIGHTS).

Between 1923 and 1936 the first HAWK family of fighters was developed, one of the early PW-8 models in 1924 being used to cross the United States in 21 hours 48 minutes. Racing land-planes and seaplanes followed, several times capturing the world speed record and winning the Schneider Trophy. Fastest of all was the Curtiss R3C, using a descendant of Curtiss' own D-12 engine, eventually developed into the Conqueror employed in many later

F9F-8 Cougar

fighters and bombers. In parallel with Army and Navy Hawk fighters, Curtiss built the Falcon two-seat observation aircraft in many versions, some of them after the 1929 merger between Curtiss and its rival Wright Aeronautical Company, which resulted in the formation of the giant Curtiss-Wright Corporation (see WRIGHT CORPORATION).

In addition to a vast range of CONDORS, HELLDIVERS, Falcons, Sparrowhawks, Shrikes and Seagulls, the Curtiss-Wright plant built the first monoplane HAWK 75 in 1935. The Hawk went into production for the French and other foreign air forces and as the P-36 for the U.S. Army; later development included the P-40 (see WARHAWK). Other major wartime products were the SB2C/A-25 Helldiver, and C-46 COMMANDO. Despite development of the SC-1 Seahawk, several piston and piston/jet fighters, and the XF-87 four-jet night fighter, Curtiss-Wright failed to secure significant postwar aircraft sales and closed its airplane division in 1947.

D

D.VII

Generally considered the best operational German fighter used in quantity during World War I, the Fokker D.VII was selected for production after a fighter design competition held at Adlershof in January 1918. Of strong, welded tubular-steel construction and powered by a 160-hp Mercedes or 185-hp BMW engine, the D.VII was easy to handle even at high altitude and highly maneuverable. It had a wingspan of 29 ft. 3 in. (8.84 m) and length of 23 ft. (7.01 m). Top speed was 117 mph (188

Curtiss P-40

km/h).

The first D.VIIs reached front-line units in April 1918, and by November of that year at least 42 *Jagdstaffeln* were D.VII-equipped. In all, some 1,000 D.VIIs were produced. Such was the Allies' regard for this aircraft, with its twin Spandau machine guns, that the Armistice Agreement gave it special mention, demanding that all first-line D.VIIs be surrendered. However, parts and engines for a considerable number of new D.VIIs were smuggled into various neutral countries, and the aircraft continued on active service in various air forces for many years.

Daimler-Benz

Famous German automobile and aircraft-engine manufacturer, formed in 1926 by the merger of two companies, Daimler and Benz, that had been pioneers in the development of the internal combustion engine.

Even before World War I Daimler and Benz were producing engines for powered flight. Until 1911, all the gasoline engines used by Count Zeppelin's airships were of Daimler manufacture, and in 1913 Daimler and Benz were both successful in the Kaiserpreis-Wettbewerb competition for aircraft engines.

During World War I, 6-cylinder water-cooled engines were developed by both Daimler and Benz. The D.I-D.IIIA Mercedes (Daimler) series, which was used in the Fokker D.VII, the Albatros scouts, and many other aircraft, eventually produced 200 hp; Benz power plants turned out up to 230 hp (Bz.IV).

In 1926 Daimler-Benz AG was formed by the merger of the two companies, and the new firm began producing both airship diesel engines and the DB 600, an inverted V-12 water-cooled gasoline engine (1937). This began a series of engines used in many of the Luftwaffe's wartime aircraft. The 1,175-hp 12-cylinder DB 601 powered the Me 209 record-breaker of 1939, and was fitted to the Bf 109E and F, early Bf 110s, the Do 215, and other aircraft. The V-12 DB 603, producing up to 1,900 hp, powered the Me 410, Do 217, and Ta 152. The DB 605 was a development of the 601 eventually yielding 2,000 hp and used by later Bf 110s and the Bf 109G. The DB 606, a pair of 601s coupled together, was evolved for use by the He 177, and the 24-cylinder DB 610 (3,000 hp) was installed in the Ju 288, Do 317, and other aircraft.

Experimental engines under development during World War II included the 16-cylinder DB 609 (2,660 hp), the 36-cylinder Double-W DB 630 (4,000 hp), and the 007 turbojet.

In 1956 the DB 720 and 721 gas

Fokker D.VII

turbines appeared, followed by the DB 730 turbofan. The Daimler-Benz gas turbine program was taken over by Turbo-Entwicklung (an associated company) in 1968.

Dambusters

Early in 1943 the Royal Air Force proposed to destroy the six most important dams in western Germany (Mohne, Eder, Sorpe, Ennepe, Lister, and Schwelme), thus flooding large areas and cutting off hydroelectric power supplies to industry.

A special LANCASTER squadron (No. 617) was formed on March 21, 1943, with handpicked crews. Their aircraft were modified to carry a special mine that was the brainchild of Barnes WALLIS. The mine was designed to be rotated by a drive off a hydraulic motor before release from an altitude of 60 ft. (18 m). It would then skip or bounce across the water to the unprotected top wall of the dam, sink to its base, and explode. To maintain the precise height required for dropping the mine, two spotlights mounted below each aircraft were aligned so that their beams coincided on the surface of the water when the altitude was exactly 60 ft.

The operation took place on the night of May 16/17, 1943, and involved 19 aircraft. The Mohne and Eder dams were both breached, but eight Lancasters failed to return. The squadron commander, Wing Commander Guy Gibson, received the Victoria Cross but was later killed in a Mosquito when returning from a raid on Germany (September 19/20, 1944).

After this exploit 617 Squadron became known as the Dambusters. It operated as a special precision bombing unit for the rest of the war and in 1977 was flying Vulcans from its original base at Scampton.

Dassault-Breguet

Founded by Marcel BLOCH, who adopted a Resistance code-name (Char d'Assault) as the new family name, spelling it Dassault, the firm of Dassault-Breguet is France's largest privately-owned aircraft manufacturer. Though the French government was trying to nationalize the aircraft industry, he set up Société des Avions Marcel Dassault soon after World War II and was soon the only French company to be manufacturing jet fighters (the Ouragan of 1949, the MYSTÈRE, and the Super Mystère). In 1956 Dassault flew the first Étendard naval attack and reconnaissance aircraft, a new version of which went into production in 1976.

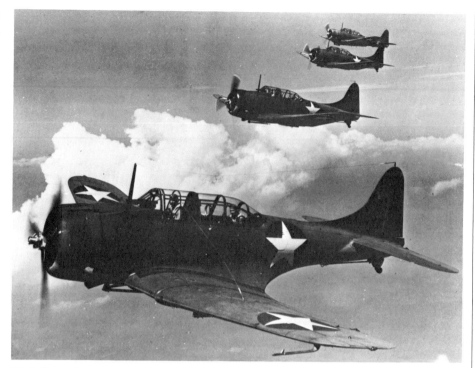

SBD Dauntlesses

By far Dassault's greatest success was the MIRAGE, the original version of which entered production in 1959. Capable of twice the speed of sound, this simple delta was produced in interceptor, all-weather fighter, all-weather attack, reconnaissance, and trainer versions; it led to the simplified Mirage 5, in which all-weather electronics are replaced by additional fuel and weapons. By 1977 more than 1,900 of all versions had been delivered, including 62 much larger Mirage IV bombers, while advanced Mirage F and G variants have also been developed.

In partnership with what was then Sud-Aviation, work began in 1962 on a business jet, the Mystère 20. In 1963 this was adopted by the Business Jets division of Pan Am as the Fanjet Falcon, and by 1977 sales exceeded 400, together with a further 170 of the smaller Falcon 10 version, with production beginning on the enlarged long-range Falcon 50 trijet. The Mercure jetliner was a disappointment, only 10 being sold, but Dassault has been trying to build the improved Mercure 200 in partnership with McDonnell Douglas and Aérospatiale.

Société Louis Breguet, the oldest aircraft manufacturer in France, merged with Dassault in December 1971 to form Dassault-Breguet. The main programs brought in by Breguet were the maritime reconnaissance Atlantique, the JAGUAR, and the ALPHA JET, an advanced trainer being developed with DORNIER.

Dauntless

The chief U.S. Navy dive-bomber in the first years of World War II, Dauntlesses were involved in combat from the time of Pearl Harbor onward, and they played a major role in destroying Japanese warships during the Battle of the CORAL SEA, at MIDWAY (where they sank three carriers and severely damaged a fourth), and in the Solomons. The Dauntless was originally procured by the Navy as the Northrop BT in 1936, but after Northrop became a full division of Douglas in 1938 work was continued under the new designation SBD. There were eventually six major SBD models, and the last of 5,936 aircraft produced was completed in July 1944.

The Dauntless was powered by a Wright R-1820 radial engine giving a top speed of 250 mph (400 km/h). A two-seat low-wing monoplane, it had a defensive armament of four .50 machine guns and featured perforated wing flaps which could be operated as dive brakes. It had a wingspan of 41 ft. 6 in. (12.65 m) and length of 33 ft. (10.06 m).

DC-3

Probably the single most significant aircraft in the history of civil aviation, the twin-engined Douglas DC-3 combined new standards in safety, speed, comfort, and reliability with operating costs low enough to make air travel as we know it today a commercial reality.

The DC-3 was produced and entered service alongside the DST (Douglas Sleeper Transport) essentially a DC-3 but with berths for 14 passengers. The DC-3, originally seating 21 (later increased to 28), was a direct development of the 14-passenger twin-engined DC-2 of 1934, itself an extremely advanced aircraft with variable-pitch propellers and a cruising speed of 170 mph (274 km/h).

The DC-3 made its maiden flight on December 17, 1935, and entered service with American Airlines, which had been using Curtiss CONDORS, the following June. Together with the DC-2 (flown by TWA) and the BOEING 247 of 1933 (in service with United), the DC-3 represented a radical break with the slow and ungainly biplanes and trimotors then in use in civil aviation. By the time of Pearl Harbor a total of some 450 DC-3s (including 38 DSTs) had been delivered to commercial operators, including 100 to airlines outside the United States.

During World War II military versions of the DC-3 (the C-47 and C-53) saw intensive service in almost every possible role and every combat zone; a Russian model, the Lisunov Li-2, was even license-built in the Soviet Union. After the end of the war the DC-3 remained for many years the mainstay of airlines throughout the world. By the time production ceased in 1946, a total of 10,929 machines had been built (more than any other transport airplane), of which 10,123 were originally produced for military use.

Several hundred DC-3s were still flying in the mid-1970s (one aircraft is known to have amassed 84,000 hours of flying time). Some were even brought back into military service as AC-47

Dragonfly gunships for use in Vietnam. They were equipped with three rapid-firing (6,000 rounds per minute) "miniguns" aimed through the port-side fuselage door and windows.

Early versions of the DC-3 used 850-hp Wright Cyclone engines. The substitution of the 1,200-hp Pratt & Whitney Twin Wasp raised the aircraft's cruising speed from 192 mph (307 km/h) to about 207 mph (331 km/h), and increased the range from the original 1,400 mi. (2,240 km) to as much as 2,100 mi. (3,360 km).

DC-4

The first four-engined Douglas airliner, the DC-4 was one of the earliest of a new generation of four-engined transports that pioneered long-range routes in the late 1940s. It was one of the first distinct signs that postwar civil aviation would use land-based transports rather than flying boats. The 52-seat DC-4 prototype, with four 1,150-hp Pratt & Whitney Hornet engines, was first flown in June 1938. It had triple tail fins and joined United Air Lines in 1939 as the Super Mainliner. It was one of the first large aircraft to have a nosewheel undercarriage. Eventually fitted with a conventional single tail, it proved too large for the airline requirements of the day and was sold to Japan, where it crashed.

The design was scaled down and re-emerged as the 44-passenger DC-4A, with a maximum speed of 280 mph (450 km/h), a length of 93 ft. 11 in. (28.63 m), wingspan of 117 ft. 6 in. (35.81 m), and range of 2,140 mi. (3,444 km). Before any deliveries could be made to commercial customers, the United States entered World War II

and production was diverted to the armed forces. The aircraft served as the C-54 Skymaster and U.S. Navy R5D, using 1,450-hp Pratt & Whitney engines.

The DC-4A went on to fly with most of the world's major airlines in the early postwar years (the majority of aircraft were converted military models, since Douglas soon went on to concentrate on the DC-6); a Merlin-engined version built by Canadair (the North Star) was operated by Trans-Canada Airlines. The 1,242nd and last DC-4 was delivered to South African Airways in August 1947, but many were still in commercial service in the early 1970s (mostly as freighters). The basic DC-4 design was one of the first to undergo the stretching process for greater payload and speed, now a familiar feature in airliner design, and the DC-6 and DC-7 were the results.

DC-6

A stretched and faster development of the DC-4, the four-engined DC-6 was the first Douglas airliner to possess a pressurized fuselage and the first designed for postwar use. The initial version of the DC-6, powered by 2,100-hp Pratt & Whitney R-2800 engines, had a wingspan (like the DC-4's) of 117 ft. 6 in. (35.81 m); length was 100 ft. 7 in. (30.65 m). It cruised at 313 mph (504 km/h) carrying up to 68 passengers and entered airline service (with United Air Lines) in April 1947.

The DC-6A freighter had its fuselage stretched to 105 ft. 7 in. (32.18 m), and its passenger version, the DC-6B seating up to 92, joined American Airlines in April 1951. Powered by four 2,500-hp

DC-3

R-2800s, it had a top speed of 360 mph (519 km/h). DC-6 airliners saw extensive service with U.S. and foreign airlines during the 1950s, and were operated by the U.S. Air Force (as the C-118/C-118B military transport) and U.S. Navy (R6D-1).

DC-7

One of the last and fastest of large piston-engined transports, the Douglas DC-7 entered airline service in 1953. Seating as many as 95 passengers, it was powered by four 3,250-hp Wright R-3350 turbocompound engines and cruised at about 365 mph (587 km/h). The long-range DC-7B had increased fuel capacity and featured improved flap operation for heavily-loaded takeoffs from short runways. The DC-7F was an all-freight version.

The last of the series of propeller-driven Douglas airliners was the DC-7C Seven Seas, which began nonstop North Atlantic services in 1956. With a fuselage stretched to 112 ft. 3 in. (34.21 m) and a wingspan of 127 ft. 6 in. (38.86 m), the DC-7C carried up to 105 passengers. It was powered by 3,400-hp Wright turbocompound engines and had a maximum speed of 405 mph (652 km/h). Range was in excess of 4,000 mi. (6,437 km).

DC-8

Second (after the Boeing 707) of the first generation of large four-jet American airliners, the Douglas DC-8 was one of the most successful jet transports of the 1960s. By the time the prototype made its first flight on May 30, 1958, over 130 aircraft had been ordered, and when production ceased in 1972 McDonnell Douglas had completed a total of 556.

The initial types (Series 10, 20, 30, 40, and 50) all seated up to 179 passengers, were 150 ft. 6 in. (45.87 m) in length with a wingspan of 142 ft. 5 in. (43.41 m). Series 10 and 20 were designed for domestic services, while Series 30, 40, and 50 were long-range intercontinental versions. To reduce drag, Series 40 and 50 had an improved wing shape and leading-edge slots; in 1961 a DC-8 Series 40 became the first jet airliner to fly supersonically (it reached 667 mph, 1,073 km/h—or Mach 1.012—in a shallow dive). Series 50 aircraft, with four turbofans of 18,000 lb. (8,165 kg), had a maximum cruising speed of 579 mph (932 km/h) and range of 5,720 mi. (9,205 km).

The basic DC-8 was stretched in the Super Sixty Series, starting in 1966 with the Super 61, which seated 259 and was

DC-7C

DC-9

187 ft. 5 in. (57.12 m) long. The Super 62 (also introduced in 1966) was 6 ft. 8 in. (2.03 m) longer than the Series 50. It incorporated aerodynamically improved wings and repositioned engine pods, and had an increased range. The final version, the Super 63, combined the size of the 61 and the aerodynamic improvements of the 62, entering service (with KLM) on July 27, 1967.

DC-8F Jet Trader freight versions were developed from both the Series 50 and Super Sixty types.

DC-9

The short-range twin-jet Douglas DC-9 made its maiden flight in 1965, and 90-passenger Series 10 aircraft entered service in December of that year. This initial version of the DC-9 has a wingspan of 89 ft. 5 in. (27.26 m) and length of 104 ft. 5 in. (31.82 m) and is powered by Pratt & Whitney turbofans of 12,250 lb. (5,557 kg) thrust. Series 20 aircraft have longer-span wings for hot climate/high altitude use; the Series 30 (1966) is powered by uprated engines of 14,000 lb. (6,350 kg) thrust and seats 125. The fuselage of the Series 30 was lengthened to 119 ft. 3 in. (36.35 m), wingspan is 93 ft. 5 in. (28.48 m). Further stretching produced the Series 40 (1967), with a length of 125 ft. 7 in. (38.28 m), and the Series 50, 134 ft. 7 in.

(41.02 m) long and seating up to 139 passengers. Cruising speed of all versions of the DC-9 is in excess of 560 mph (901 km/h); range varies between 995 mi. (1,601 km) and 1,484 mi. (2,388 km).

Passenger, cargo (DC-9F), convertible (DC-9CF), and passenger-cargo (DC-9RC) versions exist in all DC-9 series. There are also several military variants (C-9A Nightingale ambulance, C-9B Skytrain II Navy transport, and VC-9C for the Air Force's Special Air Missions Wing).

DC-10

Initially produced to meet a 1966 American Airline's requirement for a large-capacity airbus capable of operating from runways of normal length, the three-turbofan Douglas DC-10 first flew on August 29, 1970. AA and United began operating the Series 10 in 1971. This version has a wingspan of 155 ft. 4 in. (47.35 m) and length of 181 ft. 5 in. (55.30 m). It is powered by General Electric engines of 41,000 lb. (18,570 kg) thrust and was intended for domestic routes of up to 3,600 mi. (5,780 km). The Series 30 (1972) has General Electric engines of 51,000 lb. (23,100 kg) thrust and a wingspan increased by 10 ft. (3 m); the Series 40 (originally known as the Series 20), which was first flown in February

DC-10

1972, is powered by Pratt & Whitney turbofans with water injection and exhaust nozzles. All versions are designed to carry a maximum of 390 passengers. Cruising speed is in excess of 600 mph (965 km/h).

De Bellevue, Charles

Officially credited as the top American fighter ace of the Vietnam War, "Chuck" De Bellevue flew as weapons systems operator in F-4 PHANTOMS that shot down six MiGs. Since the Phantom was a two-seater fighter, the aircraft commander (pilot, front seat) and WSO (rear seat) were deemed equally responsible for a "kill" and both so credited. All De Bellevue's victories were obtained in 1972, while flying with the 555th Tactical Fighter Squadron, 432nd Tactical Reconnaissance Wing, out of Udorn in Thailand. The first was achieved on May 10, with Capt. Richard S. Ritchie as aircraft commander. On July 8 two MiG-21s were destroyed with the same pilot, as was another MiG-21 on August 28. The two final victims, both MiG-19s, fell on September 9 when Capt. John Madden was pilot.

Capt. Ritchie was the highest-scoring Air Force pilot of the war with five victories. Another WSO, Capt. Jeffrey S. Feinstein, and the U.S. Navy team of Lt. Randy Cunningham and Lt. Willie Driscoll, were also credited with five victories each.

Decca navigator

This radio navigation aid employs low frequency radio signals (70–130 kHz) generated by chains of transmitting stations consisting of a Master station and two or three Slave stations situated 50–80 mi. (80–130 km) distant from the Master. The unmodulated radio waves create a pattern of hyperbolic position lines along which the phase difference between the outputs of the Master and the Slave stations is constant. Instruments in the aircraft identify these position lines by phase comparison carried out at a common frequency. A flight log is an integral part of the system, which can also augment other navigational aids by connection to a computer (see also NAVIGATION; RADIO).

de Havilland

A major British manufacturer, de Havilland Aircraft Co. Ltd was formed in 1920 by Geoffrey de Havilland (1882–1965). De Havilland had flown his first biplane as early as 1909 and was chief designer at the Royal Aircraft Factory (1911–1914), and the Aircraft Manufacturing Company (1914–1920), where he built the DH 1 pusher biplane and the DH 4 single-engined bomber (produced under license by the U.S. and used during the war by U.S. pilots), as well as the DH 9A.

After the war the company concentrated on plywood civil airplanes and built eight-passenger DH 18s for early services on the London–Paris route. These were succeeded by the larger DH 34, which flew cross-Channel services between 1922 and 1926. The DH 27, DH 42, and DH 56 military prototypes all failed to win acceptance; these were followed by the very fast DH 65 Hound in 1926, and the DH 77 interceptor in 1929, but de Havilland's military aircraft ceased to be important for another decade.

The DH 53 Humming Bird, a low-powered, single-seat monoplane of 1923, convinced de Havilland that aircraft of less than 50 hp were of little practical value. He then produced the remarkably successful DH 60 MOTH, which first flew in 1925. DH 50 single-engined five-seat biplanes and DH 66 Hercules trimotor transports were built for British and Australian airlines (1923–1929), and de Havilland factories were established in Australia, Canada, South Africa, and New Zealand. In the 1930s the company built Tiger Moth and Moth Minor trainers; Dragon, Dragon Express, and Rapide biplane transports; the DH 88 COMET racer; and the Albatross and Flamingo airliners. It eventually went almost entirely over to MOSQUITO production during World War II.

The first VAMPIRE jet fighter flew in 1943, followed in 1946 by the DH 108 (the first British airplane to exceed Mach 1) and the Sea Vixen. The Dove all-metal Rapide-replacement appeared in 1945, and the world's first jet airliner, the COMET, in 1949.

American-designed Hamilton propellers were manufactured under license from 1935, and de Havilland Propellers Ltd., formed in 1946, produced a considerable range of guided weapons. The DH 125 (HS 125) executive jet was the last DH aircraft produced before de Havilland became part of HAWKER SIDDELEY in 1960. The TRIDENT jet airliner was eventually flown under the Hawker Siddeley name.

Apart from its airplane production, de Havilland was an important manufacturer of engines, the first of which was the four-cylinder Gipsy (1927). In 1930 this was modified to operate inverted (Gipsy III), thus lowering the nose line of single-engined aircraft. As the Gipsy Major, this engine eventually developed 150 hp. The four-cylinder Gipsy Minor produced 90 hp (Moth Minor), and the six-cylinder Gipsy Queen (also an inverted engine) powered the Rapide, Dragon Express, Vega Gull and DH 88 Comet, among others. In supercharged form the Gipsy Queen developed over 200 hp. The inverted V-12 Gipsy King (up to 525 hp) was used in the DH 93 Don and the Albatross.

De Havilland jet engines included the centrifugal Goblin (1941) of 3,500 lb. (1,585 kg) thrust, used in the Vampire and Saab J 21R, and its successor, the Ghost, of 5,000 lb. (2.265 kg) thrust, which powered the Comet and Sea Vixen. De Havilland's engine activities eventually became a part of ROLLS-ROYCE.

Delta

When an aircraft's wings and tail are merged into one large flying surface (resembling an isosceles triangle) with a swept-back leading edge, the design is known as a delta (from the shape of the Greek letter "delta" Δ).

Because of its wide chord (see WINGS), a delta can have an aerodynamically thin wing section while retaining sufficient depth for the incorporation of strong spars and room for fuel, engines, and equipment. In addition, the control surfaces are located well behind the aircraft's center of gravity, where they can exert the most force.

The structural strength of a delta-wing resists distortion caused by the stresses and strains of flight. Drag is reduced by burying the engines within the wing and by eliminating the tail-plane, while the large wing area reduces takeoff and landing speeds without the need for FLAPS.

Disadvantages of the delta layout are the nose-high takeoff and landing attitude necessary to obtain lift from a high angle of attack, maintenance problems incurred by burying the engines within the wing, an excessive increase in wing area if a long span is specified for high-altitude flying, and inferior spin recovery.

The first operational delta-wing aircraft was the Gloster Javelin (1951); others were the Convair F-102 DELTA-DAGGER and F-106 Delta-Dart, the Dassault MIRAGE and the SAAB-37 Viggen.

Delta Air Lines

Delta Air Lines is one of the United States' largest domestic operators, ranking third among the world's airlines, excluding Aeroflot, in the number of passengers it carries annually (over 26,000,000). The airline traces its origins to what was the world's first commercial crop-dusting service, established in 1925. As Delta Air Services, the company began carrying passengers between Atlanta and Birmingham in

June 1929, with an extension to Dallas that August. In 1934, as Delta Air Corporation, the airline was awarded the Atlanta–Fort Worth and Atlanta–Charleston mail contracts. Delta's route network eventually grew to extend far beyond the southeastern states. In 1953 it acquired Chicago and Southern Air Lines and in 1972 Northeast Airlines. Delta now has some 35,000 route miles (56,000 km), predominantly in the eastern half of the United States, but with transcontinental services to Los Angeles, San Diego, and San Francisco. There are also flights from Boston to Bermuda; Boston and Miami to Montreal; New Orleans and Los Angeles to San Juan; New Orleans to Caracas, Maracaibo, and Montego Bay; and from Boston and New York to the Bahamas.

Delta's present fleet consists of 62 DC-9s, 32 DC-8s, 3 Boeing 747s, 18 TriStars, and 74 Boeing 727s.

Delta-Dagger

The Convair F-102 Delta-Dagger jet fighter was the first operational U.S. delta-wing aircraft. It owes its origin to research over a period of many years by the German-born designer Alexander LIPPISCH into tailless and delta-wing aircraft. Convair's experimental Model 7002 led to construction of a test aircraft, the XF-92A, which was successfully flown in February 1949. From it came the F-102 project for a supersonic all-weather interceptor armed with air-to-air missiles. The first example flew in October 1953 but was not capable of supersonic speeds until AREA RULE had been applied to reduce drag. Deliveries of F-102As to the Air Defense Command began in July 1955, and the 875 completed by April 1958 equipped 25 squadrons. The Delta-Dagger was powered by a J57 turbojet of 17,200 lb. (7,802 kg) thrust, giving a

maximum speed of 825 mph (1,327 km/h). Wingspan was 38 ft. 1 in. (11.61 m) and length 68 ft. 4 in. (20.38 m); armament consisted of six AIM-4 Falcon missiles. The Delta-Dagger also served with Air National Guard units but was withdrawn by the end of 1975.

From the F-102 was developed the F-106 Delta-Dart (1956), powered by a J75 turbojet of 24,500 lb. (11,100 kg) thrust. This 1,500-mph (2,413-km/h) all-weather interceptor (F-106A), also produced as a two-seat operational trainer (F-106B), is armed with up to six air-to-air missiles and (since 1973) with an M-61 multibarrel cannon. Wingspan of the F-106 is 38 ft. 3 in. (11.66 m) and length 70 ft. 8 in. (21.54 m). Production (277 F-106As, 63 F-106Bs) was completed in 1960.

Deperdussin

French aircraft manufacturer of the years immediately before World War I, Deperdussin became famed for its racing monoplanes that used highly advanced methods of construction. Armand Deperdussin began business in 1910, and his first products were conventional monoplanes generally similar to those of Blériot. Deperdussin soon began to concentrate on high-speed aircraft, however, and developed smaller and smoother wings and carefully faired landing gears, and fitted new and more powerful engines such as the 100-hp Gnôme. By 1912 Jules Védrines had raised the world speed record to 108.18 mph (174.06 km/h) in one such streamlined Deperdussin.

With the help of Louis Bechereau, Deperdussin developed the world's first airplanes with MONOCOQUE fuselages. These 1913 Deperdussin racers used multiple layers of thin tulip wood giving a thickness of about 0.16 in. (4 mm) for the fuselage; the pilot was

TF-102A trainer (foreground) with F-102A Delta-Dagger

seated internally; the wings had no external bracing; and the 160-hp Gnôme 14-cylinder two-row rotary engine was carefully cowled. Marcel Prévost set new world speed records three times as pilot of a monocoque Deperdussin, finally reaching a figure of 126.67 mph (203.81 km/h); a floatplane Deperdussin also won the first contest for the Schneider Trophy.

Armand Deperdussin was later convicted of frauds involving 28 million francs and received a suspended prison sentence. His aviation interests were taken over by SPAD and he eventually committed suicide by shooting himself in a Paris hotel in 1924.

Dewoitine

The aircraft manufacturer founded by Émile Dewoitine in the early 1920s produced some of the best French fighters and transports of the years between the wars. Dewoitine's first aircraft, the D1 monoplane fighter, had made its maiden flight in 1921, and it set a pattern for the company's subsequent designs. It was a clean and simple machine with a parasol wing (carried on struts above the fuselage), constructed of the then new light alloy Duralumin, with fabric covering. The D1, powered by a 300-hp Hispano water-cooled engine, found a ready market among the air forces and navies of Europe, and 126 were built by Ansaldo in Italy. The D9 of 1924, powered by a Gnôme-Rhône (Bristol) Jupiter radial engine, also sold widely and was license-built in other countries. It carried the exceptional armament of four fixed guns, as did a number of later fighters that included the D21, an aircraft again very successful outside France, being built in Argentina and Czechoslovakia.

Finding no sales in his own country, Dewoitine transferred operations to Switzerland in 1928, where he produced the D27, one of which was still flying in 1960. After manufacturing several more fighter types, he returned to France and formed Société Aéronautique Française, building the D371, D373, and D376 for the French air force and navy, and also gaining Air France orders for large commercial transports such as the D333 of 1934, a cantilever monoplane powered by three 570-hp Hispano radials that had its main landing wheels carried in large "trouser" fairings. With a four-member crew and seating eight passengers, the D333 could reach 188 mph (302 km/h). The later D620 carried 30 passengers at 200 mph (322 km/h), and featured retractable landing gear.

DH 4

Dewoitine's most famous products were single-seater fighters of a series that began with the D500 of 1932. These low-wing aircraft were among the most successful fighters of the transitional period between the fabric biplane and stressed-skin monoplane. All were powered by water-cooled Hispano-Suiza engines (690–860 hp), and had prominent fixed landing gear usually faired with spats. Capable of reaching speeds of up to 250 mph (400 km/h), and highly maneuverable, they were armed with two or four machine guns, and often mounted a cannon firing through the propeller hub. Large numbers were exported, and the 352 sold to the Armée de l'Air and French navy between 1934 and 1937 represented 60 percent of French fighter strength.

In 1936, Société Aéronautique Française was nationalized as SNCAM. Two years later it produced the completely new D520, a stressed-skin fighter with retractable landing gear and 910-hp engine. Generally rated the best French interceptor of the war, the D520 was only available in very limited numbers in 1940 but still gained more than 100 confirmed victories over Luftwaffe aircraft. The Vichy government kept the D520 in production, 740 being built by the time the Germans occupied southern France in 1942 and seized 411 of them.

DH 4

Designed for bombing and reconnaissance duties, the Airco de Havilland DH 4, a single-engined two-seater biplane, began to reach British Royal Flying Corps and Royal Naval Air Service units in 1917. Initially, its armament consisted of a Lewis gun for the observer and a fixed Vickers gun firing through the spinning propeller. Later the complement of weapons was doubled. With a wingspan of 42 ft. 4½ in. (12.91 m) and a length of 30 ft. 6 in. (9.30 m), it could carry a bomb load of

460 lb. (205 kg). A shortage of the 250-hp Rolls-Royce engines originally specified for the DH 4 led to the use of the 200-hp Siddeley Puma, the Galloway Adriatic, the V-12 RAF 3a, and the Fiat A.12.

The high performance of the DH 4 (its maximum speed with a Rolls-Royce Eagle VIII engine was 133.5 mph, 215 km/h) made it a successful bomber serving in France, Italy, Russia, and the Aegean. Employed as a home-defense fighter, a DH 4 shot down the Zeppelin L70 during the last German airship raid on England in World War I (August 5, 1918).

In addition to British production, more than 4,000 Liberty-engined DH 4s ("Liberty Planes") were license-built in the United States, of which 200 saw action in France during 1918. Another 283 went to the U.S. Navy and Marine Corps. The DH 4B was a modified "Liberty Plane," and the DH 4M (of which 180 were constructed) was Boeing-built with a welded steel-tube fuselage.

After the Armistice, the DH 4 found various commercial uses with pioneer civil carriers and was built under license in the United States where it operated as a mailplane. Chile, Greece, Iran, and Spain also used the DH 4, and a few examples remained in service with the Belgian air force until 1932.

DH 9A

The DH 9A, which came into use at the very end of World War I as what would now be called a tactical bomber, became one of the mainstays of the RAF. It was descended from the 230-hp DH 9 single-engined bomber, which began to enter service with the British Royal Flying Corps in December 1917. Although highly maneuverable, the DH 9 proved vulnerable because of an unreliable engine and a cruising speed of a mere 78–85 mph (120–136 km/h)

A two-seat DH 9A of the RAF in India, 1929, carrying its own spare wheel

in formation at 10,000 ft. (3,050 m).

The improved DH 9A, with a larger wing area and either a 360-hp Rolls-Royce Eagle or a 400-hp Liberty engine, began to replace the original DH 9 in June 1918. By the end of World War I the "Ninak," as it was called, had dropped 10½ tons of bombs on German cities, each aircraft being capable of carrying a 450-lb. (200-kg) load at a maximum speed of 114 mph (183 km/h). The DH 9A could stay in the air for 5¾ hours and had a service ceiling of 16,500 ft. (5,030 m). Armament consisted of one forward-firing Vickers gun and a swiveling Lewis gun. Its wingspan was 45 ft. 11½ in. (14.00 m) and length 30 ft. 3 in. (9.22 m).

After the war, the DH 9A served as the RAF's standard British-based day bomber until Fairey Fawns began to arrive in 1925. It saw service in operations against the Bolsheviks in Russia (1919–20), and in Iraq, India, Egypt, Palestine, and Aden it was replaced by the Westland Wapiti (which incorporated as many DH 9A parts as possible) only in the early 1930s.

Over 4,000 DH 9A aircraft were built, and surplus airplanes were thus readily available for civil use. Many were in fact specially converted after World War I (two-passenger DH 9B and C). Substantial numbers also found their way onto the inventories of over a dozen air forces.

Dive-bomber

One of the most hated—and least enduring—of all types of warplanes, the dive-bomber did not appear until the late 1920s. German interest began with the Junkers K.47 (1928), a low-wing monoplane developed from the Junkers A.48, which served in small numbers with the embryo Luftwaffe and the Chinese air force. German interest in the dive-bomber lapsed until Ernst UDET of the Luftwaffe visited the United States in 1931 and saw the U.S. Navy's HELLDIVERS. The Navy and Marine Corps had been experimenting with dive-bombing techniques in Central America and the Caribbean for several years, and Udet was considerably impressed with the aircraft. He ordered two of the Curtiss machines and arranged for further trials in Germany, using a Focke-Wulf 56 equipped with bomb racks. The results were so impressive that a requirement for a Luftwaffe dive-bomber was formulated. From four designs (He 118, Ar 81, Ha 137, Ju 87) the Ju 87 STUKA was chosen and appeared with the Condor Legion in Spain in December 1937.

During the German attacks on Poland (1939) and France (1940), the Stuka was a successful adjunct to Blitzkrieg tactics. With the Luftwaffe in control of the air it proved highly successful at destroying pinpoint targets, and it also had a considerable demoralizing effect on opposing troops and civilians.

Employed against the RAF in the Battle of Britain, however, Stukas suffered heavy losses and had to be withdrawn from the main battle. For the first time the Luftwaffe was not in immediate and full control of the air, and the vulnerability of dive-bombers to enemy fighters was clearly demonstrated.

In the war in the Pacific the U.S. Navy and Marine Corps employed the SBD DAUNTLESS and the SB2C Helldiver (principally against shipping), and the RAF used the Vultee Vengeance in limited numbers in the China–Burma–India theater of the war. British manufacturers also constructed a handful of dive-bombers, but the Barracuda alone saw extensive service, while the French Navy made only limited use of its Loire–Nieuport 40 series of aircraft in 1939–40.

The role for which the dive-bomber had been produced came to be filled toward the end of World War II by fighter-bombers (rocket- and bomb-carrying Thunderbolts, Spitfires, Typhoons, Fw 190s, Bf 109s), aircraft that had the performance and armament to defend themselves in aerial combat.

Do 17. See FLYING PENCIL

Doolittle, James Harold (1896–)

American racing pilot and military aviation strategist, Doolittle enlisted as a flying cadet in the U.S. Army Signal Corps in 1917, was later commissioned and was involved in special flying activities. On September 4, 1922, he made the first flight across the United States in one day, taking 22 hours 35 minutes to cover 2,163 mi. (3,481 km) in a DH 4B. He won the 1925 Schneider Trophy race flying a Curtiss R3C floatplane at a speed of 232.57 mph (374.27 km/h) over the Baltimore course.

Doolittle resigned from the Army Air Corps in 1930 and during the next few years made a number of record-breaking flights in racing aircraft, setting a new world speed record for landplanes in 1932. In 1940 he returned to military service and when war came organized and led the first strike against Tokyo on April 18, 1942, flying B-25 MITCHELLS from the carrier *Hornet*. Although the bombs did little significant damage, the raid gave a much needed boost to Allied morale, and Doolittle was later awarded the Congressional Medal of Honor for his heroism.

He commanded the Twelfth Air Force and its Bomber Command in North Africa during 1942–43, moving to Britain at the end of 1943 to take command of the EIGHTH AIR FORCE. Doolittle remained in this command until hostilities ceased, then transferred his headquarters to Okinawa in the Pacific. After his retirement he chaired the Air Force Scientific Advisory Board.

Dornier

Claude Dornier (1884–1969) began his career in aviation with Zeppelin Werk Lindau during World War I. In 1922 the company became Dornier Metallbauten. Its early products included the Libelle and Delphin flying boats; the Duralumin-fuselaged four-passenger Komet II and its larger development of 1924, the six-passenger Komet III; the Spatz 80-hp trainer (1923); and the single-engined eight/ten-seat Merkur (1926), of which there was also a floatplane version.

In the late 1920s and 1930s Dornier established a considerable reputation with its large flying boats (which included the WAL and DO X), but it also built the Do C2 single-engined floatplane for mail services, the Do F twin-engined freighter with 92-ft. (28-m) wings having parabolically curved leading edges, and the 10-passenger Do K high-wing monoplane powered by four 240-hp engines (two pushers and two tractors). The Do 27 bomber-transport, with twin 750-hp BMW engines, was

one of the aircraft operated by the Luftwaffe in its earliest days, and Dornier designs went on to become well known during World War II, with the Do 17 (FLYING PENCIL) and its derivatives used widely as Luftwaffe bombers and night fighters.

The Do 24 trimotor flying boat (1937) was originally designed for use in the Dutch East Indies, but it also found extensive employment with the Luftwaffe and (in 1944–53) with the French navy, while 12 were sold to Spain in 1944. The Do 18, a twin-engined flying boat based on the Wal design, was intended for transatlantic mail services, but spent most of its career as a Luftwaffe reconnaissance machine.

Also designed for North Atlantic mail routes, the sleek Do 26 four-engined flying boats were similarly taken into service by the Luftwaffe at the beginning of World War II and continued to fly until spares were no longer available.

Although most of Dornier's wartime aircraft were flying boats and bombers, it did produce one extraordinary fighter, the Do 335. Ready at the close of the war to enter service, it was a twin-engined aircraft with both tractor and pusher propellers capable of 480 mph (771 km/h), which made it one of the fastest piston-engined airplanes of all time.

Ater the war, Dornier temporarily transferred operations to Madrid, where it produced the Do 25 single-engined high-wing monoplane. It was succeeded by the Do 27 (powered by a single 270-hp Lycoming engine), which was Dornier's first new postwar design, and with which production was moved back again to Germany. It was followed by the Do 28D Skyservant, one of several twin-engined derivatives of the Do 27 design, capable of seating up to 12. Dornier later built the V/STOL Do 31, the Do 132 helicopter, and was the German partner in the Franco-German ALPHAJET project. Dornier also undertook license-production of Lock-

heed F-104G STARFIGHTERS and Fiat G.91s for the Luftwaffe.

Douglas

Founded at Santa Monica, Cal., in 1920, Douglas Aircraft had among its early products the Liberty-Engined DT-1 and DT-2 biplane torpedo-bombers. The DT-2 formed the basis for the design of the famed WORLD CRUISER and the DT-4 650-hp Wright-powered bomber. The 0-2 series of single-engined reconnaissance biplanes continued in production throughout the 1920s, and among other aircraft of the period were the M-4 mailplane, the C-1 Army transport/ambulance machine, and the twin-engined T2D-1. In the early 1930s Douglas produced the PD-1 flying boat (a modified version of the U.S. Naval Aircraft Factory's PN-10 and PN-12 machines), together with the Dolphin six-seat twin-engined amphibian and the 640-hp 0-38 observation biplane. The experimental DC-1 airliner of 1933 led directly to the DC-2 and DC-3, which marked the beginning of Douglas' outstanding record in the field of commercial aviation. A bomber/transport version of the DC-2, the B-18 Bolo (DB-1), was built in small numbers for the Army Air Force in 1939 and also saw service with the Royal Canadian Air Force as the Digby (B-18A); a comparable development of the DC-3 was known as the B-23 Dragon (C-47).

The TBD-1 Devastator (1937) was the first monoplane torpedo bomber to enter service with the U.S. Navy and took part in the Battle of MIDWAY, but its 825-hp Twin Wasp engine gave a maximum speed of only 225 mph (362 km/h), and its defensive armament of a single fixed forward-firing .30 machine gun and one swiveling .50 in the rear cockpit proved extremely inadequate. The SBD-1 DAUNTLESS dive-bomber (1940) could attain a speed of 255 mph (410 km/h) and saw successful Navy and Marine Corps service.

Dornier Do 18

Douglas C-124 Globemaster

The twin-engined DB-7 series (A-20, P-70 Havoc) established itself as one of the most successful U.S. medium bombers of World War II. With a top speed of over 300 mph (480 km/h) and range of 1,000 mi. (1,600 km), it could carry a 2,000-lb. (900-kg) bomb load. Aircraft served with the RAF (as Bostons) and with Soviet forces, even achieving some success as night fighters. To replace the DB-7 Douglas produced the A-26 INVADER in 1942. Commercial airliner development meanwhile continued through the high-wing twin-engined DC-5 of 1938, which was constructed only in small numbers. The four-engined DC-4 Skymaster was, however, a highly popular design leading on to the DC-6 and DC-7. The DC-8, DC-9, and DC-10 jetliners found a good airline market in the 1960s and 1970s.

Successful Douglas combat planes of the postwar years included the Navy F4D Skyray delta-wing fighter (1951), which set a world speed record of 752.94 mph (1,211.48 km/h) in 1953, the A4D Skyhawk attack bomber, the B-66 Destroyer/A3D Skywarrior twin-jet attack bomber/reconnaissance plane, and the F3D Skyknight all-weather night fighter.

The SKYRAIDER of 1945, with a Wright R-3350 piston engine, was a highly versatile aircraft originally intended as a replacement for the Dauntless; it remained in production for 12 years. The C-124 Globemaster military transport (1949) had a wingspan of 174 ft. (53.03 m); powered by four 3,500-hp Pratt & Whitney R-4360 engines it could carry a 50,000-lb. (22,650-kg) payload. Even larger was the C-133 Cargomaster (1956), which had four 6,000-ehp T34 turboprops and carried a 100,000-lb. (45,300-kg) payload over 1,300 mi. (2,100 km). The X-3 Stiletto supersonic research aircraft of 1952 made relatively few flights, but the earlier SKYROCKET and Skystreak had been very successful.

Douglas Aircraft became part of MCDONNELL DOUGLAS in 1967.

Douhet, Giulio (1869–1930)

Italian general famous for his advocacy of strategic air power. After training as an artillery officer, he became commander of the Italian Aeronautical Battalion before World War I. A passionate believer in a separate military air arm, distinct from the army and navy, he lobbied politicians so strongly in support of his beliefs, openly criticizing the conduct of the war, that he was court martialled and jailed for a year for making false statements and disclosing confidential information.

By 1918 his thesis had been proved plausible by the battles of World War I and he was appointed to the General Aeronautical Commission, publishing in 1921 *The Command of the Air*.

In this influential book he argued that air power would be the key to all future wars, and that the role of ground and sea forces would be limited to defense. He called for the creation of a massive and independent air force, which by crippling enemy aircraft and by strategic bombing of centers of population and industry would ensure a rapid victory in any future war. Douhet's views were not entirely novel, having been voiced by others who had knowledge of aircraft potential (e.g. the designers F. W. Lanchester in Britain and Giovanni CAPRONI in Italy), but they did find a ready audience among key figures in the nations who were destined to fight World War II and therefore had a considerable influence.

Do X

Constructed at DORNIER's Altenrhein factory in Switzerland, the Do X made its first flight in July 1929. At the time it was the world's largest aircraft, with a wingspan of 157 ft. 6 in. (48.00 m) and length of 133 ft. (40.54 m). The interior was luxuriously appointed with a dining room nearly 60 ft. (18.3 m) long.

Initially powered by 12 525-hp Siemens Jupiter motors mounted back-to-back above the wing, it was re-equipped with 600-hp Curtiss Conquerors and made a number of experimental flights, performing one 50-minute trip with 169 people aboard. It had a maximum speed of 134 mph (214 km/h) and cruised at 118 mph (189 km/h).

On November 2, 1930, the Do X left Friedrichshafen, Germany, to fly to New York via Amsterdam, Calshot in England, and Lisbon. Fire damaged a wing in Portugal, the hull was damaged taking-off from the Canary Islands, and New York was not reached until August 1931.

A second Do X, with 12 550-hp Fiat engines, was supplied to Italy as a heavily-armed bomber. But it proved difficult to operate, due to its high ratio of weight to wing area. The first Do X was destroyed by Allied bombing of Berlin during World War II.

Draken

The delta-wing SAAB J 35 Draken (Dragon) was designed to meet a Swedish air force requirement for a Mach-1.5 fighter that would replace the subsonic J 29. The Avon-powered Draken prototype first flew on October 25, 1955.

Production J 35A aircraft, with an

Dornier Do X

afterburning Avon RM6B engine of 15,000 lb. (6,800 kg) thrust and two 30-mm Aden cannon, began to reach operational squadrons at the end of 1959. These machines were capable of Mach 1.8 and could carry as many as four Sidewinder air-to-air missiles. The J 35B of 1959 had a longer, aerodynamically more efficient rear fuselage (first introduced on late J 35As) and a new collision-course gun- and missile-sight. The SK 35C was a two-seat trainer. A more powerful Avon engine of 17,635 lb. (8,000 kg) thrust gave the J 35D of 1960 a speed in excess of Mach 2; the J 35E of 1963 is a photo-reconnaissance version of this model mounting seven cameras. The J 35XD, with wingspan of 30 ft. 10 in. (9.40 m) and length of 50 ft. 4 in. (15.35 m), has been supplied to Denmark.

By the mid-1970s the numerically most important Draken variant in Swedish air force service was the J 35F, armed with license-built Falcon air-to-air missiles. An export version, the J 35X (supplied to Denmark), has greater internal fuel capacity, provision for external ventral tanks extending the combat range to over 1,200 mi. (1,900 km), and attachment points for external weapons under the wings and fuselage. J 35X aircraft in service with the Finnish air force are designated F-35 (fighter), RF-35 (photo-reconnaissance), and TF-35 (trainer). A special all-weather Draken, the J 35XS, was ordered by Finland in 1970.

J 35B Draken on temporary service with the Finnish air force

TF-15A Eagle trainer

Drone

A pilotless aircraft that may be used as a practice target for guns or missiles, for research purposes such as air and fall-out sampling, or for reconnaissance and surveillance. Drones are controlled from a ground station, or by apparatus installed in another aircraft. In strict usage only conventional aircraft types adapted to remote control are termed drones; aircraft designed from the outset for remote control are classed as remotely piloted vehicles (RPVs). Drones do not generally carry warheads, but the U.S. Navy has experimented with drone helicopters capable of launching torpedoes.

E

Eagle

American air superiority fighter. Designed with a particular emphasis on maneuverability and acceleration, and incorporating modern metals such as titanium and boron composites, the McDonnell Douglas F-15 Eagle has two Pratt & Whitney afterburning turbofans of 27,000 lb. (12,400 kg) thrust; giving a 2:1 thrust/weight ratio. Maximum speed is above Mach 2.5 at over 36,000 ft. (11,000 m); Mach 1 is attainable in a climb. Wingspan is 42 ft. 9½ in. (13.04 m) and length 63 ft. 9½ in. (19.44 m).

Delivery of production F-15A single-seaters and TF-15A two-seat trainers to the U.S. Air Force began in 1974.

Eagle Squadrons

Before the United States entered World War II, a number of American volunteers joined the RAF, and eventually three special American-manned RAF squadrons called Eagle Squadrons were created. The first, 71 Squadron, initially equipped with HURRICANES, became operational early in 1941. A second squadron, 121, followed in July of that year, and a third, 133, in Sep-

tember. All three were eventually equipped with SPITFIRES, and in September 1942 formed a unit that began to operate from Debden, Essex, in the east of England. In the same month the Eagle Squadrons officially became U.S. units, forming the 334th, 335th, and 336th Fighter Squadrons of the 4th Fighter Group. During operations with the RAF they had been credited with destroying over 70 enemy aircraft, the majority by 71 Squadron.

In their new guise the Eagle Squadrons continued to operate Spitfires, but in early 1943 they received P-47 THUNDERBOLTS. As part of the Eighth Air Force they flew escort operations, but were not notably successful with the P-47. Late in 1943 one of the 4th Group's senior officers, Lt. Col. Don Blakeslee, was assigned to lead the first P-51 MUSTANG unit on its initial operations. His delight with the aircraft, which had much in common with the Spitfire, led him to persuade the commanding general of VIII Fighter Command to reequip the 4th Group with Mustangs. In the period from the end of February through May 1944, the 4th Group became the most successful fighter group operating from England. Foremost pilot was ex-Eagle Don GENTILE, who claimed some 15 victories during this period. By late in 1944 few of the old Eagle Squadron fliers remained at Debden; most had returned to the United States. But the 4th Group went on to amass the highest combined total of air and ground-strafing victories of all Eighth Air Force groups (1,016).

Eagleston, Glenn T.

Highest-scoring fighter ace of the U.S. 9th Air Force in World War II, Eagleston was sent to England with the 354th Fighter Group, the first American combat unit to operate the Merlin-engined P-51 MUSTANG. Eagleston was credited with the Group's first "probable" when he attacked a Bf 110 Zerstörer near Kiel on December 13, 1943. On February 10, 1944, Eagleston's P-51B was severely damaged when mistakenly attacked by a P-47 Thunderbolt. He was able to fly the aircraft back from Germany but had to bail out near his home base. During the first five months of 1944 Eagleston ran his score up to $14\frac{1}{2}$ victories, and further service added another four enemy aircraft to the total.

Eagleston was given command of the 4th Fighter Interceptor Wing at Suwon in Korea during May 1951. Flying with this F-86 Sabre unit he destroyed two MiG-15 jet fighters to bring his final career score to $20\frac{1}{2}$.

Earhart, Amelia (1898–1937)

The first woman to cross the Atlantic by air (June 17–18, 1928, as a passenger in the Fokker *Friendship* flown by Wilmer Stultz from Newfoundland to Wales), Amelia Earhart later became the first woman to pilot an airplane across the Atlantic when she flew a Lockheed VEGA solo from Newfoundland to Ireland on May 20–21, 1932. Amelia Earhart competed as pilot of a Vega in the Women's Air Derby, Santa Monica–Cleveland, August 1929, and established a women's autogiro altitude record of 19,000 ft. (5,800 m) in April 1931.

In September 1932 she set a women's transcontinental record, Los Angeles–Newark, N.J., in 19 hours 4 minutes. In January 1935 she made a solo flight from Hawaii to California, and on June 1, 1937, left Miami in a twin-engined ELECTRA to circumnavigate the Earth at the equator. Her copilot was Fred Noonan. They were last heard of heading for Howland Island in the South Pacific on July 2, 1937.

Early aviation

No one knows how much time elapsed between the earliest fantasies of flight and the first practical airplane designs. But we do know that once the accurate principles of heavier-than-air flight had been established by the Englishman Sir George CAYLEY, who worked during the years 1798–1853, there was a slow but continuous build-up of theoretical and practical knowledge in Europe and the United States. Fifty years later the Wrights achieved sustained powered flight, and in little more than another decade high-flying airplanes were maneuvering brilliantly over the battlefields of Europe. Between Cayley's remarkable insights and the mass-production of World War I were the work and daring of scores of engineers, test pilots, inventors, and entrepreneurs. Many were inspired amateurs, some were methodical technicians, and still others were farcically—sometimes tragically—wrongheaded in their approach.

Sir George Cayley had worked on the basic problems and set forth the configuration for the classic airplane: fuselage, wings (with dihedral, each wing sloping up for lateral stability), and a rear fixed tail (with an angle between the setting of the wings and horizontal tail for longitudinal stability). Notable attempts were made as early as 1847, by W. S. Henson and John Stringfellow in England, using a steam-driven twin-propeller monoplane launched down a ramp. Later unsuccessful attempts using the same general technique were made by Félix du Temple in France (1874), Alexander MOZHAISKI in Russia (1884), Clément Ader in France in 1890 (leaving the ground, but without any system of control), and Sir Hiram Stevens Maxim in England in 1894 (lifting off in a giant biplane but restrained by a railed track). However, the more theoretical German Otto LILIENTHAL began using what we would call hang-gliders in 1891 to determine how powered flight might be achieved. He was killed in 1896, after he had made over 2,000 glides and laid down a basis for both design and piloting. Percy PILCHER in Britain and Octave CHANUTE in the United States followed with better gliders producing more lift and constructed of stronger materials. When the WRIGHT BROTHERS began their methodical glider experiments in 1901, leading to the first powered and sustained heavier-than-air flight in 1903, they could build on a considerable body of experience and theory. The American Samuel P. LANGLEY failed to fly his 52-hp gasoline-powered craft in 1903, but in 1902–04 Capt. Ferdinand Ferber and Ernest Archdeacon successfully became airborne with Wright-inspired gliders, and in 1905 another Frenchman, Gabriel VOISIN, flew the Archdeacon floatplane glider off the Seine while towed by a motorboat. Voisin used boxkite wings based on the kites flown in Australia from 1893 onward by Lawrence Hargrave. In 1906 the Romanian Trajan Vuia made short hops in a tractor propeller monoplane, and in France an extraordinary Brazilian balloonist named SANTOS-DUMONT made a series of short powered flights (considered the first sustained flights in Europe) in October–November 1906. Santos-Dumont's curious biplane with the tail in front covered a distance of 720 ft. (219 m).

By this time improved engines were becoming available for aircraft propulsion, notably the water-cooled Antoinette and the air-cooled rotary Gnôme first marketed in 1907, the year in which the Voisin brothers, BLÉRIOT, Santos-Dumont, and Henry FARMAN, all working in France, produced and flew outstanding new monoplanes and biplanes that were ultimately to prove more successful than the layout adopted by the Wrights. This was especially the case with the monoplanes of Blériot, the Blériot VI and the 50-hp VII of 1907 being the first such aircraft to fly successfully.

Also in 1907, two Frenchmen made the first tentative hops by a helicopter, the first heavier-than-air VTOL device. Paul Cornu just managed to get off the ground on November 13 near Lisieux, and the Breguet brothers built the Breguet-Richet helicopter, which flew with another 50-hp Antoinette engine on September 29 near Douai, although it lacked a means of control and had to be held by four men.

In North America Alexander Graham Bell's Aerial Experiment Association began to produce results in 1908, when U.S. Army Lt. Thomas E. Selfridge's *Red Wing* flew in March piloted by Canadian F. W. Baldwin. Later that year Selfridge, when flying with Orville Wright, whose machine suffered structural failure, was to be the first person killed in an airplane crash. Baldwin's own machine was *White Wing*, flown in May, and in August Glenn Curtiss' *June Bug* began a long flight program flown by Curtiss—the second American to fly after Wright—(see CURTISS; WRIGHT) and Canadian J. A. D. McCurdy. McCurdy himself built and flew *Silver Dart* in Canada in 1909.

Meanwhile in Britain A. V. ROE, Geoffrey DE HAVILLAND, and S. F. Cody built and flew successful machines, as did Hans Grade in Germany, Jacob Christian H. Ellehammer in Denmark (he had flown tethered as early as September 1906), and Robert Esnault-Pelterie and Koechlin-de Pischoff in France. Esnault-Pelterie deserves particular credit for building an excellent monoplane, with a seven-cylinder air-cooled engine of 30 hp, and reaching speeds of the order of 55 mph (88.5 km/h) under full control in June 1908.

Two events stand out in 1909. The first was the crossing of the English Channel by Blériot in July. The second was the first great aviation meeting in history, organized by a champagne company at Reims in August. The impact of the Reims meeting was very great, both on the general public and on influential politicans and high-ranking military officers. By August 1909 aircraft could be reliably designed, built, and flown. Aviation had now reached its second period of development. See PIONEER FLIGHTS.

Eastern Air Lines

Eastern Air Lines is one of the "Big Four" trunk airlines of the United States. In fleet size and passengers it ranks second among the world's airlines, excluding Aeroflot. Eastern's origins can be traced to Pitcairn Aviation, which began operations in the mid-1920s. Pitcairn was awarded the New York–Atlanta mail contract, and at the end of 1928 it went on to take over Florida Airways' Atlanta–Miami route. In January 1930 the company's name was changed to Eastern Air Transport, and in June it extended its routes north to Boston. Eddie RICKENBACKER became general manager of EAT, and the airline was renamed Eastern Air Lines in 1934. Rickenbacker bought control in 1937 (Eastern had been owned by General Motors) and he remained at the head of the airline until 1963.

A notable innovation in air transport, their air shuttle, was introduced by Eastern in April 1961, when it inaugurated its now famous no-reservation guaranteed-seat New York–Boston and New York–Washington services.

The airline now serves more than 100 cities, with an extensive network covering the eastern half of the United States, and with services to Denver, Los Angeles, Portland, and Seattle. International operations include routes to Canada, the Caribbean, and Mexico. Eastern's present fleet consists of 15 Lockheed Electras, 30 TriStars, 113 Boeing 727s, 3 DC-8s, and 81 DC-9s.

Nearly 28,000,000 passengers are carried annually.

Eckener, Hugo (1868–1954)

Pioneer German airship captain, Eckener was one of the most notable champions of lighter-than-air flight in the years between the wars. Born at Flensburg, Germany, Eckener was educated in Berlin, Munich, and Leipzig and became a journalist and critic of Count ZEPPELIN's rigid airships. A flight at Berlin/Staaken converted him and, in 1906, his yachting skills assisted him in becoming a prominent Zeppelin captain. He established the German Airship Transport Company (DELAG) and operated pre-1914 domestic Zeppelin services. During World War I he was adviser to the German Naval Air Service, training airship crews and directing Zeppelin construction. Immediately after the war he attempted to restart commercial airship travel, and in 1924 commanded the Zeppelin *ZR-3* on its flight from Germany to Lakehurst, N.J. (the *ZR-3*, later the U.S. Navy *Los Angeles*, was a war reparation payment to the United States). Four years later he became captain of the GRAF ZEPPELIN, which made over 100 commercial Atlantic crossings. Eckener also flew the *Graf Zeppelin* to Egypt in March 1929; around the world (starting from Lakehurst in August 1929); on South American services from May 1930; and on an Arctic survey flight in 1931. Eckener refused to cooperate with the Nazis, who relieved him of his airship command shortly before the destruction of the *Hindenburg*—which heralded the end of commercial airship travel.

Eighth Air Force

In terms of men, equipment, and scale of operations, the Eighth Air Force was the largest of the 16 air forces contained within the U.S. Army Air Forces during World War II. It was based in Britain from the spring of 1942 until the summer of 1945, and was the supreme realization of the doctrine of daylight precision bombing. Intended to partner the RAF's Bomber Command in a combined campaign of strategic bombing against German war industry, its first heavy-bomber operations were undertaken in August 1942, but the bombing offensive was then curtailed by the need to build up the Twelfth Air Force to support the North African landings of November of that year.

During the winter of 1942/43 the few B-17 FLYING FORTRESS and B-24 LIBERA-TOR units then involved in operations pioneered the high-level precision bombing techniques that remained basically unchanged to the end of hostilities. Heavy-bomber reinforcements arrived in Britain in the spring and summer of 1943, allowing the campaign to gather strength. During the unescorted daylight missions deep into Germany some of the most intense air battles of history took place. Enormously heavy losses during the fall of 1943 forced the Eighth Air Force to develop ways of giving its fighter units the range needed to accompany bombers to their targets.

During the winter and spring of 1943/44 the Eighth Air Force more than doubled in size, even though a substantial portion of designated reinforcements was diverted to Italy to form a second strategic bombing force, the Fifteenth Air Force. With the advent of the Merlin-powered P-51B MUSTANG fighter, the Eighth could supply fighter escort to its most distant targets. The superiority of the American fighter forces, both in equipment and numbers, wrested air supremacy from the Luftwaffe even over Germany by the spring of 1944. Strategic bombing was diverted to invasion-support operations in the summer of 1944 and did not resume its stride until the fall of that year, delivering the most telling blows in the early spring of 1945, after an extremely bad winter had limited the offensive.

During the final months of the war the Eighth Air Force launched as many as 2,000 bombers and 1,000 fighters on a single day's operations—the largest air striking force ever committed to battle. By the end of the war in Europe the Eighth had delivered 691,470 tons of bombs in 459 operational days of flying. A total of 4,162 heavy bombers were lost, and 42,013 men killed or missing in action with them; 2,053 fighter pilots were killed or missing and 2,222 fighters lost. Fighters claimed 5,222 enemy aircraft shot down in combat and over 4,000 destroyed on the ground by strafing. Bomber gunners downed over 6,000 enemy fighters, and the ammunition used by both bombers and fighters amounted to more than 100 million rounds. Sixteen men were awarded the highest American decoration for bravery, the Congressional Medal of Honor.

Eindecker

Developed from a series of unarmed, single-seat monoplanes, the Fokker E.I *Eindecker* monoplane achieved instant fame on July 1, 1915, when Lieutenant Kurt Wintgens destroyed a French airplane with his E.I's syn-

Fokker E.I Eindecker with Anthony Fokker in the cockpit

chronized machine gun, which was able to fire between the spinning blades of the propeller (see INTERRUPTER GEAR). His success was soon repeated by such German pilots as Oswald BOELCKE and Max IMMELMANN, and the nimble little monoplane quickly achieved a deadly reputation over the western front. The E.III variant was the most widely used, some 258 being built; some aircraft had experimental two- and three-machine gun installations. Wingspan was 31 ft. $2\frac{3}{4}$ in. (9.50 m) and length 23 ft. $7\frac{1}{2}$ in. (7.16 m). Relatively low-powered (Oberursel engines of 80–160 hp) and with a mediocre performance (maximum speed was only 87 mph, 140 km/h), the Fokkers still proved far superior in maneuverability to the slow Allied two-seaters. The advent of the DH 2 and NIEUPORT scout early in 1916 diminished this superiority, at last putting an end to the "Fokker Scourge," and soon after Immelmann's death the Eindecker was withdrawn from operations and replaced with ALBATROS and Halberstadt biplanes. By August 1916 only a few were in front-line use, mainly in the Middle East.

Ejection seat

During World War II a number of fighters were developed from which successful escape in emergency, by the traditional method of bailing out, was hazardous or impossible. The Consolidated-Vultee XP-54, for example, had a propeller placed behind the pilot, so that an ejection seat, which hinged sharply down, was necessary to throw the pilot clear. In the German Dornier Do 335 the tail and rear propeller could be jettisoned, so that the pilot could bail out safely. The Heinkel He 219 V4 and V6 both tested compressed-air ejection seats, and the He 162 jet had a simple cartridge-ejected seat fitted as standard equipment.

By 1946, when the first live air ejection took place with a Martin-Baker seat, peak acceleration had been reduced by using two cartridges in series, while drogues to stabilize the seat were combined with a face blind to protect the occupant. During the 1950s fully automatic seats were introduced, which jettisoned or broke the canopy, fired the seat, released the occupant (even if unconscious) at a safe height and then released his parachute. The rocket seat introduced much greater propulsive power, to cover use on the ground without forward speed.

A very few seats ejected downward, while in some aircraft the whole cockpit was arranged to be jettisoned, bringing the crew down by multiple chutes and thereafter serving as a shelter.

El Al Israel Airlines

El Al, Israel's international airline, was formed in November 1948, and began operating to Europe in the following year, flying DC-4s. Lockheed Constellations were soon acquired for African and transatlantic services. Its first jet aircraft, Boeing 707s, went into use in 1961. El Al flies regular passenger and cargo services between Tel Aviv and 13 European cities, with additional routes to New York, Montreal, Mexico City, Johannesburg, and Nairobi.

Although a number of terrorist attacks have been made on El Al aircraft, the airline has only suffered one successful hijacking. Security arrangements surrounding El Al's operations are considered probably the best in the entire air transport industry.

The current fleet consists of 4 Boeing 747s, 8 Boeing 707s, and 3 Boeing 720Bs. The airline's subsidiary, Arkia, operates smaller aircraft on domestic services. El Al carries some 675,000 passengers annually.

Electra

Name of two American airliners, both built by Lockheed. The original Electra was the Lockheed 10 of 1934, a 10-

passenger twin-engined airliner designed by H. L. Hibbard. It was flown by commercial operators throughout the world on routes up to 900 mi. (1,450 km) and used by the Army and Navy. Powered by two Pratt & Whitney Wasp Junior engines, the Electra cruised at about 200 mph (320 km/h). Its wingspan was 55 ft. (16.76 m) and length 38 ft. 7 in. (11.76 m).

The Lockheed 188 Electra was a medium-range aircraft first flown on December 6, 1957. It was designed to meet the specific requirements of American domestic airlines and was the first (and only) U.S.-built turboprop airliner. Normal passenger accommodation was 74, but some of the 172 Electras that were eventually built were equipped to carry 98. Serious crashes in 1959 and 1960 caused Lockheed to put all aircraft through a modification program, which included strengthening parts of the wing structure. The Electra was powered by four Allison 501 turboprops giving a top speed of 448 mph (720 km/h). Its wingspan was 99 ft. (30.17 m) and length 104 ft. 6 in. (31.85 m).

The maritime reconnaissance P-3 Orion, which was developed from the Electra, was used extensively by the U.S. Navy during the 1960s.

Elevators

Hinged control surfaces which govern the pitch (the nose-up or nose-down attitude) of an aircraft. They are normally part of the tailplane or stabilizer. Moving the control column forward hinges the elevator downward, increasing the lift imparted by the stabilizer and raising the tail so that the aircraft dives. Pulling the control column back raises the elevator, which reduces the lift of the stabilizer and causes the tail to drop, thus enabling the aircraft to climb.

Elevons

Aircraft that have no separate tailplane as such, have instead control surfaces that combine the functions of both ELEVATORS and AILERONS. Such control surfaces are called elevons, and are linked to the control column so that they act in unison (as elevators) when the controls are moved longitudinally, and operate in opposition (as ailerons) if lateral movements of the column are made.

Emily

The Japanese Navy's four-engined KAWANISHI H8K (code-named "Emily" by the Allies) was the fastest flying boat

to see combat in World War II. The H8K was a high-wing cantilever monoplane with nonretractable stabilizing floats and a crew of 10. It made its maiden flight in 1941, but the design needed to be reworked to counteract the airplane's tendency to porpoise on the water. H8K1 production aircraft, with improvements that included a deeper hull, modified planing surfaces, and an enlarged tail fin, were operational by the beginning of 1942.

The 18th machine to leave the factory was fitted with 1,850-hp engines, and it and subsequent aircraft were designated H8K2s. They had a top speed of 290 mph (464 km/h) with a remarkable range of 4,450 mi. (7,160 km). The H8K2 had a wingspan of 124 ft. 8 in. (38.00 m) and length of 92 ft. 3½ in. (28.12 m). Armament consisted of bow, dorsal, tail, and beam turrets, each mounting a single 20-mm cannon, with four hand-held machine guns in lateral positions; each aircraft could carry a bomb load of 4,400 lb. (2,000 kg) or two 1,764-lb. (800-kg) torpedoes. The fuel tanks were fully protected; armor plate surrounded crew positions; and search radar was carried. "Emily" was a very formidable warplane and gained the respect of Allied fighter pilots.

A transport version (H8K2-L Seiku or "Clear Sky") had its armament reduced to only two guns; the 36 transports built could carry up to 64 passengers and supplies. The H8K3, with a retractable dorsal turret and retractable wing-tip floats, failed to achieve production status; only two were built. A total of 167 H8K flying boats of all types were constructed.

F

F-4. See PHANTOM

F4F. See WILDCAT

F4U. See CORSAIR

F.VIIB-3m

F-5. See FREEDOM FIGHTER

F6F. See HELLCAT

F.VII

The Fokker trimotor was one of the most widely used airliners of the 1920s and early 1930s. It was derived from the Fokker F.VII, a single-engined 360-hp high-wing monoplane carrying 8 passengers, which made its maiden flight in 1924. Five of these machines were built for KLM, and a cleaned-up version, the F.VIIA with a 480-hp engine, was employed by a number of European airlines. When a pair of additional engines were fixed below the wings of an F.VIIA for the 1925 Ford Reliability Tour, the result was the first Fokker trimotor, the F.VIIA-3m. Produced with either Wright Whirlwind or Armstrong-Siddeley Lynx engines, F.VIIA-3ms saw extensive service throughout the world. Wingspan of the F.VIIA-3ms was 63 ft. 4 in. (19.30 m) and length 47 ft 10 in. (14.58 m). The 10-passenger F.VIIB-3m had a wingspan increased to 71 ft. 2 in. (21.69 m) for longer services and was also used on many pioneer flights.

F8F. See BEARCAT

F9F. See COUGAR

F-14. See TOMCAT

F-15. See EAGLE

F-16

U.S. air superiority fighter now being built for a number of European countries by General Dynamics. Following a close and bitterly fought industrial competition with the French MIRAGE, Holland, Belgium, Denmark and Norway signed agreements for 348 F-16s to be built in a joint European/U.S. effort. The design of the F-16 goes back to a competition set in motion by the U.S. Air Force in 1972 when GENERAL DYNAMICS and NORTHROP were each authorized to build an experimental fighter demonstrating the effectiveness of new structural materials, methods of control and electronic equipment. It was not clear that either would become operational since there appeared to be no operational requirement. But the European countries that had been flying F-104 STARFIGHTERS for a number of years had begun to seek a successor from European and American companies. The USAF then boosted the prospects of the General Dynamics design by choosing the F-16 as its new light fighter in January 1975. A few months later the four European countries also chose the F-16, and their choice has since been augmented by orders from Iran and Spain.

The F-16 is a single-seat fighter powered by a Pratt & Whitney afterburning turbofan engine of the type fitted to the USAF's new F-15 EAGLE fighter. It has a distinctive engine intake under the fuselage. Deceptively simple in appearance, the F-16 has a performance possibly higher than that of any other fighter. The thrust of its engine is some 50 percent greater than the weight of the craft itself and in combination with sophisticated wing aerodynamics, provides very great maneuverability. The F-16 will be able to out-turn any adversary, for example. Perhaps the most remarkable feature is that the point of balance is so far back that the F-16 would be impossible to fly without a special electronic system that confers artificial stability. The advantage of this seemingly undesirable characteristic is that it makes the tailplane (which normally does no more than provide stabilizing and control forces) bear a part of the lift. The wing can therefore be made smaller, resulting in a substantial saving in weight. Maximum speed is only around Mach 1.6, but the emphasis is on air superiority, with its requirement of extreme maneuverability.

F-80. See SHOOTING STAR

F-84. See THUNDERJET

F-86. See SABRE

F-100 Super Sabre. See SABRE

F-102. See DELTA-DAGGER

General Dynamics YF-16

F-104. See STARFIGHTER

F-106 Delta Dart. See DELTA-DAGGER

F-111

The GENERAL DYNAMICS F-111 fighter was the first operational aircraft in the world to feature a variable-geometry, or SWING WING DESIGN. It serves with the U.S. Air Force as a long-range attack fighter, strategic bomber, and reconnaissance aircraft.

The F-111 was originally designed as an attack fighter to replace the Republic F-105 Thunderchief. It was to be supersonic at all altitudes in order to deliver nuclear weapons below the radar screen. At the same time the F-111 was being designed for the Air Force, the U.S. Navy outlined its need for a new long-range carrier fighter with considerable endurance.

The two sets of requirements were totally different, but in 1961 the U.S. Secretary of Defense called for the choice of a common design. Conflicting needs could only be reconciled through the use of a swing wing. General Dynamics was selected to build the Air Force plane (F-111A), GRUMMAN the slightly modified Navy fighter (F-111B). The Air Force version is powered by two Pratt & Whitney engines of 20,000 lb. (9,072 kg) thrust and has a wingspan of 63 ft. (19.20 m) and length of 73 ft. 6 in. (22.40 m). The F-111B was far too heavy for naval operations, and the order was canceled. The F-111A experienced severe aerodynamic and engine problems, and several were lost through structural failures. There were also serious shortcomings in performance. The F-111A and its improved versions eventually became successful fighters and bombers, flying operationally in Vietnam. But the aircraft has never lived down the early failures and will never be produced in great quantity.

FAA

Control of civil aviation in the United States is vested in the Federal Aviation Administration (FAA), which traces its origin to the Aeronautics Branch of the Department of Commerce, which was formed by the Air Commerce Act of 1926.

An independent Civil Aeronautics Authority came into being in 1938, to be superseded in 1940 by the Civil Aeronautics Administration. The Federal Aviation Agency dates from 1958, becoming part of the Department of Transportation (as the Federal Aviation Administration) in 1967.

Almost half of the FAA's 50,000 staff are engaged on air traffic control duties, manning some 400 airport control towers, 25 air route traffic control centers, and over 300 flight service stations.

Certification of aircrew and the certification of airplane airworthiness are both FAA responsibilities. The Airport and Airway Development Act of 1970 established the Airport Development Aid Program (ADAP) and the Planning Grant Program (PGP), intended to improve the 13,000 facilities within the United States, only about a third of which were publicly owned. Some 4,500 airports have paved runways, 3,900 of them lighted, but only about 60 possess runways exceeding 10,000 ft. (3,000 m) in length. Under ADAP, the FAA was authorized to allocate funds on a cost-sharing basis for airport improvements and planning projects.

Other FAA responsibilities include the alleviation of noise and air pollution by aircraft, airport security to prevent or deter skyjacking or sabotage, and operation of Washington's two airports (Washington National and Dulles International).

FAA development projects include further automation of the air traffic control system by expanding computer

Fairchild C-119 Flying Boxcar

capability, introducing an MLS (Microwave Landing System) to supplement and eventually replace the existing instrument landing system, producing a wake vortex avoidance system, and developing a Discrete Address Beacon System (DABS) to upgrade the present air traffic control radar beacon surveillance system.

Fagot. See MIG 15

FAI

The Fédération Aéronautique Internationale was established on October 14, 1905, as the result of efforts by the Aero-Club de France to form an international body which could supervise the establishment of aviation records and control sporting flying. Representatives from Belgian, German, French, Italian, Spanish, Swiss, British, and American organizations met in Paris in October 1910 to bring the FAI into being.

By 1920 there were 18 affiliated groups, and meetings were being held approximately once a year. Today the FAI includes national bodies from over 50 countries, and supervises balloon, sailplane and parachute events, as well as those for powered airplanes.

The FAI's codes govern record attempts by formulating categories, stipulating the margin by which an existing performance figure must be surpassed, and outlining the procedures that the sponsoring national association must follow to ensure ratification of the new record.

Fairchild-Republic

Today a part of Fairchild Industries, Fairchild-Republic is the result of the acquisition of Republic Aviation by Fairchild-Hiller in 1965. Fairchild was established in 1923, and began to manufacture the FC-1A 200-hp high-wing monoplane. In the years before World War II it produced a series of single-engined lightplanes, notably the models 42, 71 (six passengers), 22, and 24—all high-wing monoplanes—and the F-45 low-wing cabin machine.

World War II aircraft included the M-62 Cornell single-engined trainer (1939; PT 19, PT 23, PT 26), the M-62A Cornell (PT 19A), the Forwarder (RAF Argus) communication aircraft, and the type 91 six/eight-passenger single-engined amphibian (1936), used by the RAF in the Mediterranean for air/sea rescue work.

After the war Fairchild developed the C-82 Packet military transport (1944), 280 of which were built. The C-119 (1947) was a development of the C-82 with a flight deck relocated to give improved visibility and the maximum permissible gross weight raised to 77,000 lb. (35,000 kg). The C-123 Provider twin-engined military transport, originally a Chase design, was produced in the 1950s, and the Fokker F-27 FRIENDSHIP was built under license.

Republic Aviation was founded as Seversky Aircraft Corporation in 1931. Its early products included the SeV-3

F-111

three-seat amphibian and two-seat trainer. In 1937 Alexander de Seversky set a new New York–Havana record in the Seversky Racer, on which the P-35 (EP-1, Model 100) fighter was based. This Twin Wasp-powered 300-mph (480-km/h) interceptor, with a partially retractable undercarriage and twin machine guns, served in some numbers with the Army Air Corps between 1938 and 1942, and was also built as the two-seat Model 2PA trainer. A few EP-1s were still flying as photo-reconnaissance aircraft with the Swedish air force in 1948.

In 1939 the title Republic Aviation Corporation was adopted, and the P-35 design was developed through the XP-41 to the turbosupercharged 375-mph (603-km/h) P-43 Lancer (1939), with four .50 machine guns, which had a notably good performance at altitude. FLYING TIGER pilot Robert L. Scott flew a Lancer over Mount Everest in 1942.

The powerful THUNDERBOLT fighter first flew in 1941, and was followed in postwar years by the THUNDERJET and Thunderstreak. The heavy F-105 Thunderchief fighter-bomber (1955) could attain Mach-2 speeds and saw service in Vietnam.

Only two prototypes were produced of the XF-91 (1949), powered by a J47 turbojet and a tail-mounted rocket motor, which had inverse taper wings (wider at the tips than at the root). The four-engined XF-12 Rainbow photo-reconnaissance aircraft (1946) also failed to achieve production status either for military purposes or as a high-speed airliner. The Seabee single-engined amphibian (1944) proved, however, to be a popular light aircraft. In 1965 it was experimentally fitted with twin 180-hp Lycoming engines to achieve STOL performance.

Fairchild-Republic activities in the 1970s included development of the AU-23 Peacemaker (a single-engined counterinsurgency version of the Pilatus Porter), and the A-10A twin-jet close-support aircraft program.

Fairey

Charles R. Fairey (1887–1956) established Fairey Aviation in July 1915, initially to produce Short 827 seaplanes and Sopwith 1½-strutters under license. The first successful Fairey-designed aircraft was the two-seat Campania (1917). Over 60 of these single-engined float-planes served both from shore bases and from the carriers *Campania* (hence their name), *Nairana*, and *Pegasus*.

A Fairey rework of the Sopwith Baby produced the Hamble Baby, with adjustable trailing-edge flaps. The N9 and

Fairey Battles of the RAF

N10 single-engined seaplanes of 1917 led to the Fairey IIIA and B (limited production in 1918); the IIIC, which was flown by the British, based in Archangel, against the Russian Bolsheviks; the IIID, 207 of which were built for the British services (four of them carried out a formation flight from England to South Africa and back in 1926); and the immensely successful Napier Lion-engined IIIF (1926), which served with the RAF and the Fleet Air Arm from 1927 until 1941. The Gordon bomber and the Fleet Air Arm's Seal were radial-engined developments of the IIIF. Notable among Fairey's other early aircraft were the Flycatcher carrier-borne fighter (1922), 195 of which were built, and the Fawn light bomber (1923).

When the Curtiss-engined Fox two-seat daytime bomber appeared in 1925, it was faster than most contemporary fighters, with a maximum speed of 156 mph (265 km/h). Kestrel engines were later installed, but production was limited to 28 machines. A descendant of the Fox was the Firefly fighter (1925), which had to be extensively altered before it was ordered by Belgium.

Two special Long-Range Mono-planes were also built by Fairey. The first, J9479 (1928), crashed fatally in Tunisia on December 16, 1929, en route to Cape Town, but the second, K1991 (1931), set a new world distance record of over 5,000 mi. (8,000 km; England to Southwest Africa) in February 1933.

The twin-engined Hendon heavy bomber (1930) served with only one RAF squadron, but it was one of the world's first monoplane bombers. The single-engined Battle monoplane light bomber (1936) was produced in considerable numbers despite being poorly armed and capable of only 257 mph (413 km/h). Battle squadrons were badly mauled over France in 1940, and the aircraft was taken off operations shortly afterward.

From the Battle stemmed the P4/34 prototype bomber and the Fulmar two-seat eight-gun fighter (1940). The Fulmar equipped carrier-borne squadrons until the Griffon-engined Fairey Firefly, armed with four 20-mm cannon, entered service in 1943. Already obsolete at the outbreak of World War II, the Swordfish biplane torpedo-bomber of 1934 still had a distinguished combat career and was retained in service until 1945— longer than its intended successor, the Albacore of 1938.

The Barracuda torpedo dive-bomber joined the Fleet Air Arm in 1943 and saw considerable action, but the post-war Spearfish (1945) did not enter production.

After World War II Fairey produced the turboprop Gannet antisubmarine and airborne early warning aircraft (1949), which flew with the Royal Navy into the 1970s, the FD1 research delta-wing (1951), the Gyrodyne (1947) and Rotodyne (1957) compound (rotor and propeller) helicopters, and an ultra-light helicopter (1955).

Among Fairey's last products were the Fairey Delta 2 research aircraft, one of which (WG 774) set a new world speed record of 1,132 mph (1,821.39 km/h) in 1956.

Farman

Henri and Maurice Farman, the sons of an English journalist living in Paris, were among the pioneers of aviation in Europe, and founders of one of the largest French aircraft manufacturers in the years before World War II. In 1907 Henri (1874–1958), flying a Voisin pusher biplane, raised the world speed record to 32.73 mph (52.66 km/h) and the distance record to 2,530 ft. (711 m), although the Wright Brothers had of course far exceeded these records unofficially. On January 13, 1908, Farman won a prize of 50,000 francs for the first circular flight of 1 km (0.6 mi.) in Europe. He went on to design his own

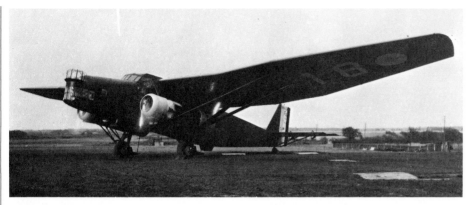

Farman F-222 bomber

pusher biplane with ailerons and the new 50-hp Gnôme rotary engine, and flew it in 1909 for a record 112.5 mi. (181.04 km). In 1912 he joined his brother Maurice (1877–1964), who was also an aircraft designer, to form Avions H. and M. Farman. Two Maurice Farman pusher designs, the "Longhorn" and "Shorthorn" (the nicknames refer to length of their landing skids) were trainers and reconnaissance aircraft in the early stages of World War I. The Henri Farman HF-20, another pusher reconnaissance machine, was replaced by the larger, faster F-40 or "Horace" Farman of 1915, a two-seat biplane fighter powered by a 160-hp pusher engine, with a top speed of 84 mph (135 km/h) and a gunner seated ahead of the pilot. The F-50 twin-engined night bomber appeared toward the end of World War I, soon after which Avions Farman produced the twin-engined GOLIATH 12-passenger transport for the scheduled services of Lignes Farman and other pioneer airlines.

During the next 20 years Farman built numerous military and civil aircraft in quantity. They included single-engined F-70s for Casablanca–Dakar and Algiers–Biskra services as well as for the Polish company Aero; F-121

Jabiru four-engined transports; F-190 single-engined five-seat monoplanes; F-220 and F-2220 airliners designed for Air France's South Atlantic mail routes (1935–38); F-300 eight-passenger trimotors; F-430 light twins; and F-2231 long-range mailplanes. Farman aircraft often appeared ungainly and boxlike, but they were renowned for carrying heavy loads over great distances. The company also designed and manufactured a number of successful 12- and 18-cylinder aircraft engines and did valuable research on cabin pressurization.

When the French aircraft industry was nationalized in 1936, Henri and Maurice Farman retired, and the firm was merged with Hanriot to form Société Nationale de Constructions Aéronautiques du Centre (SNCAC). One of its last notable products were the three Farman F-2234 four-engined monoplanes built in 1938 for experimental high-altitude North Atlantic flights by Air France. In October 1939 an F-2234 flew from France to Rio de Janeiro to join the search for the German pocket battleship *Graf Spee*; later another bombed Berlin. Eventually all three went into civilian use between Dakar and Porto Natal. After the fall of France German planes were built.

Fiat

During World War I the aviation branch of Fiat (then known as the Società Italiana Aviazione) produced the SIA 7B, a 260-hp two-seat reconnaissance biplane (1917). This aircraft proved to be structurally weak, as did the SIA 9B bomber reconnaissance machine of the same year, which had a 700-hp engine and carried 772 lb. (350 kg) of bombs.

The Fiat R.2 (1918) was a smaller, sturdier development of the 7B. It served with the Italian air force as a reconnaissance and bomber aircraft until 1925. Early postwar Fiat designs included the C.R.20 400-hp single-seat biplane, the A.120 high-wing reconnaissance monoplane, the R.22 two-seat reconnaissance biplane, and the B.R.2 single-engined biplane bomber.

In 1932 the C.R.32 fighter made its maiden flight. This 600-hp biplane, with its twin .50 machine guns, was still numerically the leading Italian interceptor in 1939, although it was replaced the following year by the highly versatile 840-hp C.R.42 Falco biplane. The G.50 Freccia monoplane (1937), with the same Fiat A.74 radial engine as the C.R.42, saw extensive service in 1940–1941, but its twin machine guns were inadequate and the MACCHI M.C.200 was generally preferred. A redesign of the G.50, the G.55 Centauro, had a license-built DB 605 V-12 engine and mounted three cannon in addition to its two machine guns. It proved an outstanding fighter, capable of 385 mph (620 km/h). A further development with a DB 603 engine (the G.56) did not enter production. The B.R.20M Cigogna twin-engined bomber (1936) was used by the Italian air force in some numbers, and the trimotor G.12 transport (1940) formed part of Alitalia's fleet when the airline began operations in 1946.

Several trainers were built by Fiat after World War II (the piston-engined

Fiat G.12 of Alitalia

71

Three generations of classic fighter aircraft: Sopwith Triplane of World War I, P-51 Mustang of World War II, Royal Navy F-4K Phantom of the 1970s

G.46, G.49, and G.59; the Nene-powered G.82 jet), and STARFIGHTERS and F-86K SABRES were constructed under license. The G.91 fighter-bomber, trainer (G.91T), reconnaissance aircraft (G.91R) and strike fighter (G.91Y) first flew in 1956 and was subsequently built under license in West Germany, serving with the Italian and Portuguese air forces and the Luftwaffe.

In 1969 Fiat's aerospace activities (other than aircraft engines) were transferred to Aeritalia, established as Italy's principal airframe manufacturer. Fiat's G.222 twin-turboprop military transport (1970), which was ordered by the Italian air force, underwent its development under the new company's direction. Aeritalia also became a partner in the production of the TORNADO.

From almost the earliest days of aviation Fiat was also engaged in the production of aircraft engines. The first Fiat engine was built in 1908, and during World War I 15,000 aircraft power plants were produced. Liquid-cooled V-12 engines manufactured during the interwar years included the 410-hp A.20 for fighters, the 550-hp A.22 developed into AS.2 and 3 racing engines, the geared A.22R used by the SAVOIA-MARCHETTI S.55 and S.56, and 970-hp A.25 fitted to bombers. The 24-cylinder 3,100-hp AS.6 of 1933 powered the Macchi M.C.72, holder of the world speed record; the V-12 A.24R was used for both military and commercial aircraft; and the A.30 powered the C.R.32. Radial engines included the 7-cylinder 140-hp A.50 and 150-hp A.70, the two-row 14-cylinder A.74 and A.76, and the 18-cylinder A.80 (1,000 hp) and A.82 (1,400 hp). Since World War II Fiat has been engaged largely in production of engines under license from other manufacturers (engines such as the Ghost, J35, Orpheus, J79), as well as in component manufacture and overhauls.

Fighter

During the early months of World War I, the pilots and observers of opposing reconnaissance aircraft began to use rifles and revolvers to fire at each other. Although pushers gave a better field of fire for a machine gun mounted in the front cockpit, the more maneuverable tractor was at an advantage in a dogfight even though its machine gun had to be fitted above the center section of the upper wing to fire at an upward angle.

To make the best use of the maneuverability of an aircraft with a front-mounted propeller, it needed forward-firing guns so that the pilot could simply aim his aircraft at the target, but the spinning propeller blades were always in the way. The French experimented with armor-plating the propeller blades of some of their aircraft in order to deflect the bullets from a fuselage-mounted gun, but FOKKER developed an INTERRUPTER GEAR for the Germans, which enabled the gun to fire efficiently between the blades. With this equipment, the Fokker EINDECKER became preeminent along the western front in 1915–16, although the more agile ALBATROS D.II and D.III biplanes with their twin machine guns eventually supplanted it by the summer of 1916. The continuing search for maneuverability led to the appearance of Sopwith and Fokker TRIPLANES in 1917, but biplanes had reestablished their ascendancy by the end of World War I (Sopwith CAMEL, Fokker D.VII).

Wartime designs continued in use during the early postwar years, and the fighters of the late 1920s and early 1930s were merely more powerful, aerodynamically refined developments of the basic two-gun biplane type. A revolution in aircraft design came in the mid-1930s and established the monoplane as the standard fighter configuration, with an armament of up to eight machine guns (as in the SPITFIRE and HURRICANE), a 20-mm cannon and two machine guns (BF 109D), or two 20-mm cannon and two machine guns (RATA).

The emphasis on overall performance was maintained through World War II, although wingspans were sometimes reduced, or "clipped," to increase maneuverability at low level. The high rate of fire of the .303 machine gun was found insufficient to compensate for its small caliber, and 20-mm cannon were adopted by Britain, although the United States continued to rely largely on .50 machine guns. By the end of the war piston-engined fighters were operating at over 30,000 ft. (9,000 m) and attaining speeds of 450 mph (720 km/h) or more, with laminar-flow wings and engines developing 3,000 hp. Jets first appeared in 1944–45 (ME 262, METEOR), and within 10 years supersonic speeds were being regularly achieved in dives. AREA RULE configuration made possible level flight at speeds over Mach 1, and cannon guns entirely supplanted machine guns.

Performance became for a period virtually the sole criterion in the design of fighter aircraft, exemplified by the STARFIGHTER (the "missile with a man in it"), with a six-barrel rotary cannon and provision for air-to-air missiles. Two-seater day fighters (such as the Bf 110 ZERSTÖRER) had previously been largely unsuccessful due to their size, but the advent of complex, electronically-governed weapon systems brought a renewal of interest in multiseat fighters (F-4 PHANTOM, Tu-28 "Fiddler"), with a tendency to replace guns entirely by missiles. Experience over Vietnam indicated that dogfights were not entirely things of the past, and in the mid-1970s there was a renewed interest in maneuverability (as in the F-16) at the expense of outright performance, while cannon guns again became standard armament.

Night fighters (also used as INTRUDERS) have always tended to be larger and heavier than interceptors, partly to give maximum endurance and partly to accommodate a second crewman for operating electronic equipment. In this case maneuverability is not the prime consideration.

Ground-attack fighters are generally adaptations of interceptors. They carry heavy cannon guns, rockets, and bombs, and are usually protected by extra ventral armor against ground fire.

Fishbed. See MIG 21

Flaps

Flaps are fitted to the inboard section of the trailing edges of an aircraft's wings in order to improve handling qualities at low speeds. Lowering or extending the flaps increases the camber of the wing and thus the lift the wing is able to provide (see WINGS). As a result, the STALLING speed of the aircraft is lowered and the landing speed reduced. The use of flaps also increases the drag exerted on the airplane by the SLIPSTREAM, so that the landing approach follows a steeper angle of descent to the runway. The use of flaps on takeoff again augments the lift, and permits a steep climb.

Simple flaps are merely hinged surfaces at the back edge of a wing. To preserve smooth airflow over the upper part of the wing, split flaps fitted only to the lower surface may be used. A variation of this pattern is the Fowler flap, which extends backward and has the additional advantage of increasing the actual wing area.

Slotted flaps have a gap at the hinge line between the wing and the flap through which the airflow is speeded up, thus delaying the onset of the TURBULENCE that leads to stalling. Leading edge slats perform a similar function but are located at the front of the wing. At low speeds they open automatically and the airflow is speeded up through the gap between the slat and the wing to delay the onset of a stall. At high speeds, they are held shut by air pressure.

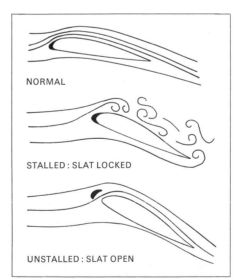

Automatic leading edge slats, like slotted flaps, increase the airflow over the upper surface of the wing. This delays the onset of turbulence and consequent stall. In normal flight the slots lie flush along the leading edge of the wing. As the angle of attack increases the slat moves forward, and a smooth airflow can be maintained over the wing. If the slat remains locked (e.g. for aerobatics) turbulence and stalling will occur

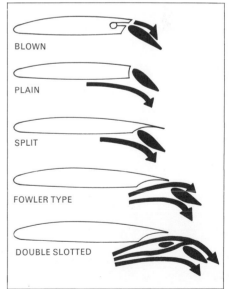

Typical types of flaps, devices that increase the lift provided by the wings and the drag exerted by the slipstream. Slotted flaps increase lift still further by providing a smooth airflow over the upper surface of the wing; blown flaps, often using air bled from an engine compressor and expelled at the hinge line, perform a similar function

Flight simulator

One of the earliest flight simulators was the Link Trainer, first sold by Edwin Link to the U.S. Army Air Corps in 1934 to help improve the ability of mail pilots to fly on instruments. The Link Trainer, often called "the blue box," was a simple representation of an aircraft cockpit, fitted with all primary flight controls and instruments, and surrounded by a crude "airframe." Mounted on pivots about all axes, the device could be positioned by mechanical and pneumatic actuators so that its motion obeyed the commands of the pilot inside. In particular, all instruments faithfully responded as they would in a real aircraft (except that aerobatic maneuvers were not possible). By 1939 an optional extra was a "crab," a small wheeled device driven by electrical signals from the Trainer to crawl slowly across a map in front of an instructor, showing the student's accuracy of navigation. In World War II more than 700,000 student aircrew, nearly all of them pilots, received Link Trainer instruction. Pupils could learn instrument flying much more cheaply than if they were actually in the air, with greater safety and reliability, and at any time of day or night, while at the same time aircraft were released for other purposes; later it became possible for the instructor to inject simulated faults or malfunctions, to test the pupil's ability to handle emergencies.

Today there are numerous "general aviation trainers," not unlike the classic Link in principle, used to teach pupils to fly lightplanes (especially on instruments) and handle emergencies. But a far bigger market is concerned with advanced transport and combat aircraft. Here the simulator is a vital part of the crew training and ongoing pro-ficiency testing. Advanced simulators are designed to reproduce the exact entire flight deck (sometimes, in maritime reconnaissance planes, for example, the whole fuselage) of a particular aircraft. The flight deck may be coupled to a powered motion simulator to produce both linear and rotary motions. The whole system is connected to a computer, usually of the digital type, which can introduce any conceivable in-flight emergency. Through the flight deck windows can be seen an exact representation of the surroundings of an airport, right down to the runway in use, or a simulated air or ground target, or even an outline of another aircraft for combat training. Flight simulators are also used as tools to aid aircraft design.

Floatplane

Floatplanes are aircraft intended for operation off water which are equipped with planing surfaces that do not form an integral part of the fuselage structure (see SEAPLANE).

Normally a pair of long floats or pontoons are mounted below the fuselage or wings; subsidiary stabilizing floats toward the wing tips are unnecessary. Some floatplanes have only a single central float, however, and are equipped with auxiliary stabilizing floats.

Floats can also be fitted to some landplanes in place of a wheeled undercarriage, and various warplane types have been modified in this way, particularly fighters (Spitfire, Zero) and torpedo bombers (Swordfish, Albacore). The adverse effect of the floats on performance and the availability of carrier-borne aircraft led to the abandonment of these and similar projects.

Before the advent of helicopters, small floatplanes were often carried on warships (Vought OS2U Kingfisher, Nakajima E8N "Dave," Arado 196), and even on submarines (Yokosuka E14Y "Glen," Arado 231).

Present-day manufacturers of light aircraft occasionally market floatplane versions of their models for use from lakes or rivers in areas where conventional landing strips are unavailable.

Flyer

The *Flyer* was the WRIGHT BROTHERS' first airplane. On December 17, 1903, the *Flyer* reached a speed of 8 mph (13 km/h) traveling into a 21 mph (34 km/h) headwind, and took-off. This flight lasted just 12 seconds, but the fourth and last flight that day was 59 seconds long and covered 852 ft. (260 m). These were the world's first sustained and controlled man-carrying powered flights.

The *Flyer*, a biplane with a wingspan of 40 ft. 4 in. (12.29 m), a wing area of 510 sq. ft. (47.38 m²), and an overall length of 21 ft. 1 in. (6.43 m), had two counter-rotating pusher propellers driven by bicycle chains and powered by a 12-hp gasoline engine designed and built by the Wrights. Control was by double rudder and elevator structures, and by wing-warping (a system of twisting the wing tips to achieve controlled banked turns). The pilot lay on the lower wing beside the engine; the loaded weight of the *Flyer* was only some 750 lb. (340 kg).

Later designated *Flyer I*, this historic airplane was followed by *Flyer II* and in 1905 by *Flyer III*, generally considered the world's first fully maneuverable and "practical" airplane, rather than a powered glider. On October 5, 1905, the *Flyer III* covered a distance of over 24 mi. (38.6 km) in 38 minutes 3 seconds.

On the 45th anniversary of its first epoch-making flight, *Flyer I* was formally given a permanent home in the Smithsonian Institution (it is now exhibited in the National Air and Space Museum).

Flying Bedstead

The name "Flying Bedstead" was given to two Rolls-Royce aircraft-engine thrust-measuring rigs, looking somewhat like four-poster beds. Powered by two Nene jet engines, the first rig made tethered flights in July 1953, followed by free flights from August 1954. These constituted the first VTOL aircraft flights that did not use rotating wings.

Arado 196 floatplane

The Wright Flyer *making its historic flight on December 17, 1903; Orville Wright piloting, Wilbur Wright on foot*

631 six-engined flying boats were flown only briefly by Air France in 1947 (on services to the West Indies). Another giant, Howard HUGHES' Hercules, the aircraft with the largest wingspan ever built, made only one short flight. Military flying boats were still used in the 1970s by the Soviet Union (BERIEV Be-8 "Madge" and Be-12 "Mail") and Japan (SHIN MEIWA), but attempts to produce jet-powered flying boat fighters (Convair Sea Dart, 1953; Saunders-Roe SR/A1, 1947) were unsuccessful. The high drag of the bulky flying-boat hull, the presence of well-equipped airports and long runways in almost every part of the world, and the reliability of modern aircraft engines over the longest flights are all factors that have worked against the revival of the golden age of the flying boats.

Flying boat

A flying boat is a SEAPLANE in which the principal planing surface is an integral part of the fuselage structure (or hull). Subsidiary stabilizing floats (sometimes retractable) are usually fitted toward the wing tips, or instead sponsons may be attached to the fuselage sides. Since even the largest flying boats do not require the expense of airport construction, and given that such aircraft could cross the ocean in stages and enjoy a certain added safety factor in the event of an emergency in mid-ocean, it was natural that they should play a major role in pioneering long-distance civil aviation in the years between the wars.

The American aviator and designer Glenn CURTISS is credited with the design of the first flying boat (1912), and large numbers of ocean patrol flying boats were built by Curtiss during World War I. One aircraft, the NC-4, crossed the Atlantic (in several stages) in 1919 (see PIONEER FLIGHTS). During World War I the British Felixstowe flying boats were also successful in both maritime reconnaissance and bombing roles, and their example led to a series of postwar RAF flying boats that included the Supermarine Southampton (1925) and SHORT Rangoon (1930) and Singapore (1926–34).

Lufthansa used the WAL and its successors, also built by DORNIER, on transatlantic mail services, and Britain's Imperial Airways employed flying boats on many of their long-distance routes in the late 1920s and early 1930s, using aircraft such as the Short Calcutta and Kent. In the years immediately before World War II, American flying boats pioneered transpacific routes (SIKORSKY S-42, MARTIN 130) and shared the first commercial flights across the North Atlantic with British Short C-class "Empire Boats" (normally used on routes to Africa, India and the Far East).

Maritime reconnaissance flying boats were widely used in World War II by Britain (the Short Sunderland), the United States (CATALINA and MARINER), and Japan (KAWANISHI H6K and the outstanding H8K EMILY). German-designed flying boats included the Dornier Do 18 and Do 24 and the BLOHM UND VOSS Bv 138 and huge six-engined Bv 222 Wiking.

After 1945 flying boats never recovered their prewar eminence. Long-range land-based troop carriers developed during the war led to a new generation of four-engined airliners on intercontinental routes, although the BOEING 314 and civil derivatives of the Sunderland (Sandringham and Solent) saw a few years' more service. The giant SAUNDERS-ROE Princess never became operational, and the French Latécoère

Flying doctor services

A number of organizations providing airborne medical attention and ambulance services now exist, but the idea of a flying doctor service was pioneered in Australia. The first such service was initiated in May 1928 by the Aerial Medical Service of Australia, using a DH 50 chartered from Qantas. Alfred Traeger, an electrical engineer, developed a radio transmitter/receiver powered by a pedal-operated generator, and patients and doctor were able to communicate in Morse code over a range of 300 mi. (480 km).

After the withdrawal of Australian government support during the Depression, the administration of Queensland stepped in to assist, and in 1933 the service was extended to Victoria. In 1934 radio telephony sets replaced the Morse code units, simplifying the task of communications, and in 1942 the organization was officially designated

Imperial Airways' Short "Empire Boat" Caledonia *over Manhattan, July 1937, after the airline's first North Atlantic survey flight*

B-17G Flying Fortress

The Flying Doctor Service. By the mid-1970s it had 13 bases, was equipped with sophisticated radio equipment and modern airplanes such as the Beagle 206, and covered roughly two-thirds of Australia. The service is free; finance is provided principally by government grants and private donations.

Similar services have been developed in other parts of the world. In 1947 the International Grenfell Association and the Saskatchewan Air Ambulance Service began operations in Canada, and in 1961 the African Medical and Research Foundation established the first flying doctor service on the African continent.

Flying Fortress

World War II U.S. heavy bomber, probably the most famous of all American military aircraft. The Flying Fortress was the world's first all-metal four-engine monoplane bomber, appearing first in 1935. The relatively high cost restricted U.S. Army Air Corps orders to only a few dozen examples of the prewar models, but in 1941 a complete redesign of rear fuselage and tail produced the B-17E, which was put into quantity production and was soon followed by the basically similar but more combat-worthy B-17F.

By mid-1942 three plants were producing this model which was replaced on the production lines by the "chin" turreted B-17G the following year. The later versions were heavily armored and carried as many as 20 machine guns, to make up for the lack of fighter escorts. By the time production was terminated in 1945, a total of 12,731 Fortresses of all kinds had been made.

Early Fortresses were first used in combat by the RAF in 1941 and at the end of that year by USAAC units in the Pacific. Thereafter the majority of B-17s went to equip the U.S. EIGHTH AIR FORCE in Britain; together with a smaller number with the 15th Air Force in Italy, they played an important part in the strategic bombardment of Germany. Powered by four 1,200 hp Wright R-1820 radials, the B-17G had a top speed of 295 mph (474.6 km/h). Its wingspan was 103 ft. 9 in. (31.6 m) and length 74 ft. 9 in. (22.7 m).

Flying Pencil

So slender was the twin-engined Dornier Do 17 mailplane of 1934 that it earned the nickname "Flying Pencil." Rejected by Lufthansa because of its cramped passenger accommodation, it was taken into use by the Luftwaffe as a medium bomber and began to reach operational squadrons in 1937 (Do 17E and F, with glassed-in nose cap and twin fins). The Do 17Z (1939) had an enlarged nose to accommodate a four-man crew. The Do 215 was an export version of this variant.

The Do 217 was a faster and larger development of the original Do 17. It was stressed for greater loads and initially had air brakes for dive-bombing. Wingspan was 62 ft. 4 in. (19.00 m) and length 59 ft. 9 in. (18.20 m); maximum speed was 320 mph (515 km/h). The aircraft began to reach the Luftwaffe in 1941 (Do 217E), and by the following year had also been adapted for night fighting (Do 217J). Before production ceased in 1943, the Do 217K and M variants had been evolved, with very bulbous noses. The Do 217N was a night fighter based on

Do 17 Flying Pencil

the 217M.

The Do 317 did not proceed beyond the prototype stage.

Flying Tiger Line

The all-cargo Flying Tiger Line is the world's third-largest airfreight carrier, surpassed only by Aeroflot and Pan American. It was formed in 1945 as National Skyway Freight Corporation. Its founders had flown with Claire Chennault's "Flying Tigers" over China during the war, and the airline assumed its present name in 1946. A cargo-only scheduled airline, Flying Tiger was soon equipping itself with war-surplus DC-3s, Curtiss Commandos, and DC-4s. It began a scheduled coast-to-coast service in 1949 and started a route to the Far East in 1969. The company also operates worldwide airfreight charters and military air-transport contracts. The present fleet consists of 3 Boeing 747 freighters (with more on order) and 16 DC-8 freighters.

Flying Tigers

Popular name for the American Volunteer Group formed to assist China before America entered World War II. The project was the brain-child of Claire L. Chennault, a retired American officer acting as an air adviser to the Chinese. In 1941 he obtained unofficial sanction to recruit 100 pilots from the American military services, together with a contingent of ground personnel. One hundred Curtiss P-40 fighters (later marked with the distinctive shark's-mouth decoration that gave the group their name) were shipped to Burma, and the force of three squadrons began training during September 1941, its original purpose being to provide air defense for Chinese cities.

The first action involving the AVG took place on December 20, 1941, when its P-40Bs (see WARHAWK) made successful interceptions of Japanese bombers attacking Kunming, in southwest China. The Group also saw extensive combat over Burma and had many successes during the early months of 1942.

Although the P-40's climb and maneuverability were inferior to that of Japanese fighters, Chennault trained his pilots to make the best use of the P-40's superior diving speed and firepower. These tactics enabled Flying Tiger pilots to destroy large numbers of enemy aircraft.

In June 1942 the AVG was officially absorbed into the U.S. Army Air Force as the 23rd Fighter Group, continuing operations in the China–Burma–India

Distinctive "shark's mouth" of a P-40B of the Flying Tigers

theater of operations until the end of the war.

Focke-Wulf 190

Designed around the BMW 139 radial engine, the Fw 190 is generally considered the best German fighter of World War II. It made its first flight in July 1939, and was eventually used in a variety of combat roles, some 20,000 Fw 190s being eventually built. Production Fw 190A-2s, with 1,600-hp engines and an armament of two 20-mm cannon and four machine guns, began reaching operational squadrons in May 1941. These were followed by progressively more powerful versions with four cannon and two machine guns. "Tropicalized" aircraft were sent to North Africa, and fighter-bomber versions could carry up to 3,970 lb. (1,800 kg) of bombs (Fw 190F). There were also Fw 190 night fighters and Fw 190G ground-support aircraft. The Fw 190B had a DB 603 engine but did not enter production, due largely to the outstanding performance of the 426-mph (685-km/h) Fw 190D interceptors powered by a Jumo 213 power plant, which entered service in the winter of 1943–44. The Fw 190D had a wingspan of 34 ft. 6 in. (10.50 m) and length of 33 ft. 5 in.

(10.20 m). Final development of the highly successful Fw 190 series was the Ta 152, which was designated in recognition of its designer Kurt TANK. A few Ta 152E photo-reconnaissance aircraft and Ta 152H high-altitude fighters were delivered to the Luftwaffe, but the only variant to see real operational service was the Ta 152C, which had a DB 603 engine with methane/water and nitrous-oxide injection. Maximum speed was 472 mph (760 km/h) at 41,000 ft. (12,500 m).

Focke-Wulf 200. See KONDOR

Fokker, Anthony Herman Gerard
(1890–1939)

Born in Java of Dutch parents who returned to the Netherlands in 1894, Fokker built his first aircraft in 1910 and went on to produce a number of monoplanes in the years immediately preceding World War I, some of which were purchased by the Germans.

His M5 design became the German E.I in military use. Equipped with Fokker's revolutionary INTERRUPTER GEAR that enabled a machine gun to fire through the propeller arc, the Fokker EINDEKKER became the most

Focke-Wulf 190G, with drop tanks under the wings for added endurance and a 1,110-lb. (500-kg) bomb under the fuselage

formidable fighter on the western front. In 1917 the Fokker Dr.I TRIPLANE that was to be made famous by Manfred von RICHTHOFEN made its appearance in combat, and the following year witnessed the debut of both the superb D.VII and the parasol-wing D.VIII monoplane.

After the Armistice, Fokker escaped back to his native Holland with six train loads of aircraft and spares to reestablish himself as an airplane designer at Amsterdam.

Realizing the potential of civil aviation he produced a number of successful passenger transport designs—the single-engined F.II and F.III, which were both used by the newly-formed KLM, followed in 1924 by the first F.VII (a single-engined machine carrying 8 passengers). The following year the three-engined F.VII-3m appeared, 116 examples of this type eventually being built. In addition to its success as a commercial airliner, the F.VII-3m completed a number of long-distance flights in the hands of Van Lear Black, Sir Charles Kingsford Smith (*Southern Cross*, the transpacific record-breaker), Lt. Cdr. Richard BYRD (to the North Pole), and Amelia EARHART (*Friendship*, which flew the Atlantic).

The Fokker trimotor airliners evolved through the F.IX, F.XII and F.XIV to the five F.XVIIIs (of which *Pelikaan* and *Snip* were long-distance record-breakers).

Military designs were still being produced, and the 1920s saw the appearance of the D.XIII fighter, together with the C.IV and C.V reconnaissance planes.

The American Fokker company, founded originally as a sales and information bureau, but later a manufacturing concern, came under the control of General Motors in 1929, but following the crash of a TWA Fokker trimotor in 1931 doubts were cast on the composite wood-and-metal construction practised by Fokker. The all-metal airplane was now favored, and Fokker was forced out of the General Aviation Corporation (which the U.S. company had become in May 1930).

Back in Europe, Fokker produced the four-engined F.XXII and F.XXXVI airliners, but the all-metal Douglas DC-2 was now available to KLM and other airlines, and neither of these big Fokkers went into production. The last Fokker biplane fighter was the D.XVII of 1931, the D.XXI monoplane appearing in 1936 together with the twin-boom G.I, followed by the pusher/tractor D.XXIII of 1939. Thirty-four T.V twin-engined bombers were built from 1937 onward, while examples of T.VII-W

seaplanes reached both the RAF and the Luftwaffe after the occupation of the Netherlands in 1940.

Fokker himself died in 1939, but after World War II his company produced the S-11 primary trainer, the S-14 Mach Trainer, the F-27 FRIENDSHIP airliner and the faster medium-range F-28 Fellowship. The concern is now a subsidiary of Zentralgesellschaft VFW-Fokker mbH, located in Amsterdam.

Fonck, René Paul (1894–1953)

With a score of 75 officially credited victories (he privately claimed 127), the Frenchman René Fonck was the leading Allied fighter pilot of World War I. He joined the French army in 1914 and in February 1915 achieved a transfer to the air service for pilot training. With his first squadron, *Escadrille* C.47, he flew Caudron two-seaters on general reconnaissance and bombing duties, scoring his first aerial victory on August 6, 1916. Several further victories led to his transfer to the élite Cigognes ("Storks") Group, in which he was assigned to *Escadrille* Spa 103 on April 15, 1917. Renowned for his marksmanship, Fonck rapidly added to his tally and by February 1918 he was credited with 21 victories, to which he added 10 more by May 9. On that one day he scored a remarkable 6 victories, and he continued to achieve many doubles and triples during the remaining months of the war.

After the war Fonck became a stunt and aerobatic pilot. He attempted a transatlantic flight from New York in a Sikorsky S-35, but he crashed on takeoff and his two companions were killed. He later helped reorganize French fighter defenses in the 1930s.

Ford Trimotor

Nicknamed the "Tin Goose," the Ford Trimotor was an all-metal, high-wing monoplane airliner with a corrugated fuselage and spartan interior. It was a sturdy and extremely popular aircraft on heavily-used routes throughout the United States and in a number of foreign countries, and was only superseded by the new generation of "modern" airliners, typified by the Boeing 247 and DC-2 and DC-3.

The Trimotor was manufactured in two major versions by Stout Metal Airplane Division of Ford Motor Co. at Dearborn, Mich. The 4-AT (2 crew and 11 passengers), powered by three 300-hp Wright Whirlwind engines, had a wingspan of 74 ft. (22.56 m) and length of 49 ft. 10 in. (15.19 m); 86 were built between 1926 and 1933. The 14-passenger 5-AT (wingspan 77 ft. 10 in., 23.72 m, length 50 ft. 3 in., 15.31 m) used 420-hp Pratt & Whitney radial engines; a total of 116 were built in the period 1928–33.

The specially adapted 4-AT named *Floyd Bennett* and used by Richard E. BYRD on the first flight ever made to the South Pole is preserved at the Henry Ford Museum, Dearborn.

Foxbat A/B. See MIG 25

Freedom Fighter

The first in the series of NORTHROP F-5 tactical air-superiority fighters were known as Freedom Fighters (later variants of the F-5 appeared under the name Tiger). First flown in 1959, the F-5 was specifically developed for small air forces needing a versatile lightweight fighter that was relatively inexpensive to operate and maintain. The F-5 has proved highly successful in these aims, and some 15 Nato and Seato countries purchased it during the 1960s. The single-seat F-5A was the most widely used model; it was powered by two General Electric J85 turbojets giving a top speed of 945 mph (1,520 km/h).

Ford Trimotor

Armament consisted of two 20-mm cannon together with seven external stations for guided missiles or bombs. Its wingspan was 25 ft. 3 in. (7.70 m) and length 47 ft. 2 in. (14.38 m).

The later F-5E model had an improved performance with greater maneuverability and was used by U.S. Air Force squadrons to play the part of MiGs in training exercises.

French Air Force

Established in 1910 with five aircraft (a Blériot, two Farmans, and two Wrights), France's Aviation Militaire had grown to 25 squadrons (4 in the French colonies) by the outbreak of World War I in 1914. At the Armistice four years later, the Aviation Militaire totaled 3,500 aircraft organized in 255 squadrons, among them many scout units flying Spads and Nieuports. In the years after World War I the Aviation Militaire was the world's largest air force, and it was equipped almost entirely with French-designed aircraft. A major problem was, however, that service aircraft were manufactured by more than 30 different companies, though in the 1920s the general use of the Gnôme–Rhône (Bristol-license) Jupiter did something to reduce the wide variety of engines employed.

In 1933 a major reorganization established a separate Air Ministry, and the air force became known as l'Armée de l'Air. Although still the most powerful air force in the world in 1935, with active squadrons numbering 160, it suffered from poor direction and what continued to be an extremely fragmented manufacturing industry. Nationalization and regrouping of most of the companies in 1936 created chaos, and swift expansion in the late 1930s did not take place as planned. Vast orders were placed in the United States in 1938–39, but in 1940 the Armée de l'Air flew aircraft generally inferior to those of the Luftwaffe. Chief fighter types were the Morane-Saulnier 406, Curtiss Hawk 75, Bloch 151, and Dewoitine 520; principal bombers were the Potez 63, Breguet 691, Amiot 143 and 350, Farman 222, and LeO 45. After the fall of France a few aircraft and crews escaped to fight with the FAFL (Free French Air Force); most stayed with the Vichy Air Force flying on the side of the Axis.

In 1945 the Armée de l'Air was re-established, and soon became heavily engaged in wars in Indochina (until the French withdrawal in 1954) and Algeria (until independence in the 1960s). Today the Armée de l'Air is based almost wholly in metropolitan France. It is well

F-5 Freedom Fighters

The French Air Force's supersonic strategic bomber, the Mirage IVA, capable of delivering nuclear weapons

equipped with the MIRAGE III, 5 and F1 fighters, JAGUAR attack aircraft, Mirage IVA bombers (with KC-135 tankers), and a wide range of transports, helicopters, and trainers.

The Aéronavale (French naval aviation) was formed in 1912. It divided into carrier- and shore-based groups in 1925, and by 1940 was a large but poorly-equipped force, most of the aircraft of which managed to reach Allied territory after the French defeat. Today the Aéronavale is one of the largest naval air arms in the world, with two active carriers, squadrons of Crusader fighters, Étendard attack aircraft, Alizé and Super Frelon antisubmarine machines, and Neptune and Atlantic patrol and antisubmarine planes, backed up by numerous helicopters and a helicopter carrier. ALAT (Aviation Légère de l'Armée de Terre) was formed in 1952 and is equipped mainly with lightplanes and helicopters. ALAT was one of the

pioneers in the use of armed helicopters in guerrilla warfare; among the chief helicopter types it employs today are the Gazelle, Puma, and Alouette III.

Friendship

A safe, proven and economical turbo-prop airliner designed by FOKKER, the F-27 Friendship was intended as a replacement for the DC-3 for short-haul services. The plans for the Friendship date back to the early 1950s, when the DC-3 had already been in service longer than any previous transport, and a number of manufacturers, with efficient and reliable gas turbines at their disposal, set out to challenge the famous piston-engined airliner. In 1955 Fokker flew a high-wing airliner, powered by two 2,050-hp Rolls-Royce Dart turboprops, which it called the Friendship. The aircraft has been in continuous production ever since, testifying to the soundness of the original formula, and some 650 have been built for both civil and military application. A typical F-27 weighs about 45,000 lb. (20,385 kg) fully loaded, some 80 percent more than the DC-3, and cruises at nearly 300 mph (482 km/h), almost twice as fast. With a wingspan of 95 ft. 2 in. (29.00 m) and a length of 77 ft. 3 in. (23.56 m), it can accommodate 40–44 passengers against the DC-3's 21–28. The Friendship's greatest advantage is its much higher productivity, the cost per seat-mile of flying the F-27 is a fraction of the cost of running a DC-3 or similar aircraft.

Fuel

Aviation fuel is basically either gasoline or kerosene. High octane gasoline is used for piston engines, while jet engines, or the basically similar gas turbines that power turboprop aircraft, use either kerosene or gasoline. The calorific value, or amount of heat energy in each pound of fuel, varies little between the two types; other properties are more important.

Piston-engine fuel is called Avgas (Aviation gasoline), and its most important characteristic is its resistance to "knock" or detonation—the violent and premature explosion of fuel that can occur in place of normal smooth burning and can quickly wreck a power unit. The degree of knock resistance is defined by octane numbers—the higher the number the greater the resistance. For many years an 87-octane fuel was available for low-powered piston engines, but as the result of rationalization and standardization, this has recently been replaced by a 100-octane fuel given the designation 100L. High-performance engines require fuels specified by two octane numbers—one for the lean running used during cruising and another for the rich running employed during takeoff at full power. Typical grades are 100/130 (lean/rich) and 115/130. Other fuel characteristics such as volatility are important, but to a lesser degree.

In jet engines and gas turbines the fuel burns continuously and at substantially constant pressures, without fluctuations or peaks, so that knock resistance is of no importance, though other properties such as specific gravity and calorific value are. Many jets and turboprops run on kerosene, known variously as JP1, Avtur or ATF (Aviation Turbine Fuel) and ATK (Aviation Turbine Kerosene). This is essentially the same as domestic heating kerosene and has the merit of being relatively uninflammable in the event of a crash. Some aircraft, however, use Avtag

(Aviation turbine gasoline) or the much more inflammable JP4. This is a cheap-to-produce "wide-cut" gasoline, that is, a gasoline containing a wide range of the various grades of fuel extractable during the refining process. Among the characteristics of jet fuel that may have to be held within tight limits is thermal stability. This is particularly important for supersonic aircraft. Such high-performance aircraft are heated rather than cooled by their passage through the air, and the fuel may be at 392°F (200°C) by the time it reaches the engine.

Fuel tank

In all aircraft constructed until roughly 1944, fuel tanks were separate from the airframe structure; they were generally made of aluminum and could be removed. One major advance, made in the mid-1930s, was the development of the first of several types of self-sealing tanks. The walls of such tanks have a sandwich-like cross-section, with a layer of material that quickly swells on contact with the fuel. Any holes made by bullets or shell splinters rapidly close themselves and the escape of most of the tank's contents is prevented.

Present-day fuel tanks are generally of two types. The flexible bag tank or bladder is made of multiple layers of nylon and neoprene rubber. The integral tank, developed during World War II, does not exist as a separate entity but is a specially sealed portion of the airframe. Modern transport and military aircraft have integral wing tanks, and often integral tanks in the fuselage and even the fin or flaps. For

extended range military aircraft sometimes carry external drop tanks, jettisoning them when they are empty.

Explosions in tanks can be prevented by purging the space above the fuel with nitrogen or another inert gas, and by surrounding the tanks (and sometimes filling their interior) with low-density inert material such as reticulated foam, which barely affects the amount of fuel that can be carried but prevents leaks or fires.

Fury

The original HAWKER Fury was an elegant biplane fighter with a wingspan of 30 ft. (9.14 m), which first flew (as the Hornet) in 1929. It was the first RAF aircraft to exceed 200 mph (320 km/h) and was developed from the Hart single-engined bomber, one of the first major designs of Sir Sydney CAMM. Fitted with a Rolls-Royce Kestrel engine and twin machine guns, the Fury operated with the RAF from 1931 until 1939 and also equipped the air forces of Yugoslavia (some with Lorraine or Hispano engines), Iran (initially with Hornet, later Mercury engines), Portugal, and Spain (Hispano engines). The final version of the Fury, with a Kestrel XVI engine, was capable of 242 mph (389 km/h). Other derivatives of the Hawker Hart were the Nimrod naval fighter, Demon two-seat fighter, Osprey naval reconnaissance plane, Audax army liaison aircraft, Hardy general-purpose machine, Hartbees (used by South Africa), Hind bomber/trainer, and Hector (an Audax replacement).

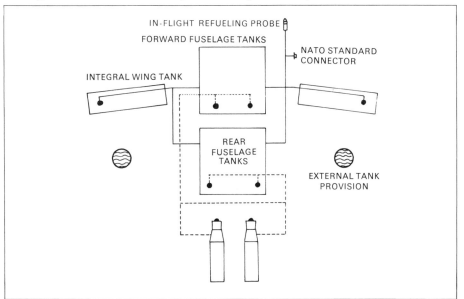

The fuel system of a modern high-performance military aircraft (like that of the twin-engined Panavia Tornado outlined here) employs integral fuel tanks in both wings and fuselage, as well as provision for drop tanks and in-flight refueling

The name Fury was also applied to a Centaurus-engined Hawker fighter, developed from the TEMPEST, which made its maiden flight on September 1, 1944. This aircraft was used in single-seat and two-seat trainer form by the Royal Navy (1947–57) as the Sea Fury, as well as by the Netherlands, Pakistan, West Germany (as a target tug), Burma, Cuba, Australia, Canada, and Egypt; a number of non-"navalized" Fury aircraft (some two-seaters) went to Iraq. The Royal Navy's aircraft saw extensive active service in the Korean War.

The Sea Fury had a wingspan of 38 ft. 5 in. (11.71 m) and a length of 34 ft. 3 in. to 34 ft. 8 in. (10.43 m–10.55 m) depending on the variant. Maximum speed was 460 mph (740 km/h) at 18,000 ft. (5,500 m), and the standard armament was four 20-mm cannon.

Hawker Fury of 1931–32

G

Gabreski, Francis S. (1919–)

Top-scoring American fighter pilot in Europe during World War II, Gabreski was credited with 28 air victories (all German fighters) and $2\frac{1}{2}$ strafing successes. Gabreski achieved all his victories while flying P-47 THUNDERBOLTS with the 56th Fighter Group of the Eighth Air Force from bases in England. An original member of the Group, which was the first U.S. formation to receive the P-47, Gabreski was assigned to the 61st Fighter Squadron and later became its commanding officer.

Gabreski was taken prisoner when forced to crash-land his P-47 on German territory while on a ground-strafing mission (July 20, 1944). During the Korean War, Gabreski flew F-86 SABRES with the 4th and 51st Fighter Wings and was credited with a score of $6\frac{1}{2}$ MiG-15s.

Galaxy

The LOCKHEED C-5A Galaxy military transport is currently the world's largest airplane. The Galaxy made its maiden flight on June 30, 1968, and over 80 have entered U.S. Air Force service since 1970. Equipped with an upward-hinging nose and rear clam-shell doors and ramp, the aircraft can carry a maximum payload of 265,000 lb. (120,200 kg) or 345 fully-equipped troops.

Four specially designed General Electric TF39 turbofans of 41,000 lb. (18,600 kg) thrust give a maximum speed of 571 mph (919 km/h) over a range of as much as 6,500 mi. (10,460 km). Wingspan is 222 ft. 8 in. (67.87 m) and length 247 ft. 10 in. (75.54 m).

Galland, Adolf (1912–)

One of Germany's most important fighter tacticians and a heavily decorated World War II pilot. Galland was a Lufthansa pilot who acquired fighter experience with the Italian air force. A

Fuselage

The fuselage is the main body of an airplane (excluding its wings, tailplane, and undercarriage), the hull of a flying boat, or the main body of a helicopter. Fuselages may be constructed either as a braced box framework that is a basis for *formers* and *stringers* over which a covering is placed, or as a shell-like MONOCOQUE.

Early aircraft like the Wright *Flyer* lacked anything that could be called a fuselage, but after sustained flight made protection for the pilot from the slipstream necessary, the fuselage evolved into the basic component of the airplane, to which wings for lift and tail surfaces for control were fixed.

Until 1915 fuselage construction was almost wholly in wood, but in that year the German firm JUNKERS built the J-1, generally recognized as the first all-metal airplane. Duralumin, an aluminum alloy combining strength with lightness, was widely used by the 1930s; in recent decades plastic and fiberglass have been employed for lightplanes and special metals such as titanium for high-speed aircraft.

A conventional airplane has a single fuselage, but a few planes of the past have had two (the Blackburn TB of 1915, and the P-82 Twin Mustang of 1945). In a flying wing, such as the prototypes of the NORTHROP B-35, there is no fuselage as such.

Lockheed C-5A Galaxy

crash in 1935 while stunting in a Focke-Wulf Stieglitz damaged his left eye, but he obtained a commission in the new Luftwaffe, fought with the Condor Legion in Spain (1937–38) flying Heinkel He 51s as ground-attack fighters, took part in the Polish campaign (September 1939) and flew a BF 109 on the western front in May 1940.

Galland was a Battle of Britain fighter pilot, taking over command of the crack JG26 "Schlageter" *Geschwader* in August 1940. He succeeded MOELDERS as Inspector of Fighters in December 1941, but was relieved of his post in January 1945 because of his criticism of the inept GOERING. For the last few months of the war General Galland commanded Jagdverband 44, flying ME 262 jet fighters, and when taken prisoner on May 5, 1945, he had 104 victories to his credit.

In the 1950s he became involved in civil aviation as a director of Air Lloyd.

Garros, Roland (1888–1918)

A leading French World War 1 fighter ace, Roland Garros was a pilot for the French manufacturer MORANE-SAULNIER in the years immediately before the war. He established several records for long-distance and endurance flights, including the first crossing of the Mediterranean, from Saint-Raphael on the coast of France to Bizerte in Tunisia (September 23, 1913). In August 1914 Garros enlisted in the French air service, joining *Escadrille* MS.23, and flying Morane-Saulnier two-seat monoplanes on general reconnaissance duties. On April 1, 1915, in a Morane Parasol Type L fitted with a forward-firing machine gun and with steel deflector wedges on its propeller, Garros succeeded in shooting down a German aircraft. At least two more victims fell to his novel but dangerous gun arrangement before April 18, when a faulty engine forced him to land behind the German lines, where he was taken prisoner (see INTERRUPTER GEAR). Three years later, Garros escaped and returned to France, where he joined *Escadrille* S.26 on August 20, 1918. He was killed in action on October 5, when his aircraft propeller was shattered during combat and he crashed near Vouziers.

General Dynamics

One of the world's largest industrial corporations, with 14 principal operating units in the fields of aerospace, shipbuilding (in particular, nuclear submarines), electronics, mining, and other areas of industry. Formed by John Jay Hopkins in 1947, its aerospace operations were dramatically increased by the acquisition of a controlling interest in CONVAIR in 1953. General Dynamics aerospace activities are centered at four divisions: San Diego (Convair), Fort Worth, Pomona (Cal.), and Orlando (Fla.).

During the 1950s the main aircraft programs were those taken over from Convair, the B-36, F-102 (DELTA-DAGGER), F-106 (Delta-Dart), B-58 (Hustler) and 240-series of airliners (METROPOLITAN). The Models 880 and 990 jetliners produced at the end of the decade were technically advanced, but commercially unsuccessful. In 1960 the corporation's Fort Worth division won a contract to supply the U.S. Air Force with a new tactical fighter (TFX) intended to replace virtually all the previous fighters and attack aircraft of the Air Force, Navy and Marine Corps. The resulting aircraft, the F-111, combined a variable-geometry SWING-WING DESIGN, an afterburning turbofan engine, titanium construction, and several other major advances. Unfortunately, it suffered from excessive weight and drag, and eventually a much smaller number than planned were built (none for the Navy). The next Fort Worth aircraft, the F-16 lightweight fighter (LWF) first flown in January 1974, has been ordered in large numbers by the Air Force, four Nato countries, Iran, and Spain. With a thrust/weight ratio higher than unity in the clean condition, the F-16 has a high combat performance for close-range dogfighting and is also being developed to fly attack and reconnaissance missions carrying external weapons. Features include "fly by wire" electronically-signaled controls, a cockpit with reclining seat and side-stick controller, reduced static stability in pitch to increase combat maneuver power, a tail made largely of advanced composites, a wing with variable camber (but little sweep) blended into the fuselage, and long forebody strakes to provide vortex control.

In 1948 Convair flew the first ballistic rocket ever to have a gimbal-mounted thrust chamber, providing control of the trajectory throughout flight. Over the succeeding years it has produced many missiles and space launchers, of which the best known is the Atlas. In 1959 this became the first ICBM (intercontinental ballistic missile) to enter service, and hundreds remained operational until 1965. The E and F series Atlases were then rebuilt by Convair to fly as SLV (standardized launch vehicle) space boosters, carrying Agena or Centaur upper stages, and by 1977 over 480 space missions had been flown.

Convair also produces the Centaur upper stage, with high-energy liquid hydrogen propulsion.

In 1977 the San Diego plant was producing the fuselage of the DC-10 and mid-fuselage of the Space Shuttle Orbiter. The Pomona division began operation in 1953 by developing and mass-producing the Terrier ship-to-air guided missile. Terrier, and its successor Tartar, have served with many navies, and today Pomona produces the Standard ARM anti-radiation missle.

General Electric

The American industrial giant, General Electric, entered the aeronautical field with the production of superchargers for the piston-engined aircraft of the day. In 1918, the first GE turbo-supercharger was successfully tested at high altitude, and many years of further development led eventually to the highly successful turbos of World War II, which conferred outstanding high-altitude performance on such aircraft as the B-17 FLYING FORTRESS, B-24 LIBERATOR, P-47 THUNDERBOLT, and P-38 LIGHTNING. The B-29 SUPERFORTRESS had two turbosuperchargers feeding each engine, and larger turbos were fitted to such later machines as the enormous B-36 and the B-50 (a reworked B-29).

In 1941 GE began the development of aircraft gas turbine engines. The Steam Turbine Division at Schenectady developed the TG-100 (later designated XT31) turboprop. The Turbosupercharger Division near Boston was assigned the task of "Americanizing" the British WHITTLE-designed engine. The result—the GEC Type 1 turbojet—ran after 28 weeks of development on March 18, 1942, and two Type 1A engines (later designated I-16, and finally J31) took the first American jet aircraft, the P-59 AIRACOMET, into the air on October 1, 1942. The much more powerful I-40 (later J33) was the first jet engine to be mass-produced. The axial J35 followed, and was turned over to ALLISON to manufacture; the J47 was made in very large quantities (over 36,700), and was succeeded by the advanced variable-stator J79, which has made possible American Mach-2+ combat aircraft of many types and is still in production with 18,000 units so far delivered. From these basic turbojets has come the world's greatest range of aircraft engines. Among them are the J85 lightweight turbojet and its CJ610 commercial version, the CF700 aft-fan transport engine derived from the CJ610, the CJ805 civil turbojet and CJ805-23 aft-fan engine, the T58 turboshaft (producing about 1,000 hp), the T64 turboshaft and turboprop in the 2,700–4,500 hp class, the T700 advanced turboshaft of 1,500 hp, the TF34 and TF39 military turbofans and the commercial CF34, the large CF6 (power plant of the DC-10, A.300B Airbus, and late-model Boeing 747s), the F101 (engine of the swing-wing B-1 bomber), and the CFM56 civil turbofan in the "ten ton" class being developed in partnership with France.

General Electric is also one of the world's largest producers of aircraft electric systems, airborne armament (it developed the airborne multibarrel "Gatling" cannon), and other aerospace and defense equipment.

Gentile and Godfrey

Two World War II American fighter aces credited with destroying or damaging 30 enemy aircraft in a month, while flying as a team. Members of the 4th Fighter Group in England, they had both previously served with the RAF. In late February 1944 their Group became the first in the Eighth Air Force to convert to the long-range P-51B MUSTANG, and during the next six weeks it was assigned as escort for bombers on daylight raids over Germany. In this period John Godfrey usually flew as wingman for Don Gentile, the engine cowlings of their Mustangs being painted with a red-and-white checkerboard to enable each pilot to recognize the other in combat.

Gentile returned to the United States in April 1944 with 23 victories in the air and 7 from strafing; he did not fly again in combat was was killed in a flying accident in January 1951.

Godfrey also returned to America on leave but returned to fly a second tour. When shot down and taken prisoner in August 1944 he had 18 air and 12½ ground victories. He died in 1958.

Geodetic construction

In the years between the wars the Englishman Barnes WALLIS began to develop a unique form of aircraft construction, which had its origins in methods used on a number of large airships. In essence geodetic construction is a form of MONOCOQUE construction in which all members lie along the surface of the three-dimensional curved form of the final body—an airship hull or airplane fuselage. The resulting structure almost resembles basketwork, and is made up of intermeshing light-alloy sections, held together by rivets and/or bolts. The fabric skin is attached over the outside.

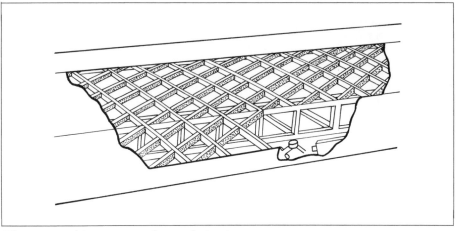

The intermeshing pattern of a geodetic wing structure

The first geodetic airplane was the Vickers-Armstrong Wellesley bomber of the late 1930s. In its successor, the WELLINGTON, geodetic construction was used for wings and tail as well as fuselage, and allowed many badly damaged aircraft to return home for repair when conventional structures would have either failed in flight or been unrepairable. The later Windsor bomber showed the limitations of geodetic construction in aircraft capable of speeds above 300 mph (480 km/h), and no geodetic aircraft have been built since 1946.

Gliding

After the attainment of powered flight in 1903, interest in gliding declined, although there was a gliding club in England in 1909 where Jose Weiss flew his monoplane (which influenced the early designs of HANDLEY PAGE).

Germany became a major gliding center, an early pioneer being Frederic Harth, who began to study the problems of non-powered flight in 1910, the same year that two students, Hans Gutermuth and Berthold Fischer, discovered the Wasserkuppe in the Rhön Mountains of western Germany and realized its potential as a gliding venue.

The Armistice terms of 1918 severely restricted German aviation after World War I, and gliding enjoyed a resurgence under the guidance of Oscar Ursinus, editor of *Flugsport*. A meeting was held at the Wasserkuppe in 1920, and in 1921 Harth used a slope wind at Hildenstein to fly for 21½ minutes—the first pilot to achieve hill soaring. By 1922 flights of several hours were being achieved at the Wasserkuppe using slope lift, and at Combegrasse in central France, Bossoutrot became one of the first pilots to soar in a thermal. Ferdinand Schultz flew for 14 hours 7 minutes at Rossitten on the Baltic, a center where the wind was deflected by the surf and the dunes to provide lift.

In 1928 Professor Walter Georgii, director of the Rhön-Rossitten Gesellschaft research institute, began to investigate the lift attainable below cumulus clouds due to convection. One or two glider pilots had previously experienced this phenomenon, such as Wilhelm Leusch, who was killed in 1921 when his glider broke up under a storm cloud over the Wasserkuppe, but at the 1928 Rhön meeting, soaring beneath cumulus clouds became an established practice.

The very light hill-soaring gliders, sensitive to the merest breath of wind, were now replaced by sailplanes, with sufficient maneuverability to circle in the lift below a cloud, and Robert Kronfeld's Wien, designed by Alexander LIPPISCH, flew 85.5 mi. (137 km) in 1929 on the edge of a storm. Lippisch suggested the use of the variometer to find thermals in clear skies, and the 1930s witnessed great advances in German gliding, with military involvement as a source of future Luftwaffe aircrew. Towing by powered aircraft divorced sailplanes from the hitherto indispensable hills, particularly in America, France, and Poland, and the distance record was extended from 169 mi. (272 km; Gunther Grönhoff in the Fafnir, 1931) to 220 mi. (354 km; Wolf Hirth, 1934), 234 mi. (376 km; Heini Dittmar, 1934), and then 296 mi. (476 km; Ludwig Hoffman, 1935). Goal-flying to specified destinations led to sailplanes designed for greater speed, since physical fatigue set a limit to the time a pilot could remain aloft. Soaring in the standing wave generated on the lee side of hills was also investigated in the mid-1930s, and heights of over 20,000 ft. (6,000 m) were attained.

After World War II, during which troop-carrying military gliders were extensively used for airborne operations, the incentive of competition raised sailplane speeds to around 150 mph (241 km/h). Competitors learned to make steeply banked turns in order to keep within a thermal and gain height with the maximum rate of climb, flying on as quickly as possible to the next thermal. In 1962 a triangular course of 500 km (300 mi.) was set for the first time in a championship at the Polish Nationals, and high-altitude flying in standing waves eventually led to a height of 46,267 ft. (14,102 m) being reached by Paul F. Bikle flying a Schweizer 1-23E from Mojave, California, in the Bishop wave (February 25, 1961), while in New Zealand the Northwest Arch wave system was exploited for flights of up to 300 mi.

In 1964 Al Parker became the first glider pilot to complete 1,000 km when he flew from Odessa, Texas, to Kimball, Nebraska (1,036 km, 642 mi.). By 1972 the distance record had been extended to 1,460.8 km (905.7 mi.) (H.-W. Grosse of West Germany, April 25, 1972). Qualification for the international Silver gliding badge is a distance flight of 50 km in a straight line, for the Gold badge 300 km (either in a straight line or around a triangular course), and for the Diamond badge either 300 km (out-and-return or around a triangle) or 500 km distance.

For the technical aspects of gliders see SAILPLANES.

Gloster

In 1915 the newly-founded Gloucestershire Aircraft Company produced a small scout designed by Harold Boultbee, formerly chief draftsman at British & Colonial (BRISTOL). Despite the absence of a market after World War I, the company decided to carry on and appointed Harry P. Folland chief designer. He had held the same position at the British NIEUPORT Company, closed down in 1920, and so the first products were based on the Nieuport Nighthawk, an outstandingly advanced fighter. Reequipped with a superior power plant—either the Bristol Jupiter or Armstrong Siddeley Jaguar—the result was called the Mars, and versions were built for a number of countries. During the 1920s the Bamel, a Napier Lion-engined racer, and the Nightjar, a naval fighter, were followed by 23 fighter designs for 19 air forces and navies, of which the most famous were two used by the RAF, the Jaguar-engined Grebe and the Jupiter-engined Gamecock. These light biplanes were among the most maneuverable and popular combat aircraft of the mid-1920s.

In 1926 the company changed its name to Gloster, and in the mid-1930s one of its notable products was the Mercury-powered Gauntlet fighter, with a speed of 230 mph (370 km/h) and outstanding climb and maneuverability. In 1937 the RAF's last biplane fighter, the Gladiator, entered service, eventually to win fame over Norway in 1940, and to be the chief defender of Malta.

In May 1941 the first British jet aircraft made its maiden flight. This was the Gloster E.28/39, designed around the radical WHITTLE turbojet. Design also proceeded on the F.9/40 twin-jet fighter, which flew in 1943 and was fitted with four different turbojet types. Gloster Aircraft was now operating as a member of the HAWKER SIDDELEY Group (with which it had been associated since the 1930s), and part of its effort was directed to producing the Hawker Typhoon. The F.9/40 fighter entered service in June 1944 as the Gloster METEOR, the first operational jet aircraft of the Allies. It saw action against V-1 flying bombs, and in 1945 and 1946 set new world speed records at 606.38 and 615.78 mph (975.67 and 990.79 km/h). Gloster's last product was the Javelin, an all-weather night fighter that served the RAF from 1956 until replaced by the PHANTOM in 1968. Gloster was totally absorbed within Hawker Siddeley in the 1960s.

Goering, Hermann (1893–1946)

Commander of the German air force —the LUFTWAFFE—during World War II, and a World War I fighter ace, Goering was one of Hitler's closest associates, a Reichsmarschall of Nazi Germany and one of the key figures of the Third Reich. He began his World War I flying service as an observer, becoming a pilot in 1915 and taking command of a fighter squadron in 1917. His unit was incorporated in RICHTHOFEN's "flying circus" in March 1918, and in July of that year he assumed command of the entire Geschwader, both the Red Baron and his immediate successor (Reinhard) having been killed. A captain by the end of the war, he was credited with 22 victories.

After a period as a commercial pilot, he became involved in the Nazi movement and was wounded in the abortive *Putsch* of 1923. Partly as a result of drugs administered during treatment he developed an addiction to morphine, and by the time Hitler came to power in 1933 Goering was in physical and mental decline.

Appointed Minister of Aviation with the rank of General in 1933, Goering

secretly laid the foundations of Germany's new air force, and became its commander in chief in April 1935. His work bore fruit in initial successes of World War II such as the Blitzkrieg attack on Poland. But thereafter he made no real attempt to keep up with technical advances in aviation, or with the progress of the war. He failed to oppose Hitler's insistence on producing light bombers rather than fighters—an important factor in the Allied success in the Battle of BRITAIN, and in the bombing of Germany that followed. In February 1940 he had actually ordered development work on new aircraft to cease, since he considered existing Luftwaffe combat planes quite adequate for the short war he envisaged. Lack of new designs was eventually to cripple the German air force.

As the tide of war turned, Goering accused his aircrews of incompetence and cowardice. After 1942 he played little part in the direction of the German war effort. He was dismissed by Hitler at the end of April 1945 and placed under arrest, but became an Allied prisoner of war on May 8, 1945. Condemned to death by the Nuremberg war crimes tribunal in 1946, he committed suicide with poison shortly before the time appointed for his execution.

Goliath

The ungainly FARMAN F-60 Goliath was a twin-engined biplane that helped pioneer commercial air transport in Europe. It accommodated 12 passengers in two cabins, with an open cockpit for the two crew members above and between the cabins. Some 60 were built between 1918 and 1929, using Salmson, Lorraine-Dietrich, Maybach, Renault, Gnôme-Rhône Jupiter, or Siddeley Jaguar engines of between 200 and 400 hp. Wingspan was 86 ft. 10 in. (26.50 m) and length 47 ft. (14.33 m). The Goliath entered Paris–London scheduled service in 1920, and in 1921 joined Lignes Farman between Paris, Amsterdam, and Berlin, and the Brussels–London route of SNETA, predecessor of SABENA.

Goliaths were built in Czechoslovakia for the state airline CSA, and others were used in South America and by the French air force. They were all withdrawn from use by 1933; one is preserved at the Musée de l'Air in Paris.

Gotha

Large twin-engined German Gotha bombers were used for daylight bombing raids against southern England toward the end of World War I. The

Farman Goliath

main production variants, the G.IV and G.V, were issued to Heavy Bomber Squadron No. 3 in Belgium early in 1917, and from May 1917 onward, large formations were sent over England. Each aircraft was capable of carrying up to 1,100 lb. (500 kg) of bombs for about 300 mi. (480 km).

By May 1918, 22 raids had been made against England, during which the Gothas had dropped some 187,000 lb. (85,000 kg) of bombs. Their losses were high, however, and they were thereafter used on tactical night sorties along the western front. At least 24 had been brought down by the English defenses, and a further 37 machines were lost in accidents. Powered most often by twin 260-hp Mercedes engines, the Gotha had a ceiling of 15,000 ft. (4,500 m). With a wingspan of nearly 77 ft. 9 in. (23.70 m) and a maximum (loaded) speed of about 75 mph (120 km/h), the Gotha G.V. carried a crew of three or four.

After World War I Gothaer Waggon-fabrik reverted to its original business of manufacturing railroad cars and diesel engines, but it reentered the field of aviation when it produced the experimental tailless Go 147 single-engined military observation plane in 1936. During World War II the company built 3,000 Bf 110 ZERSTÖRERS and the DFS 230 assault glider, developed the flying wing designs of the Horten Brothers as the Go 229 jet fighter-bomber, and supplied the Go 242 transport glider and its unsuccessful powered derivative, the Go 244, to the Luftwaffe.

Graf Zeppelin

Probably the most successful airship in the history of lighter-than-air travel, the German rigid AIRSHIP *LZ-127*, christened the *Graf Zeppelin*, was completed in 1928. She was 776 ft. 3 in.

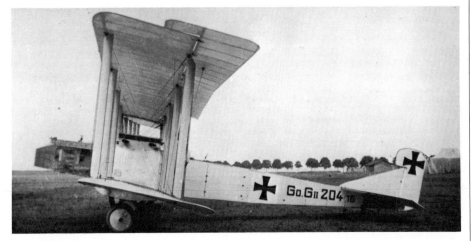

Gotha G.II

(236.60 m) long, contained 3,995,000 cu. ft. (113,000 m³) of gas, and was powered by five Maybach VL-2 560-hp engines. Instead of burning gasoline or diesel fuel, the airship's engines consumed Blaugas, a fuel that accounted for a part of the airship's total gas capacity. The crew numbered between 40 and 50, and there were 10 double cabins for 20 passengers, who were carried in a style like that of a luxury oceanliner.

After two commercial flights across the Atlantic, the *Graf Zeppelin*, under the command of Hugo ECKENER made a round-the-world flight in 1929 covering 21,255 mi. (34,200 km) in an elapsed time of 20 days 4 hours 14 minutes. The ship later made special flights to South America (1930), Russia (1930), the Mediterranean (1931), and the Arctic (1931), before being assigned to regular services across the South Atlantic. She was decommissioned on June 18, 1937, after the loss of the HINDENBURG, and was scrapped on Goering's instructions in March 1940.

In an active career of nearly nine years the *Graf Zeppelin* made some 590 flights, carried over 13,000 passengers

and 235,000 lb. (106,000 kg) of mail and freight, and covered 1,053,395 mi. (1,694,912 km) in 17,178 hours of flying time—all without incident.

Another German airship, the *LZ-130*, bore the name *Graf Zeppelin*. Planned for transatlantic service as a successor to the *Hindenburg* (which she closely resembled), and designed to be inflated with helium, the *Graf Zeppelin II* achieved only 400 hours of test and exhibition flying, inflated with hydrogen, before also being dismantled in 1940.

Ground Controlled Approach (GCA)

Radar displays in an airport's control tower can give air-traffic controllers precise information about an aircraft's position and height. In bad weather the controller can use this information to direct the pilot toward the runway for a Ground Controlled Approach (GCA).

After the aircraft has been identified on the radar screen, the controller tells the pilot what headings to fly and what height to maintain to position himself correctly for his landing approach. The pilot confirms instructions by repeating them back. About 8 mi. (13 km) from the runway a precision controller takes over. The pilot no longer repeats his instructions since any deviation from the required flight path is instantly apparent on the radar display, and the controller will issue corrections. When the runway comes into sight the pilot takes over. GCA may be used at busy airports to position aircraft for landing even in good visibility, though the pilot's own INSTRUMENT LANDING SYSTEM can allow the aircraft to make its approach, without need for control-tower guidance.

Grumman

In 1929 Leroy Grumman, who had been plant manager with Loening Aircraft, established his own company at Bethpage, Long Island. In 1931 he obtained an order for a two-seat U.S. Navy fighter. The prototype flew later that year, and led to a small production order for what became known as the FF-1, an advanced radial-engined biplane with all-metal construction and retractable undercarriage. Development of the FF-1 produced the F2F and F3F, which became the principal naval fighters of the years between 1935 and 1940. Grumman Aircraft Engineering Corporation continued to specialize in carrier-borne aircraft, particularly fighters, which included the F4F WILDCAT, the F6F HELLCAT, the twin-engined F7F Tigercat, and the F8F BEARCAT. After World War II Grumman produced its first jet aircraft, the F9F Panther, again for the U.S. Navy, which was developed into the swept-wing COUGAR. The first supersonic Grumman, the F11F Tiger, entered service in the mid-1950s. Despite the narrowing market for military aircraft, Grumman managed to secure the U.S. Navy's order for an advanced air-superiority fighter for the 1970s with its F-14 TOMCAT.

Even in its earliest days Grumman's range of products extended beyond naval fighters. Leroy Grumman's experience with Loening amphibians enabled him to produce the first Grumman amphibians, the J1F (1933), and the J2F Duck (1936). Later amphibian designs, the twin-engined JRF Goose (1937) and J4F Widgeon (Gosling, 1940), saw extensive service during World War II, as did the TBF AVENGER torpedo bomber. After the war, Grumman built the AF-2 Guardian and S-2F Tracker antisubmarine aircraft, the HU-16 Albatross amphibian, the Gulfstream turboprop executive transport and trainer, the Gulfstream II business jet, the C-1 Trader and C-2 Greyhound

Grumman aircraft of World War II: TBF Avenger (foreground) and F8F Bearcat

navy transports, the E-1 Tracer and E-2 Hawkeye airborne-early-warning aircraft, the A-6A Intruder attack aircraft, and the OV-1 for tactical battlefield observation.

Gunsight

Although the earliest aircraft guns had not been specifically designed for aerial use, some specially modified form of sight was soon found to be necessary, and by 1914 the first simple sights had been produced in many countries. Throughout World War I almost all fixed and manually-trained guns used the ring-and-bead sight. This consisted of a small ball (the bead) carried about 6 in. (150 mm) above the muzzle, and a

ring about 4.7 in. (120 mm) in diameter mounted further back along the gun, with crossed wires so that the bead could be centered easily. In actual combat allowance had to be made for the relative motion between the gunner and his target, and the drag of the slipstream; a no-deflection shot was seldom possible, and the purpose of the ring and cross-wires was to help the gunner quickly judge the right deflection and aim-off by the required amount. Many aircraft with fixed guns also used an optical sight, often carried to the right of the center line but sometimes passing through a central hole in the windshield. British optical sights were the Aldis, with a diameter of 1.8 in. (4.57 mm), used on all fighters from 1916

The basic ring-and-bead gunsight on a Lewis machine gun of World War I

until 1940, and the Neame and the Hutton, in which the rings were illuminated for night attack.

By far the greatest advance in gunsight technology came in 1940, when the British firm Ferranti began design of a gyro-gunsight (GGS). In the GGS the lead-angle (aim-off) is assessed by a combined gyro and electronic system that measures the rate of rotation of the line-of-sight to the target. In early gyro-sights the pilot looked through a sloping glass screen and watched the target ahead. He had to estimate the range from the target's apparent size and knowledge of its span, and set the range by turning a knob. This correctly positioned an aiming mark, which usually took the form of a central dot surrounded by a ring of bright diamond shapes projected on to the viewing glass by an illuminated optical system. The pilot had only to fly the fighter to keep the aiming mark on-target for him to be sure of scoring hits.

Today the task of determining range is accomplished automatically, by radar or laser carried in the fighter. The sight forms part of a more general HEAD-UP DISPLAY, which can also be used for ground-attack and other purposes.

Guppy

Giant cargo aircraft designed to carry oversized loads, the Guppies are extensively modified Boeing 377 STRATO-CRUISER airliners and C-97 military transports. The original Pregnant Guppy was modified, like later versions, by Aero Spacelines and first flew in its new form in September 1962. A massive lobe joined above the normal fuselage gave a 20-ft. (6.1-m) inside height. The Pregnant Guppy was powered by four piston engines, but turboprops were used on the later Super Guppy, which transported components for the Saturn rocket program. The nose section swings to open for easy loading.

Two Mini Guppies were followed by the Guppy 101; the latest version of the aircraft is the 201, which entered service in 1971. This has the largest freight compartment of any transport; over 25 ft. (7.6 m) in diameter, it has a capacity of 39,000 cu. ft. (1,104 m³). The Guppy 201 has a wingspan of 156 ft. 8 in. (47.75 m) and length of 143 ft. 10 in. (43.84 m); cruising speed is 288 mph (463 km/h).

Guynemer, Georges Marie Ludovic Jules (1894–1917)

French fighter ace of World War I, Guynemer was second only to René FONCK and had 54 confirmed victories

Pregnant Guppy cargo aircraft

to his credit. He began pilot training in March 1915, and on June 8, 1915, was assigned to *Escadrille* MS.3 (part of the elite "Storks" group), which was equipped with Morane Bullet monoplanes. In just over a month he had scored his first victory, but in September he was himself shot down in no-man's-land—the first of seven such incidents.

By the early spring of 1916 he had eight victories, and though a wound sustained in action on March 15 led to his absence from the front for several months, he was back in combat during July flying NIEUPORT scouts. On September 23 he brought down three enemy planes in one day before being shot down himself, although he again escaped uninjured.

In February 1917 Guynemer was promoted to *capitaine*, and notched up a triple victory on March 16 and a quadruple success on May 25 (two German aircraft were shot down in the space of one minute). On July 28 his 50th victim fell before the guns of his new Spad.

His health, never good, had now visibly deteriorated, and he was under pressure to retire from combat flying. He refused, and on September 11 failed to return from a routine patrol. Neither his body nor his aircraft was ever recovered. Guynemer fought in over 600 aerial combats and received 26 citations.

H

Halifax

The HANDLEY PAGE Halifax was the second four-engined bomber to enter RAF service in World War II (three months after the STIRLING). The Halifax

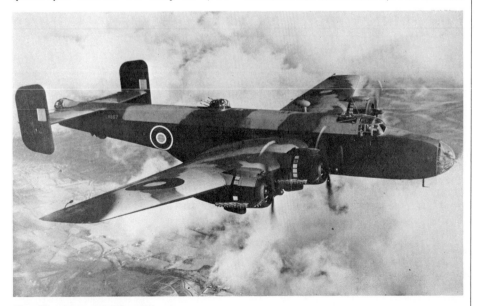

Handley Page Halifax III

Handley Page HP 42 Hannibal, *used on Imperial Airways' services between Cairo and South Africa and India*

first flew on October 25, 1939, and the first operational sorties were flown by Merlin-engined Halifax Is on March 10/11, 1941. Halifaxes continued on operations until the end of the war, flying 75,532 missions and dropping 227,610 tons of bombs.

Later models had more powerful Bristol Hercules engines; the Mark VI was capable of 312 mph (502 km/h). Nose armament was reduced to a single machine gun after the Mark II, there were tail and dorsal turrets, and the bomb load was 13,000 lb. (6,000 kg). Wingspan of the Mark III was 94 ft. 2 in. (28.70 m) and length 70 ft. 1in. (21.36 m). Variants included paratroop and glider tug versions, while specially modified Halifaxes played an important radio countermeasures role.

Handley Page

Known principally as a manufacturer of large, multiengined airplanes, the firm of Handley Page was formed by Frederick Handley Page (1885–1962) in June 1909. Early projects included a canard glider, the E50 monoplane, and the G100 biplane, all of which used an inherently stable crescent-wing configuration. During World War I, Handley Page built the O/100 twin-engined bomber for the Royal Navy. The HANDLEY PAGE O/400 bomber entered RAF service in 1918 and was used to raid strategic targets in Germany, but a four-engined development of this machine (the V/1500) was delivered too late to see action. A number of O/100s and O/400s were converted into airliners after the war, but airplanes in the W8 series were designed from the outset for civil use, although the Hyderabad night bomber of 1923 was a

later development. This series of two- and three-engined biplanes continued through the W9 and W10 airliners (1925–26), the Hinaidi bomber (1927), and the Clive bomber-transport (1928).

The Heyford bomber of 1930 represented a new biplane bomber design, over 100 being built for the RAF; while the HP 42 Hannibal and Heracles airliners became a familiar feature of Imperial Airways operations during the 1930s. Handley Page's first monoplane bomber was the Pegasus-engined Hampden of 1936 (the Hereford was an unsuccessful Dagger-powered variant of this type); the Harrow was a bomber-transport with a fixed undercarriage. About 1,500 Hampdens were eventually built, and they flew with the RAF until December 1943. By this time they had already been supplanted on the production lines by the four-engined HALIFAX, which played a major role in the bombing campaign against Nazi Germany.

In the early postwar years a number of Halifaxes were modified for civil use as Haltons, but the new Hermes airliner of 1945 proved a disappointment, although its military transport derivative, the Hastings, served with the RAF until 1968. A return to the crescent-wing layout was seen in the HP 88 research aircraft of 1951 (which crashed only two months after its maiden flight), and in the four-engined Victor jet bomber, which entered RAF service in 1957. Examples of the Victor modified as in-flight refueling tankers remained in use through the mid-1970s.

The Herald short/medium range airliner initially had four Leonides piston engines, but it was eventually produced in some numbers with twin Dart turboprops. The Jetstream was also powered

by two turboprops, but this small airliner or executive aircraft (with proposed military variants) was the last Handley Page project: the company ceased operation in 1969.

Handley Page O/400

The largest British bombers to see active service in World War I were the Handley Page O/100 and O/400. A biplane with a wingspan of 100 ft. (30.48 m) and length of 62 ft. 10 in. (19.15 m), the O/100 was powered by twin 250-hp Rolls-Royce Eagle engines that gave it a top speed of 72 mph (116 km/h). The aircraft began operations with the Royal Naval Air Service in the fall of 1916 (except for one aircraft that was inadvertently delivered to the Germans: it force-landed on the wrong side of the trenches in bad weather). The O/100 force was initially used for daylight coastal raids, but by the end of 1917 aircraft were carrying out night attacks on railroads and airfields in German-occupied France and Belgium, eventually extending their attacks to industrial targets in Germany itself. One aircraft was employed in the Aegean, but was lost due to engine failure on September 30, 1917.

A total of 46 O/100s were built. They were succeeded in production by a more powerful version, the O/400, of which over 500 were produced. The O/400 was essentially a O/100 with a rearranged fuel system (tanks in the fuselage center section and the leading edge of the upper wing), more powerful engines (360-hp Eagles), and improved performance (maximum speed 97½ mph, 157 km/h; service ceiling 8,500 ft., 2,600 m). Deliveries to the RAF began in the summer of 1918, and by the end of the war O/400s had raided industrial targets in the Saar and Rhineland, using the new 1,650-lb. (750-kg) bombs. A single machine was sent to the Middle East for use against the Turks.

After the Armistice, numerous O/400 bombers were converted for civil employment; many had been built in the United States by the Standard Aircraft Co. and were equipped with Liberty engines.

Harrier

The first fixed-wing V/STOL strike fighter to enter operational service, the Hawker Siddeley Harrier's immediate predecessors were the Hawker P.1127 and the Kestrel, aircraft that underwent extensive testing and evaluation from 1962 to 1968. The Harrier employs vectored thrust for vertical takeoff from a Pegasus turbofan, capable of giving a low-level maximum speed of 737 mph (1,186 km/h) and a dive Mach number of 1.3. With reaction jet nozzles at nose, tail, and wing tips, the Harrier can maneuver while hovering. A relatively small aircraft, it has a wingspan of 25 ft. 3 in. (7.70 m) and is 45 ft. 6 in. (13.87 m) in length.

After making its maiden flight on August 31, 1966, the Harrier began to equip RAF units in 1969, with initial variants (Marks 1, 1A, and 3) powered by a series of engines of 19,200 lb. (8,710 kg), 20,000 lb. (9,060 kg), and 21,500 lb. (9,740 kg) thrust. Armament consists of 30-mm guns (in ventral fuselage pods), together with externally carried bombs and/or rockets. Training is carried out in the two-seat Marks 2, 2A, and 4, and a maritime version for shipboard operation, the Sea Harrier, is under development for the British Navy.

The Harrier Mark 50 entered service with the U.S. Marine Corps in 1971 as the AV-8A, with provision for carrying Sidewinder missiles. The Mark 54, also ordered for the Marine Corps, has the 21,500 lb. (9,740 kg) thrust Pegasus 103 engine; the TAV-8A is a two-seat operational trainer.

McDonnell Douglas was also designing an advanced Harrier for the U.S. Marine Corps (the AV-16A), with double the range and weapons payload of the original aircraft.

Hartmann, Erich "Bubi" (1922–   )

Hartmann was the top-scoring German ace of World War II; with 352 victories to his credit he ranks as the most successful fighter pilot in the history of air warfare. Virtually all his career was spent on the Russian front,

Handley Page O/400

U.S. Marine Corps AV-8A Harrier

but his tally included a handful of American P-51 Mustangs.

Hartmann joined the Luftwaffe in 1940 and was sent to the eastern front in 1942 as a BF 109 pilot. His record was not exceptional until mid-1943 (by which time he had scored 34 victories), but on July 7 of that year he brought down 7 Russian aircraft in a single day. Within two months his score was 95. On August 24, 1944, he claimed 6 aircraft on one sortie, and 5 more before the day was out to raise his score to 301.

His final victory was achieved on the last day of the war (May 8, 1945). He had flown 1,425 sorties (engaging the enemy in 800 of them), been shot down 16 times, and on two occasions took to his parachute.

Turned over to the Russians after surrendering to American forces, the

"Black Devil of the Ukraine" was sentenced to 10 years' imprisonment as a war criminal. He returned to West Germany in 1955 and became a jet pilot in the postwar Luftwaffe.

Hawk

Two series of fighter aircraft produced by CURTISS were known as Hawks. The first Hawk was the chief U.S. Army fighter in the period from 1924 to 1933. A single-seat biplane with a wingspan of 32 ft. (9.75 m) and length of 22 ft. 6 in. (6.85 m), the initial Hawk (the PW-8 of 1923) had a 440-hp Curtiss D-12 engine, which gave it a maximum speed of 169 mph (272 km/h). A production order for 25 PW-8s in 1924 was followed by some 125 refined examples built over the next five years,

P-6E Hawk

P-36A Hawk

the official designation being changed to P-1 in 1925, and later Hawks produced as the P-2, P-3, P-5, and P-6.

Advanced trainer versions were known as the AT-4 and AT-5, 71 of which were built and many later converted into fighters. The U.S. Navy also procured a substantial number of Curtiss Hawks under the designation F6C; later models had radial engines.

The name Hawk was also applied to the P-36 monoplane fighter, Curtiss' first all-metal aircraft. It was designed in the mid-1930s as a potential replacement for the Boeing P-26A PEASHOOTER. Powered by a 1,050-hp Twin Wasp engine, the P-36A had a maximum speed of 300 mph (480 km/h) and entered service with the Army Air Corps in April 1938. Its wingspan was 37 ft. 4 in. (11.40 m) and length 28 ft. 6 in. (8.69 m). The last 31 of these aircraft served under the designation P-36C and had 1,200-hp power plants. Although

clearly obsolete by 1941, some P-36s were still operational in Hawaii at the time of the Japanese attack on Pearl Harbor.

Export versions saw considerable service with the French air force (Hawk 75A-1 and 75A-4). Other nations to use the Hawk included Britain (in the Far East until 1943), Finland (ex-French aircraft), Norway (Hawk 75A-6), Peru (P-36G, ex-Norwegian Hawk 75A-8), South Africa, Portugal, Iran (Hawk 75A-9, taken over by the RAF in 1941), and the Netherlands (in the East Indies).

Hawker

Harry G. Hawker was one of the pioneers of British aviation. In 1912 he joined T. O. M. Sopwith as test pilot. With designer Fred Sigrist the three ran Sopwith Aviation, beginning from a converted ice rink at Kingston, Surrey. Within two years Hawker had estab-

lished more than 30 records flying various Sopwith landplanes and seaplanes, and throughout World War I he tested every type of Sopwith aircraft, including the notable CAMEL. In May 1919 Hawker took-off from Newfoundland with K. Mackenzie-Grieve in an attempt to fly to England in the Sopwith *Atlantic*, but the aircraft was forced down in the ocean. In 1920 Sopwith Aviation ceased business and a new company was established and named H. G. Hawker Engineering, but in June 1921 Hawker suffered a hemorrhage while testing the Nieuport Goshawk and was killed in the ensuing crash.

In the years from 1925 onward the firm of Hawker became solidly established as a builder of combat aircraft, largely on account of the stature of its chief designer, later technical director, Sydney CAMM. Among his designs were the highly successful biplane bomber, the Hart, produced in a number of versions, and the FURY, which was the first British military plane to exceed 200 mph (320 km/h). In 1933 the company was renamed Hawker Aircraft, and continued under this name as part of HAWKER SIDDELEY, which it became part of in 1935. The HURRICANE, a durable monoplane design of 1936, was produced in great quantities until 1944. Both the Typhoon and the TEMPEST were fast piston-engined fighters (the latter exceeding 450 mph; 724 km/h) used successfully in ground-support missions. The best known postwar design proved to the the HUNTER jet fighter, which was followed by the world's first VTOL fighter, the HARRIER. In the early 1960s Hawker was integrated into an enlarged Hawker Siddeley Group.

Hawker Siddeley

British aerospace manufacturer, established in 1935, and since May 1977 part of the new state-owned industry, British Aerospace. Its origins date back to the late 1920s when Armstrong Siddeley Development Company was formed and gradually took control of numerous car and aircraft firms including Armstrong Siddeley Motors, ARMSTRONG WHITWORTH Aircraft, and AVRO Ltd. In 1935 the directors of Armstrong Siddeley and HAWKER Aircraft joined with GLOSTER to form the Hawker Siddeley Group. Under Sir Thomas Sopwith, Sir Frank Spriggs, and Sir Roy Dobson, the Group expanded to meet the needs of World War II and by 1945 had delivered approximately half of all British-built combat aircraft. The Group's factories employed nearly

500,000, and produced the HURRICANE, Typhoon, TEMPEST, FURY, Gladiator, METEOR, Anson, Manchester, LANCASTER, York, Lancastrian, and Whitley.

After 1945 the Group expanded its operations in Canada, turning Victory Aircraft into Avro Canada, forming Orenda Engines, and buying the vast Dominion Steel and Coal Group and various other Canadian firms. Subsidiaries appeared in other parts of the Commonwealth, and several other British companies joined the Group.

In 1960–1963 Hawker Siddeley absorbed DE HAVILLAND, with all its operating companies, as well as BLACKBURN (which had previously swallowed General Aircraft and itself set up an engine subsidiary) and Folland Aircraft. It also owned half of Bristol Siddeley Engines, combining the former engine activities of Bristol, de Havilland, Armstrong Siddeley, and Blackburn. For several years there were attempts to organize the company in divisions, such as Avro-Whitworth, Whitworth-Gloster, and Hawker-Blackburn, but in April 1965 the decision was taken to unify the whole Group under two main headings. Hawker Siddeley Aviation combined all the airframe activities; Hawker Siddeley Industries was responsible for all nonaeronautical products. Further rationalization took place by amalgamating many HSI units under the divisions Hawker Siddeley Electric and Hawker Siddeley Diesels. Subsidiaries include Hawker de Havilland Australia and Hawker Siddeley Canada. Hawker Siddeley Dynamics (HSD), the successor to de Havilland Propellers, became one of the world's leading designers and developers of guided missiles and infrared systems.

Today the British Aerospace products that originated in Hawker Siddeley Aviation are evenly divided between military and civil machines. The twin-jet Gnat is used by the RAF as a dual trainer, and was also produced in fighter versions. The Buccaneer is a standard tactical bomber of the RAF (and South African Air Force), differing from most other attack machines in carrying its large weapons load (16,000 lb., 7,257 kg) internally. The HARRIER is now entering production in maritime and developed AV-8B forms, and the TRIDENT remains a successful airliner. The Nimrod, still the RAF's long-range maritime patrol aircraft, is appearing in a new version carrying special surveillance radar and computers for AEW (airborne early warning) duties. Among other aircraft are the Hawk trainer and attack aircraft, the HS 748 twin-turboprop with its Coastguarder maritime reconnaissance version, and the HS 125

Hawker Siddeley HS 125 twin-turbofan business jet

business jet which is now sold with TFE731 turbofan engines.

Hayabusa

The most important Japanese Army Air Force fighter of World War II in terms of numbers was the NAKAJIMA Ki.43 Hayabusa ("Peregrine Falcon"), code-named "Oscar" by the Allies. It was available in only very small numbers at the time of Pearl Harbor (approximately 40 aircraft), but nearly 6,000 fighters were eventually built. First flown early in 1939, the Hayabusa was designed by Hideo Itokawa around the 975-hp Nakajima Ha.25 Sakae engine; armament consisted of twin machine guns. Complaints that the prototype was sluggish on the controls led to a redesign incorporating combat flaps, and the Ki.43 became a superbly sensitive and maneuverable interceptor. Combat experience soon demonstrated a need for more armor protection around the pilot and fuel tanks, and for a more powerful engine. As a result, the new Ki.43-IIa of 1943 had 13-mm head and back plates in the cockpit and a 1,105-hp Ha.115 engine.

In the summer of 1943 the clipped-wing Ki.43-IIb made its appearance. It was an aircraft capable of out-maneuvering the LIGHTNING, MUSTANG or THUNDERBOLT, but lacking the dive capability and climb performance of the American fighters. The Ki.43-IIB Hayabusa had a wingspan of 35 ft. 7 in. (10.84 m) and length of 29 ft. 3 in. (8.92 m); maximum speed was 329 mph (530 km/h). Still inadequately armored, it tended to disintegrate completely if hit.

Although the Hayabusa was already obsolete by 1944, a new version appeared over Japan as a home-defense fighter during the last year of the war.

The Ki.43-IIIa had a 1,250-hp engine giving a speed of 342 mph (550 km/h); separate exhaust stacks produced a certain amount of additional thrust. The inadequate armament of two machine guns would have been replaced by two cannon in the Ki.43-IIIb, but the conclusion of the war limited manufacture to two prototypes.

Hayate

A lineal descendant of "Oscar" (the Ki.43 HAYABUSA), the NAKAJIMA Ki.84 Hayate ("Gale") fighter first appeared in combat during 1944. In the last 14 months of the war more than 3,500 of these highly versatile aircraft were produced. Powered by a 1,900-hp Ha.45/11 18-cylinder two-row radial engine of very compact design, the Ki.84 was capable of 390 mph (630 km/h) and could out-climb and out-maneuver both the MUSTANG and THUNDERBOLT. The airframe was of sturdy construction, making the Hayate suitable for use as a dive-bomber with up to 1,100 lb. (500 kg) of bombs; armament initially consisted of two machine guns and two cannon. Wingspan of the Ki.84 was 36 ft. 10 in. (11.23 m) and length 32 ft. 7 in. (9.92 m). Code-named "Frank" by the Allies, the Hayate had its armament progressively increased, first to four 20-mm cannon (Ki.84-Ib), and then to two 20-mm and two 30-mm weapons. Trouble was experienced with the engine (which suffered persistent loss of oil pressure), the poorly-designed hydraulic system, and the undercarriage legs (inadequately hardened and prone to failure). Toward the end of the war a wooden-airframed Hayate was evolved (the Ki.106) as a means of saving light alloys. Other versions employing economies in construction were the Ki.84-II (wooden rear fuselage and wing tips)

Heinkel He 111H

and the Ki.113 (built of steel), but both these aircraft proved excessively heavy. Variants under development in 1945 included the Ki.84-III, with a turbo-supercharged 2,000-hp Ha.45ru engine, and the Ki.116, with a Mitsubishi Ha.112-II 14-cylinder radial.

He 111

Developed from the sleek He 70 single-engined airliner, the twin-engined Heinkel He 111 bomber made its first flight on February 24, 1935. Early versions were used in small numbers (under the designations He 111C and G) by Lufthansa (which found the aircraft uneconomic since it seated only 10 passengers), and it began to equip Luftwaffe bomber units in 1936 (He 111B with DB 600 engines). The He 111B first saw combat with the Condor Legion in the Spanish Civil War, in which the He 111E with Jumo engines was also tested (1938).

By 1939 the He 111P (DB 601 engines) and H (Jumo 211 engines) with fully glassed-in nose and con-

ventionally-tapered wings had become the standard Luftwaffe medium bombers. They had a wingspan of 74 ft. 1 in. (22.60 m) and length of 53 ft. 9 in. (16.39 m). Heavy losses were sustained during the Battle of Britain, but in the absence of any successful replacement for the He 111, production continued until 1944, amounting to over 7,000 aircraft. Spain was still building the He 111 under license in the 1950s as the C2111D, powered by 1,610-hp Merlin engines.

He 162. See VOLKSJÄGER

Head-up display

In 1940 British pilots and engineers began to experiment with the projection of radar displays onto the windshield of a fighter, so that the crew could watch it at the same time as they searched the sky ahead for a target. By 1944 AI (airborne interception) Mk IX radar was combined with the first presentation that could be called a head-up dis-

play (HUD), but it was not until 1955 that HUDs came into regular squadron use.

In modern combat aircraft the HUD is a central part of the system that presents the aircrew (usually the pilot) with electronically generated symbols giving all desired flight and attack information, as well as a radar display and a clear view ahead. Essential components are a bright cathode-ray tube, collimating optics, and a digital computer, usually able to be quickly reprogrammed in flight by tape. Most HUDs can be instantly switched to different modes—navigation, close-range navigation, surface attack, air attack, bad-weather landing—and typical displays present a miniature aircraft symbol, usually centered between horizon and pitch bars, with an azimuth steering arch kept centered on a vertical line lower down the display. In some HUD modes there are vertical scales for airspeed and combined altitude and vertical speed (climb/descent), while in others there are plain numbers at the top of the display giving speed in knots and altitude in feet. In surface-attack modes the computer, airspeed, radar height, and laser target range are used to control the aiming line and impact marker.

Among the aircraft equipped with advanced HUD systems are the HARRIER, JAGUAR, TORNADO, CORSAIR II, and EAGLE. Except for the last-named all have installations of British origin, and the leading English suppliers—Smiths Industries and Marconi Elliott—are now in production with the HUDWASS (HUD weapon-aiming subsystem), Flexihud (programmable

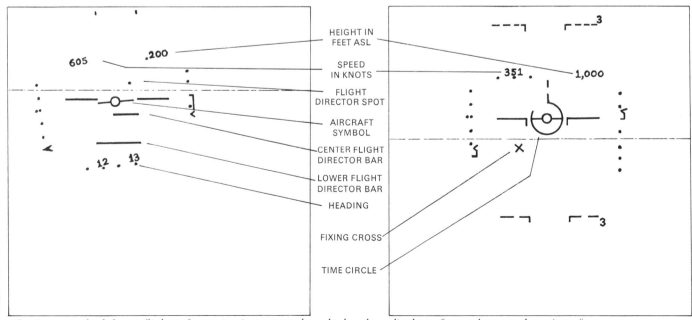

HEIGHT IN FEET ASL

SPEED IN KNOTS

FLIGHT DIRECTOR SPOT

AIRCRAFT SYMBOL

CENTER FLIGHT DIRECTOR BAR

LOWER FLIGHT DIRECTOR BAR

HEADING

FIXING CROSS

TIME CIRCLE

The form in which basic flight information is presented on the head-up display of a modern combat aircraft

by a compact tape-fed computer), and Minihud installations of various kinds, some for fitting in aircraft of earlier vintage. Complementary to HUD systems are helmet sights in which, simply by looking at a ground or aerial target (which may be out of the field of view of the HUD), the pilot can activate sensors such as radar, IR or laser rangers to lock onto it.

Heat barrier

When an aircraft is flying at speeds of Mach 2 and above, the friction between the surrounding air and the aircraft's metal skin generates considerable heat. There is no specific "heat barrier" that can be overcome in the same way that the SOUND BARRIER can be broken: the faster the aircraft travels within the Earth's atmosphere, the greater the heat produced. At speeds of Mach 6 skin temperatures of over 1,000°F can be experienced on the nose and the wing leading edges. At these temperatures aluminum alloys are significantly weakened, and special high-temperature composites including metals such as titanium must be used.

Heinkel

Employed as chief designer for the aircraft manufacturer Albatros in 1913, Ernst Heinkel (1888–1958) eventually founded his own Heinkel Flugzeugwerke AG at Warnemünde on December 1, 1922. Heinkel quickly established a reputation for his low-wing single-engined seaplanes, commencing with the He 1 and 2 of 1923 and continuing through a series of designs to the He 12 mailplane *New York* (1929), which was catapulted from the liner *Bremen* as it approached New York, and the He 58 *Atlantik* (1930), which was launched from the *Europa*. A series of trainer-type biplanes was also built, among them the popular He 72 Kadett of the 1930s, while in 1926 the first Heinkel fighter appeared (the He 23 biplane). Fighter design progressed through the He 43 and the 200-mph (320 km/h) He 49 (1932), to the He 51 biplane of 1933, which was issued to the fighter squadrons of the new Luftwaffe, fought in Spain with the Condor Legion, and was retained as an advanced trainer until about 1943.

Reconnaissance biplanes included the He 46 (1931), which remained in Luftwaffe service until 1943. The He 42 two-seater floatplane of 1931 also soldiered on into the war years, together with the He 59 (1931), a twin-engined floatplane used for air/sea rescue work. The last Heinkel floatplane to enter production

A captured Heinkel He 177A heavy bomber

was the He 115 (1938), of which over 300 were built by 1944. The prototype broke eight seaplane speed records in 1938, and Luftwaffe machines were used as bombers, minelayers, and reconnaissance aircraft.

Heinkel's first dive-bomber was the underpowered (600 hp) He 50 biplane of 1931, which served only in small numbers, the He 66 being a training/reconnaissance export version.

The single-engined two-seat He 64 streamlined monoplane of 1932 won the Europa Rundflug race of that year and was followed by the elegant He 70 4-passenger monoplane, with retractable undercarriage and 750-hp engine. This machine was capable of 223 mph (359 km/h) and in 1933 set eight world speed records before joining Lufthansa's internal routes. A number of He 70s were used by the Luftwaffe as attack bombers or for reconnaissance. Only 14 examples of the four-engined He 116 were built, five for military use, while two went to Japan as mailplanes.

The company's principal World War II product was the twin-engined HE 111 bomber, since the He 112 fighter (1935) had not found favor with the Luftwaffe (which was committed exclusively to the Bf 109) and served mostly with the Spanish Nationalists and Romania. The He 100 fighter (1938) took the world speed record at 463.92 mph (746.45 km/h) in 1939 with a special 1,800-hp engine, but the 12 production aircraft (with 1,175-hp power units) were only used for propaganda photographs to give the impression that numerous squadrons of "He 113" fighters were in service.

The He 176 rocket fighter (1939) had a very short endurance, and the outbreak of war brought an end to its development. Two examples of the He 178 were built, only one of which flew, but the aircraft made the world's first jet airplane flight at Rostock-Marienehe on August 27, 1939, in the hands of Erich Warsitz. It was powered by a Heinkel S 3B engine of 1,102 lb. (500 kg) thrust.

Many other Heinkel projects for the Luftwaffe ended unsuccessfully. The He 118 dive-bomber (1937) proved incapable of diving vertically and was abandoned; the He 119 high-speed bomber (1937), with a nose-mounted propeller driven from a 2,350-hp DB 606 engine in the fuselage, suffered from engine overheating and was also abandoned; and the He 280 fighter (1941), with twin S 8A jet engines of 1,540 lb. (700 kg) thrust, became an unsuccessful competitor for the ME 262. The He 219 twin-engined fighter-bomber (1942) did serve successfully as a night fighter, but the He 162 VOLKSJÄGER jet fighter (1944) was never fully operational.

The He 177 Greif heavy bomber (1939), which had a pair of coupled DB 601 power units in each engine nacelle, was used for both bombing and maritime reconnaissance, but it suffered constant engine fires and earned the nickname "Flying Coffin." Production of the He 277 (1943), with four separate engine nacelles, only amounted to eight machines. After the war, Ernst Heinkel undertook license-production of the Starfighter, Magister and G 91, his projected jet feederliner (He 211) and VTOL He 212 coming to nothing. In 1965 the Heinkel firm became part of VFW (Vereinigte Flugtechnische Werke).

Helicopter

A wingless aircraft that obtains lift and propulsion from overhead horizontally turning rotors. Helicopters can take-off and land vertically, they can hover, and they can move in any direction. The rotors of a helicopter fulfill the same function as the wings of an airplane in providing lift, but the helicopter's engine rotates the rotors (which are in effect rotating wings), while the plane's power unit must move forward the entire airframe in order to obtain lift from the wings.

With their VTOL capability and their maneuverability, helicopters have obvious advantages over conventional

The 18-passenger Sikorsky S-58

U.S. Marine Corps CH-53 (Sikorsky S-65) Sea Stallion salvaging a CH-46 (Boeing-Vertol 107) Sea Knight

Most helicopters are powered by gas turbine engines, and they range in size from miniature unenclosed one-man craft, through the usual three- or four-seaters, up to transports carrying 25 or more passengers, and giant "sky cranes." These include the American Sikorsky S-64, which, with a 72-ft. (21.9-m) diameter rotor, can lift over 22,400 lb. (1,015 kg), and the Soviet MI-12, produced by the MIL design bureau. The Mi-12, with 220-ft. (67-m) rotors, set a world record in 1969 when it raised a payload of 88,636 lb. (40,200 kg) to a height of 7,398 ft. (2,255 m). The greatest altitude achieved by a helicopter, a French Aérospatiale SA 315B Lama, is 40,820 ft. (12,442 m).

Helicopter designs vary. The simplest is a single rotor system with subsidiary vertical tail rotor. A tandem has two main rotors, one at each end of the machine; side-by-side rotors are displaced laterally; triple rotors usually have two set side-by-side and one carried forward. Multiple rotors may be designed to intermesh, or can be coaxial and counter-rotating. The mounting of the rotor on the drive shaft may be rigid (with only a facility for changing the pitch); set on a gimbal enabling the blades additionally to move (flap) as a unit; or provided with complex hinges so that each individual blade has a certain amount of movement both up-and-down and in the horizontal plane, together with a pitch changing facility.

For the rotor to generate lift, the pitch of the blades must be altered to an appropriate angle of attack (see WING). The available pitch variation generally ranges from about 3 to 14 degrees, and alterations in pitch are governed by the collective pitch lever in the cockpit ("collective" because the pitch of all the blades is altered simultaneously). There is a throttle control on the top of the collective pitch lever, but the inertia of the rotor blades makes it difficult to change their speed quickly, and so height is normally gained or lost by changing the pitch of the blades and therefore their lift.

To move a helicopter forward the rotors must be tilted forward so that the airflow is directed slightly to the rear. This is achieved by cyclic pitch: the pitch of individual blades is increased as they pass to the rear (producing more lift) and decreased as they rotate to the front (reducing their lift). In most cases the blades are hinged at the rotor hub, either individually or on a gimbal, so that they can rise and fall (flap) under the influence of this increasing and decreasing lift. Inclination of the rotors is governed by the pilot's control column.

aircraft in many applications. They are widely used for short-range passenger transport (up to about 100 mi., 160 km), and are valuable for spraying and other agricultural operations; for observation, survey, police, ambulance, and rescue work; for carrying supplies, for example to offshore oil rigs; for transporting heavy mining or other gear to remote areas that cannot be reached by road; and in warfare, in which a new breed of ATTACK HELICOPTERS has been developed. They cannot, however, carry as heavy loads as fixed-wing planes, and their speed is limited to a maximum of little more than 200 mph (322 km/h). (The world record stands at 220.885

mph, 355.485 km/h, achieved by a SIKORSKY S-67 Blackhawk in 1970.) The tips of the advancing rotor blades achieve supersonic velocities at quite low aircraft speeds, while at modest forward speeds the roots of the retreating blades are moving through the air relatively more slowly than the helicopter itself and generate no lift. Higher speeds (up to 316 mph, 508.5 km/h) have been reached by compound helicopters, which have a supplementary means of propulsion (usually a jet engine). Numerous experiments have been made in an attempt to build a fixed-wing plane with a helicopter's versatility (see VTOL).

The torque reaction from the spinning rotor would tend to make a helicopter rotate in the opposite direction. A vertically oriented tail rotor, driven off the main rotor gearbox, counteracts this torque. The rudder pedals change the pitch of the tail rotor blades, increasing or decreasing their thrust and enabling the helicopter to be turned. In twin rotor helicopters the rotors turn in opposite directions and the torque is therefore canceled out, directional control being achieved by differential collective pitch changes (which alter the torque and hence turn the fuselage), or (in side-by-side rotor systems) by tilting the rotors in opposite directions.

Maintaining a helicopter in position by balancing the torque of the main rotor by the thrust of the tail rotor leads in practice to a small amount of drift. This is combated by slightly inclining the main rotor, either through the medium of the cyclic pitch system or by tilting the drive shaft.

Although designs for helicopter-like devices were made as long ago as 1483 (by Leonardo da Vinci), and proposals were made to apply steam power to rotorcraft in the 19th century, the helicopter is in fact a late arrival among man's means of transportation. The gasoline engine made powered flight a practical proposition for the first time, although a helicopter in fact requires between two and three times the power of a fixed-wing aircraft to lift the same weight. The Frenchman Louis Breguet built a helicopter with quadruple bi-plane-type blades in 1907 that managed to reach 4 ft. (1.2 m) off the ground while steadied by poles, and the same year Paul Cornu constructed a machine that hovered 1 ft. (0.3 m) above the ground for 20 seconds with himself as passenger. In Russia Igor Sikorsky built two unsuccessful rotary-wing aircraft during 1909–10. None of these early experiments solved the problem of torque, despite the use of counter-rotating rotors.

Between the wars interest in helicopters declined, partly because the AUTOGIRO was being developed with some success, but the appearance of the Focke-Achgelis Fa 61 helicopter in Germany in 1936 revived enthusiasm in the powered rotating wing. The Fa 61, with twin laterally displaced rotors, was fully controllable and could descend under autorotation if engine failure occurred.

Toward the end of World War II Igor Sikorsky, now in the United States, produced the R-4, the first helicopter to enter extensive military service. This 180-hp machine carried only one passenger and could not normally hover with a full load. The Sikorsky R-5 carried two passengers in addition to its two-man crew and was developed to become the highly successful S-51 commercial helicopter. Slightly earlier appeared the small BELL 47. The first helicopter to gain approval for civilian use, it has been produced in larger numbers than any other rotorcraft. The later Sikorsky S-55 and S-58 were larger, more powerful machines that carried 10 and 18 passengers respectively, their Russian equivalent being the Mil Mi-4. A major advance since these helicopters of the early and mid-1950s has been the introduction of gas turbine engines. Helicopters gained in speed and lifting ability, and their use has widened to include specialized roles as different as skycranes and air cavalry gunships.

Hellcat

The principal U.S. Navy carrier fighter during the last two years of World War II, the F6F Hellcat was designed by Grumman as a replacement for the F4F Wildcat, and featured better climb, maneuverability, and pilot visibility. The maximum speed of 375 mph (600 km/h) was less than that of some comparable aircraft, but the Hellcat could engage the Japanese ZERO on more or less equal terms because of its superior maneuverability, the result of its broad, square-cut wings (the Hellcat possessed the largest wing area of any American single-engined fighter of the period). Its great structural strength also made it very popular with pilots, and the F6F shot down nearly 5,000 Japanese aircraft. Hellcats first saw action in the Gilbert Islands in September 1943, less than 15 months after the prototype's maiden flight. The Hellcat was normally armed with six .50 machine guns in the wings, but some later models had two 20-mm cannon and four machine guns. Power was provided by a 2,000-hp Pratt & Whitney R-2800 radial engine. Wingspan was 42 ft. 10 in. (13.05 m) and length 33 ft. 7 in. (10.24 m).

Helldiver

The Helldivers were a series of naval and Marine Corps dive-bombers developed by CURTISS in the years between the wars. The first Helldiver was derived from the Curtiss F8C Falcon, a two-seater biplane fighter-bomber that entered service in 1928. The Helldiver had a cowling around its 450-hp Pratt & Whitney Wasp engine, an upper wing of reduced span and area, and could carry either a 500-lb. (226-kg) bomb beneath the fuselage or two 116-lb. (52-kg) bombs under the wings. F8C Helldivers saw service in the Caribbean and were operated by the U.S. Navy and Marine Corps until the mid-1930s.

SB2C Helldiver

F6F-5 Hellcat

The SBC Helldiver of 1933 was also a biplane, but was powered by a 950-hp engine, had enclosed cockpits and a retractable undercarriage, and could carry a 1,000-lb. (453-kg) bomb. At the time of Pearl Harbor the U.S. Navy had nearly 200 SBC Helldivers still in service; one Marine Corps unit was also equipped with the aircraft.

The name Helldiver was finally used for the Curtiss SB2C, a radial-engined low-wing monoplane which made its first flight in 1940, and served as the standard U.S. Navy dive-bomber from 1943 to 1946. The SB2C Helldiver carried a crew of two, had four .50 wing guns or two 20-mm cannon, and carried its bomb in an internal bay. Wingspan was 49 ft. 9 in. (15.17 m) and length 36 ft. 8 in. (11.17 m). Powered by a Wright R-2600 engine, maximum speed of the Helldiver was 290 mph (467 km/h). A total of 7,194 Helldivers were built. Some 900 of these were for Air Force use (under the designation A-25A), and 1,194 were license-built in Canada. See also DIVE-BOMBER.

Henschel

Henschel was one of a number of German industrial concerns that received financial incentives to enter the aircraft industry after 1933. Its aviation subsidiary, Henschel Flugzeugwerke, had a large factory at Schönefeld, near Berlin, where Friedrich Nicolaus designed a series of military aircraft. The best-known of Henschel's prewar aircraft were the Hs 123 biplane dive-bomber and the parasol-winged Hs 126 tandem-seat military observation aircraft. The 123 served in Spain in 1937–39 and was withdrawn from wartime duty after the fall of France, only to be thrust into fighting on the eastern front between 1942 and 1944, its remarkable handling qualities suiting it eminently to night and bad-weather harassment over the battlefield. The later Hs 129 was a specialized ground-strafer and anti-armor machine, equipped with various heavy cannon and in some cases with oblique rocket guns triggered by flying over metal objects. The Hs 130 high-altitude reconnaissance machine was too complex to see service, and the Hs 132 jet dive-bomber was too late.

Henschel also produced what is sometimes claimed to have been the world's first guided missile. This was the radio-controlled and rocket-assisted Hs 293 glider-bomb, which went into production in January 1942 and was used by many types of bombers. An Hs 293 sank HMS *Egret* on August 27, 1943, and destroyed other ships and numerous land targets. Nine variants

Henschel Hs 123 of the Luftwaffe

or developments were produced for other roles; these included one of the first surface-to-air missiles, the Hs 117 Schmetterling ("Butterfly").

Hien

The KAWASAKI Ki.61 Hien or "Swallow" (code-named "Tony" by the Allies) was the only Japanese fighter with a liquid-cooled engine to see service in World War II. The Hien owed its origins to the acquisition of manufacturing rights for the German DB 601 V-12 engine (used in the Messerschmitt BF 109). This enabled Kawasaki to produce prototypes of both the Ki.61 and the smaller and unsuccessful Ki.60 (which lacked maneuverability and was abandoned).

Tested against the Bf 109E and captured examples of the Curtiss P-40E, the Ki.61 was considered to be superior to both and entered service with the Army Air Force in the latter part of 1942. The Japanese version of the DB 601, known as the Ha.40, developed 1,175 hp and gave the Hien a maximum speed of 348 mph (560 km/h). When the Ki.61 appeared over New Guinea in April 1943, it proved to have an excellent dive capability and was well able, with its twin 20-mm cannon and two machine guns, to hold its own against contemporary American fighters. Adequate armor plate and self-sealing tanks (at that time something of an innovation in a Japanese fighter) reduced combat losses. The engine's main bearings and oil system gave trouble, however, and though later versions employed 30-mm cannon (Ki.61-Id) and a 1,500-hp Ha.140 engine (Ki.61-II), and there was even-

tually a redesigned variant with a stronger airframe (Ki.61-II-KAI), the Hien was gradually abandoned in favor of the Ki.100–essentially a Ki.61-II airframe fitted with a 1,500-hp Ha.112 radial engine.

The Ki.100 came into service at the beginning of 1945 and proved immensely successful. Superior to all Allied fighters except the P-51D MUSTANG, the Ki.100 operated as a high-altitude home-defense interceptor against B-29 SUPERFORTRESSES and their escorts, building up a formidable reputation. The last variant, with a teardrop cockpit canopy, was fitted with a turbo-supercharged methanol/water-injection engine that gave a top speed of 400 mph (643 km/h) at 30,000 ft. (9,150 m). Wingspan of the Ki.100 was 39 ft. 4 in. (12.00 m) and length 28 ft. 11 in. (8.82 m).

Hindenburg

Probably the most famous of all airships was the ill-fated hydrogen-filled German-built *Hindenburg* (*LZ 129*), completed in 1936 following the success of the smaller GRAF ZEPPELIN. She was 804 ft. (245 m) long, had a capacity of 7,000,000 cu. ft. (196,000 m³) of gas, and was powered by four 1,100-hp Daimler-Benz diesels. A scheduled service was maintained between Friedrichshafen, Germany, and Lakehurst, New Jersey, carrying up to 70 passengers on each journey.

While approaching her mooring mast at Lakehurst on May 6, 1937, she burst into flames, 35 of her passengers and crew being killed. This disaster, which has never been fully explained, followed several other airship accidents during the preceding years, and effectively marked the end of the airship experiment on a large scale.

Hiryu

Unlike earlier Japanese bombers of World War II, the MITSUBISHI Ki.67 Type 4 Hiryu or "Flying Dragon" (code-named "Peggy" by the Allies) was designed in the light of combat experience. The result was the best Japanese Army bomber of the war, although fewer than 700 were produced. The prototype was completed in December 1942; it was an aerodynamically clean, twin-engined mid-wing design with the new Mitsubishi Ha.104 1,900-hp 18-cylinder two-row radial engines. The aircraft was so maneuverable it could perform a loop or a vertical turn.

After the first production batch (Ki.67-Ia), a new version was introduced, the Ki.67-Ib. This had transparent blisters at the waist gunners' positions, modified nose and tail armament, improved engine cowlings, and other detail changes. A single 20-mm cannon in a dorsal turret and four machine guns proved, however, not to be very formidable defensive armament, and the bomb load amounted to only 1,760 lb. (800 kg). But the Hiryu made up for these deficiencies with a top speed of 334 mph (537 km/h) and range of 2,360 mi. (3,800 km). Wingspan was 73 ft. 10 in. (22.50 m) and length 61 ft. 4 in. (18.70 m).

First encountered by the Allies as a torpedo-bomber during battles off Formosa and the Philippines in 1944 (the new machines had been loaned to the Japanese Navy), the Hiryu subsequently fought in China and raided American forces at Okinawa and Iwa Jima and in the Marianas. In the last stages of the war the aircraft was employed as a KAMIKAZE machine and also (in modified guise as the Ki.109 with a nose-mounted 75-mm cannon) as a home-defense interceptor. The supercharged Ha.104ru engines originally specified for the Ki.109 never became available, however, and with its standard power plant the aircraft was unable to reach the altitude at which the B-29 SUPERFORTRESSES generally attacked.

Hughes, Howard (1905–1976)

The most famous years of Howard Hughes' long association with aviation were those of the 1930s, when he was one of the world's best known aviators. Hughes had already done stunt-flying (and suffered a near-fatal crash) in his own war film "Hell's Angels," and trained, under an assumed name, as an airline pilot, before abandoning movies for flying. In 1935 he set a new world landplane speed record of 352.46 mph (567.11 km/h) in his specially designed *Hughes One* racer, and two years later flew the plane across the United States in 7 hours 28 minutes (a record unbroken until the days of jet aircraft). In July 1938, with a crew of four, he flew a special twin-engined Lockheed 14 around the world in a record 3 days 19 hours 14 minutes, returning to a hero's welcome in New York and awarded a special medal by Congress.

During World War II Hughes was involved in the design of two ultimately unsuccessful aircraft for the U.S. government. The twin-engined XF-11, a prototype of a twin-boom reconnaissance bomber and powered by eight-bladed contra-rotating propellers, crashed in 1946 on its first flight with Hughes at the controls. The eight-engined all-wood H.2 Hercules flying boat (the "spruce goose") also flew only once. The aircraft with the greatest wingspan ever built (320 ft. 6 in., 97.69 m), weighing 190 tons and powered by eight engines, in its single flight covered a distance of less than a mile at a height of 70 ft. (21.3 m) at Long Beach, Cal. (November 2, 1947), again with Hughes at the controls.

Hughes sold his 78 percent interest in TWA for over $546 million in cash in 1966; Hughes' own companies are still major defense and aerospace contractors. Since the late 1950s Hughes Tool Co. has produced several helicopters, including the lightweight piston-engined Hughes 300 (the U.S. Navy's TH-55A trainer) and the larger turbine-powered Hughes 500 (the Army's OH-6A Cayuse observation helicopter, also produced in civilian versions). The new YAH-64A armed ATTACK HELICOPTER is scheduled to go into service in the early 1980s.

Hunter

The Hunter was an outstanding British fighter that provided the backbone of RAF strength from 1954 until 1965. Although the Supermarine Swift was in fact the first swept-wing fighter to enter RAF service (February 1954), the first Hawker Siddeley Hunters were delivered in July of the same year. The prototype P.1067 flew for the first time on July 20, 1951, and set a new world speed record in modified Mark 3 form on September 7, 1953, at 727.63 mph (1,170.76 km/h).

The Avon 113-powered Mark 1s suffered engine surge when the four 30-mm Aden guns were fired, which led to restrictions on use of the armament at high altitude; the Armstrong-Siddeley Sapphire-engined Marks 2 and 5 were free of this defect. The Mark 4 had an Avon 115, from which the surging problem had been eliminated, and ammunition link containers were also fitted beneath the nose. By 1958 the RAF's standard day fighter was the Hunter 6, with an Avon engine of 10,000 lb. (4,530 kg) thrust, and an outboard extension to the wing leading edge; length is 45 ft. 10½ in. (13.98 m) and wingspan 33 ft. 8 in. (10.26 m); maximum speed is 710 mph (1,143 km/h). The FGA9 (1959) was a ground-attack variant, and the FR10 (1958) had a photo-reconnaissance function. Two-seat T7 trainers entered RAF service in 1958.

Hunters were still flying with the RAF in the mid-1970s, and have also been operated by the air forces of Sweden, Denmark, Peru, India, and Switzerland, and by many Middle Eastern nations.

Hurricane

The first monoplane fighter to enter RAF service (with 111 Squadron in 1937), the Hawker Hurricane was numerically the most important British fighter in the Battle of BRITAIN (1,300 Hurricanes compared to about 950 SPITFIRES). The Hurricane was a development of earlier Hawker biplanes such as the FURY, and made its maiden flight on November 6, 1935. By the outbreak of World War II, this Merlin-engined fighter with its eight wing-mounted .303 machine guns equipped

The RAF's Hunter 6 fighter

19 RAF squadrons. Wingspan of the Mark I was 40 ft. (12.19 m) and length 31 ft. 5 in. (9.58 m). With a maximum speed of 340 mph (547 km/h, Merlin XX-engined Mark II), the Hurricane was not an effective match for the faster BF 109, and by 1941 it was being used largely as a night fighter or fighter-bomber with two 500-lb. (226.5-kg) bombs. Its armament was increased to 12 machine guns (Mark IIB), four 20-mm cannon (IIC), or a pair of 40-mm antitank guns (IID). The last variant to see extensive service was the Mark IV (1943), intended principally as a ground-attack aircraft and capable of carrying rocket projectiles.

Hurricanes were supplied to Russia and also served (as Sea Hurricanes) as catapult and carrier-borne fighters. The type was also license-built in Canada with a Packard-Merlin engine (Marks X, XI, XII). The last Hurricane, a Mark IIC (PZ865), was completed in August 1944 and is still maintained in flying condition.

Hydraulic system

By the early 1930s the demand for higher performance had led to larger retracting landing gears, wheel-brake systems, retracting floats, flaps, gun turrets, and bomb doors, all of which could not safely or reliably be worked electrically or pneumatically and were too tiring to operate by hand. The solution was to be an airborne hydraulic system, in which a self-contained series

Hawker Hurricane IIC night fighter

of circuits containing special fluid would be energized by engine-driven pumps and controlled by electrically-operated valves. But at the time there were no appropriate materials, fluids, or seals, and there was a lack of any experience that could meet the airborne need for high performance and light weight. The only hydraulic components in mass production in 1934 were automobile brakes, which used natural-rubber seals. For aircraft mineral oil seemed the only possible fluid, but it was incompatible with natural rubber. Not until 1937 were synthetic rubbers available, and

even then most systems were energized by hand pumps and operated at pressures of 800–1,000 lb./sq. in. Two further years of intense effort led to high-precision metal parts and the O-ring seal, which allowed pressures to jump to 3,000 lb./sq. in. This in turn made the hand pump inadequate, and led to development of lightweight systems localized around the device being operated, with signals transmitted electrically.

By the end of World War II hydraulic systems were fully reliable, and often served as many as 25 items in a single

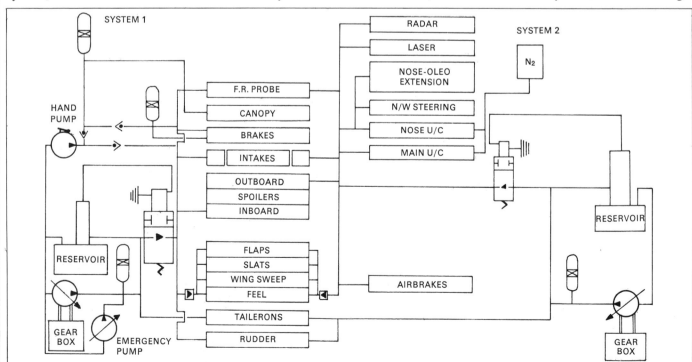

The highly complex dual-circuit hydraulic system of the Panavia Tornado multirole combat aircraft

aircraft. In the 1950s many aircraft were being developed with system pressures of 4,000 lb./sq. in., but the common pressure today is still at 3,000 lb./sq. in. In some very large aircraft, such as the Boeing 747, total hydraulic power can be rated at 1,000 hp, and the number of hydraulic components is measured in hundreds. Traditional mineral oils are still in wide use, but by 1954 the first of a growing range of alternatives (Skydrol 500) had been produced. Unlike their predecessors, these fluids were not highly flammable, but they needed a completely new range of sealing materials. Skydrol was specified for the first commercial jetliners, and it is now one of a large number of water-based or ester-based fluids used in aircraft. An even more sophisticated range of fluids and sealing materials is used in supersonic aircraft subject to high skin temperatures; engine controls and variable afterburner nozzles are operated by hydraulic rams or motors operated by the engine fuel itself. Probably the most challenging problems of all have had to be overcome in the hydraulic systems of guided missiles, in which accelerations and temperatures can be at their most extreme. Among other difficult hydraulic problems are those found in flying-control systems, where performance must be unfailingly reliable (often achieved by using two, three or four circuits in parallel) and feedback due to aerodynamic loads on the surface is unacceptable. It is not always possible to rely on the "stiffness" of an operating ram in which fluid is trapped; in some landing-gear struts the springing is achieved solely by the trapped fluid's compressibility.

I

IATA

The majority of the world's scheduled airlines are members of IATA (International Air Transport Association), which helps operators to correlate their routes, collectively negotiate fare structure, and complement each others services. International airlines are full members, domestic operators are associates.

There is a Financial Committee (currency exchange, insurance, taxation, inter-line transactions), a Legal Committee, and a Technical Committee; Traffic Conferences are normally held every two years and decide policy for passenger fares and cargo rates, other matters being referred to special conference committees.

IATA itself was founded in 1945, but owes its origins to the International Air Traffic Association which was established at The Hague in 1919. It has main offices in Montreal and Geneva.

ICAO

The International Civil Aviation Organization (ICAO) was provisionally established at a meeting in Chicago in 1944 attended by aviation representatives from 52 Allied and neutral states. A convention on international civil aviation was drawn up that established the privileges and restrictions of contracting states to develop air transport services on the basis of equality of opportunity and on sound operational and economic principles.

Also prepared at Chicago was the International Air Services Transit Agreement (providing for signatories' aircraft to overfly or land for technical reasons in each others territory) and the International Air Transport Agreement governing traffic between signatories.

On April 4, 1947, the ICAO officially came into being, with headquarters in Montreal. By 1975 there were 131 member states, whose delegates continued to uphold the original Convention and Agreements. Particular attention was paid to the encouragement of aircraft design for peaceful purposes, to safety factors, to air navigation, and to the elimination of unfair or wasteful competition. The ICAO Assembly meets at least once every three years, and there is an extensive publications program covering international conventions and agreements, ICAO procedures, air navigation, air transport, legal considerations, etc.

Icing

The accretion of ice on the airframe of a plane in flight can constitute a serious hazard. Its presence increases weight and drag, deforms the contours of airfoil sections (thus reducing lift), and may lock the control surfaces.

Icing occurs most extensively in air with a high moisture content at temperatures of between 28.4°F (2°C) and 14.6°F (−8°C), especially when a very cold airplane descends from high altitude through rain in an ambient temperature just below freezing point. It can occur in clear air as well as cloud, but it is unlikely to be heavy at temperatures below 14.6°F (−8°C) and does not generally occur at less than 0.4°F (−18°C).

Glaze forms in varying depths as a glassy coating of clear or cloudy ice. It is usually caused by rain falling on a freezing surface, or by supercooled water drops. Rime frost is encountered in freezing fogs or in stable layer clouds and consists of porous, granular ice crystals.

Hoarfrost rarely occurs in flight in temperate regions, but in the winter it must be brushed off parked aircraft before takeoff since it would otherwise increase drag.

The shear strength of ice bonded to an airframe is quite considerable. To remove it, inflatable rubber boots may be recessed in the leading edges of the wings and tailplane (only practicable in aircraft with a speed of less than 350 mph, 560 km/h), glycol-based fluids may be released onto flight surfaces to lower the freezing point of the accretion so that it falls away (this method will not usually prevent ice from forming initially), hot air may be bled from an engine to warm vulnerable parts of the structure where ice is likely to form, or electrical heating may be employed.

Ice may block small apertures, such as the PITOT TUBE (thereby rendering the airspeed indicator useless). Electrical heating is used to counteract such an eventuality, and also to prevent icing of cockpit windows, engine intakes, and propellers.

The carburetors of piston-engined aircraft are vulnerable to icing up of the choke or the throttle by a combination of low pressure and evaporation (the volatility and high latent heat of the fuel cools the air passing through the venturi). The carburetor heater must be operated whenever an aircraft with a gasoline engine begins to descend, especially through cloud. Carburetor icing can occur in temperatures as high as 59°F (15°C).

IFF

The initials IFF stand for "identification, friend or foe," the name of a device essential to every combat aircraft and any airplane flying in wartime. As first used in Britain in 1940, the IFF system normally consisted of a ground transmitter/receiver (T/R) which sent out pulses to "interrogate" all aircraft in the local airspace. Friendly aircraft carried a special IFF transponder which automatically sent back a coded reply to the ground. Hostile aircraft were unable to reply and could thus be identified and shot down.

Modern IFF systems are invariably associated with primary radar devices. They transmit directional pulses on 1,030 MHz, while cooperating targets (aircraft or ships) reply on 1,090 MHz.

The replies are processed and turned into video signals displayed on the air-traffic or air-defense displays, each target blip having its own computer-generated alphanumeric labels (in a few very simple cases the IFF reply merely lights a warning lamp). Modern systems operate in particular modes (called 1, 2, 3/A or C), while the airborne transponder sends back a complex "ident code" that can be varied subtly to prevent possible copying by an enemy. Some military IFF transponders automatically transmit this code each time the microphone switch is pressed for air/ground communication.

Il-2. See SHTURMOVIK

Il-28

The first jet bomber to enter service with the Soviet air force, the Ilyushin Il-28 (code-named "Beagle" by Nato) made its maiden flight on August 8, 1948. As the result of a priority building program, 25 were able to take part in the 1950 May Day flypast over Red Square. The Il-28 was a twin-engined straight-winged aircraft with swept tail surfaces, powered by two 5,000 lb. (2,270 kg) thrust RD-45 copies of the Rolls-Royce Nene. It carried a bomb load of up to 6,613 lb. (3,000 kg), with a maximum takeoff weight of 46,297 lb. (21,000 kg). Wingspan was 70 ft. 4¾ in. (21.44 m) and length 57 ft. 10¾ in. (17.64 m). Some 1,000 Il-28s were exported to a dozen nations, including examples of the Il-28U trainer ("Mascot").

A scaled-up version with a takeoff weight of 92,524 lb. (42,000 kg) was produced in 1952. This was the Il-46, designed to carry up to 13,228 lb. (6,000 kg) of bombs. The Il-28 had a maximum speed of 559 mph at 14,765 ft. (900 km/h at 4,500 m) and a range of up to 1,355 mi. (2,180 km). The Il-46 was marginally faster and had a maximum range of 3,055 mi. (4,920 km).

Ilya Mourometz

The Ilya Mourometz, designed by Igor SIKORSKY, was the first four-engined bomber, and it equipped what was probably the world's first strategic bomber force, created by Czarist Russia during World War I. The Ilya Mourometz was based on Sikorsky's *Le Grand* of 1913 (the world's first four-engined airplane), and established a world payload record in February 1914, when a load of 2,822 lb. (1,280 kg), consisting of 16 men and a dog, was lifted during an 18-minute flight. Several similar

The Ilya Mourometz, the world's first four-engined warplane

records were established during 1914; in June the Ilya Mourometz made a momentous flight to Kiev, covering some 800 mi. (1,300 km) in poor weather in about 18 hours' flying time. Between 70 and 80 IM bombers were built. The first IM unit was formed on August 26, 1914, and sent to the front facing East Prussia. The IM units were run on naval lines—each aircraft was numbered and replaced as necessary by a later type. The greatest number in service at any one time was 18. The "Squadron of Flying Ships," as it was called, was formed on December 10, 1914. Over 400 missions were flown, and only two aircraft were lost. A wide variety of models was built: there were five basic types of airframes, together with many alternative engines and different arrangements of defensive armament. The bomb load was carried internally in the IM's capacious fuselage.

The largest IM was the Type E2, with a wingspan of 113 ft. 1 in. (34.50 m), a maximum speed of 81 mph (130 km/h), and a range of 348 mi. (560 km). The IM-E2 was powered by four 220-hp Renault engines and carried 5,423 lb. (2,460 kg).

Ilyushin, Sergei Vladimirovich (1894–1977)

Soviet aircraft designer, Ilyushin was an engine mechanic at Petrograd from 1916 to 1918, and learned to fly in 1917. He joined the Red Army in 1919 and in 1921 became an engineering student at the Zhukovski Military Aviation Academy. While a student, he produced three gliders in the years 1923–25. He joined the Central Design Bureau in 1931 and immediately before World War II designed the famous Il-2 SHTURMOVIK ground-attack aircraft and the DB-3 (TsKB-26, TsKB-30, Il-4) bomber, which made its first flight in 1935, and in 1939 flew nonstop from the Soviet Union to North America.

The DB-3 was virtually the only long-range Soviet bomber of World War II; a total of 6,784 were built between 1937 and 1944. Normally powered by two 950-hp M-87A radial engines, it could carry up to 5,500 lb. (2,500 kg) of bombs and had a range of between 1,615 and 2,175 mi. (2,600 and 3,500 km) depending on the payload.

At the end of the war the Ilyushin design team was working on an airliner to replace Soviet-built DC-3s (Li-2s). The result was the twin-engined Il-12 of 1946 (seating 21 to 27 passengers) and its later development, the Il-14. These aircraft were used extensively in the early postwar period by Soviet bloc airlines, and the Il-14 was produced under license in East Germany and Czechoslovakia.

The IL-28 (maiden flight August 8, 1948) was the first production Soviet jet bomber. Powered by two RD-45 or VK-1 copies of the Rolls-Royce Nene, the Il-28 had a maximum speed of 559 mph (900 km/h) at 14,765 ft. (4,500 m), could carry up to 6,613 lb. (3,000 kg) of bombs and had a range of 715–1,355 mi. (1,135–2,180 km).

The Ilyushin design team subsequently concentrated on transport aircraft. The Il-18 of 1957 is a medium-sized airliner with four 4,000-hp AI-20 turboprop engines; over 600 were produced (100 for export). The Il-86 is the first Soviet wide-bodied jet airliner, designed to carry 350 passengers and with a takeoff weight of 414,470 lb. (188,000 kg).

Immelmann, Max (1890–1916)

One of the first German fighter aces, Immelmann learned to fly in February–March 1915, and joined *Flieger-Abteilung* 62 in May to fly two-seat aircraft. On July 30 he made his first flight in a Fokker EINDECKER, and on August 1 scored his first victory. By the end of the year he had seven kills, and on January

Ilyushin Il-18 "Coot" turboprop airline of Czechoslovak Airlines

12, 1916, was awarded the *Pour Le Mérite*. Immelmann developed a climbing half loop with a roll off at the top known as the Immelmann turn. By the end of March 1916 his score stood at 13, and he had received every possible honor and award, including the nickname *Adler von Lille* ("Eagle of Lille"). On June 18 he shot down his 15th victim, and that evening joined combat with two British FE2Bs of 25 Squadron. After sending one down with a dead pilot, his Fokker broke up in midair, and Immelmann was killed. German evidence stated that his gun synchronization was faulty and the Fokker's propeller was shot away, while the Royal Flying Corps gave credit to Second Lieutenant G. R. McCubbin and Corporal J. Waller.

Inertial Navigation Systems (INS)

Celestial navigation is sometimes impossible because of obscuring clouds, and radio or radar aids may be either unavailable or unreliable (in wartime they may be jammed by enemy electronic countermeasures). In such circumstances an independent navigation system contained within the aircraft and immune to outside interference must be employed: inertial navigation. German V-2 rockets embodied a simple form of guidance system incorporating gyroscopes and accelerometers. In 1945 Dr. Charles S. Draper at the Massachusetts Institute of Technology's Instrumentation Laboratory was asked by the Air Force to devise an inertial navigation system suitable for aircraft. As a result of Draper's pioneer work, INS is now available for both civil and military aircraft. An INS consists basically of three accelerometers mounted with their axes at right angles to each other on a platform that is gyroscopically stabilized to maintain itself in the same plane. A computer can interpret the readings from the acceler-

ometers in two stages to give the speed and the distance traveled. With this information almost any navigational track can be calculated.

Inertial navigation is normally accomplished by flying from one intermediate waypoint along the route to the next, details of the waypoints having been previously fed into the computer. Drift in an aircraft's INS should not exceed 1 nautical mile per hour, and the equipment's life is about 1,500 hours. An airliner normally has three INS units (one each for the A and B autopilots, together with a backup). Military use of INS requires rugged equipment, but a life of no more than 150 hours is acceptable.

In-Flight refueling

Transferring of fuel from one aircraft to another in flight has still not been adopted in commercial aviation; its principal use is to increase the range of military aircraft. The first pipeline refueling was made by two U.S. Army DH 4B biplanes in 1923. By 1939 Sir Alan Cobham was conducting transatlantic trials, and in the late 1940s the U.S. Air Force introduced in-flight refueling on a large scale. Boeing de-

veloped the "flying boom" method for the Strategic Air Command's KC-135 tanker—a basic Boeing 707. The Russian "looped hose" transfer had in fact been jointly developed by the United States and Britain. Other U.S. Air Force commands and the U.S. Navy adopted the British "probe and drogue" system, in which the receiver's probe connects into a drogue at the end of a hose trailing from the tanker. Certain helicopters are also fitted for in-flight refueling.

Instrument Landing System (ILS)

A radio aid that enables a pilot to make his final approach to the runway in bad weather, and is in fact often used as a matter of routine even in good visibility. A meter in the flight deck registers signals from "localizer" and "glide path" transmitters. The localizer antenna is in line with, but beyond, the far end of the runway, and transmits a narrow beam extending $2\frac{1}{2}$ degrees each side of the runway center line. If the plane is within this beam, the meter's localizer beam pointer hangs vertically; the pilot knows that he is on course. If the plane flies to the right of the beam, the needle swings to the left; the pilot must "fly left" to get back on course (and vice versa).

The glide path antenna is positioned near the touchdown end of the runway and in effect transmits a beam angled up at about three degrees to the horizontal. When a plane is on the glide path beam, the second meter pointer is horizontal. If the pilot deviates above the beam the pointer swings down, indicating "fly down" (and vice versa). Two additional antennae transmit vertical beacons at distances of about 5 mi. (8 km) and $\frac{3}{4}$ mi. (1 km) from the touchdown point as a position check. ILS provides an accurate and reliable guide for the approach, but not for the actual landing, which is performed visually (see AIR-TRAFFIC CONTROL).

"Probe-and-drogue" in-flight refueling: an RAF Victor tanker and two Phantoms

Instruments

The principal flying instruments (as distinct from NAVIGATION instruments and engine instrumentation) include an ALTIMETER and an ARTIFICIAL HORIZON, which are arranged in a T-shaped layout on the pilot's instrument panel, with the artificial horizon in the center.

To the left of the T is the airspeed indicator (ASI), which is graduated in knots and may have two needles (indicating tens and hundreds of knots). Inside the airtight case is an expanding metal capsule to which the needle(s) are attached. A tube leads into the capsule from the PITOT TUBE, which records the air pressure created by the aircraft's forward movement. A second tube connects the instrument case to the static vent, which measures the external atmospheric pressure. The differential between these two pressures expands and contracts the capsule, moving the needles.

The horizontal situation indicator (HSI, direction indicator) is at the bottom of the T-shaped instrument panel. It consists of a rotating compass card attached to a gyroscope that has a horizontal spin axis. Once calibrated with the magnetic compass, the HSI gives a steady reading uninfluenced by irregularities in the Earth's magnetic field or by any magnetism in the aircraft itself, and is not swung about by the aircraft's motion.

The machmeter measures aircraft speed in relation to the local speed of sound at any altitude. It contains two capsules: air from the static system is fed to the airtight case and causes the absolute capsule to expand with altitude, while air from the pitot system enters the differential capsule and expands it as the speed increases. The two capsules are both connected to the needle which records the MACH NUMBER.

Vertical speed indicators (VSIs) contain a capsule connected to the static pressure system with a capillary leak to the inside of the airtight instrument case. As the aircraft climbs (or descends), the pressure in the capsule becomes less (more if descending) than the pressure in the instrument's case and the indicator needle moves. When level flight is resumed the pressure equalizes by means of the capillary leak.

The turn and bank indicator has a gyroscope aligned across the aircraft which indicates rate of turn, together with a slightly curved fluid-filled tube containing a ball to show bank. Formerly an important instrument, the turn and bank indicator now usually occupies a subordinate position on the panel.

AUTOMATIC PILOTS, AVIONICS and engine instrumentation require their own separate data recording displays and controls in the cockpit.

Interrupter gear

As early as 1912, inventors in Britain, France, Germany, Italy, Russia, and the United States had all proposed means of firing a machine gun safely past the blades of an aircraft's revolving tractor propeller. All these ideas were ignored by governments and senior military officers, though at least two of the proposals were identical in principle to the solutions adopted in World War I. One of the experimenters was Roland GARROS, and when his Morane Type L was forced down and captured by the Germans in April 1915, they discovered the steel plates or wedges he had fitted to the propeller blades to deflect any bullets from his forward-firing machine gun. Anthony FOKKER was one of several designers to whom the German High Command showed the primitive Garros system. His engineers were not impressed and devised instead the first true interrupter gear actually to be tested. Fitted to a Fokker EINDECKER (an M.5K) in a matter of days, the device simply cut the fixed Spandau machine gun out of the firing circuit each time a blade passed through the line of fire. First appearing in combat at the end of June 1915, the Fokker monoplanes, armed with one or more guns fixed to fire directly ahead, revolutionized air warfare.

By October 1915 British teams were engaged in devising improved interrupter gears. The Royal Flying Corps received systems by Vickers, Ross, and Sopwith-Kauper, while the Royal Naval Air Service used the Scarff–Dibovsky. The Vickers F.B.19 employed a Mark II form of the Challenger gear, in which the gun was accurately synchronized with the propeller, and there were also several French and American systems.

By 1916, however, a totally new kind of interrupter gear had been devised. The Romanian Georges Constantinesco patented a method of transmitting impulses along a pipe filled with an incompressible fluid. In his C.C. Gear he linked the trigger of the gun(s) with a pump on the engine, and by adjusting the input pulses to particular positions of the crankshaft (or the whole engine, in the case of rotary engines) firing into the propeller blades could be avoided. The Constantinesco gear could be produced as a standard system, readily adaptable to any engine and almost any type of gun. It superseded mechanical interrupter gears by 1917, and remained in common use until World War II.

HOLLOW DIAPHRAGM

LINKAGE BETWEEN DIAPHRAGM AND POINTER

300

180

PITOT HEAD

WING

RAM AIR PRESSURE

ELECTRIC HEATER TO REMOVE ICE FORMATION

One of the principal instruments in any aircraft is the airspeed indicator. The Pitot tube and static vent, often (as here) incorporated in a single unit, lead to the airspeed indicator, where a capsule expands or contracts and moves a needle or pointer on the dial face

Grumman Intruder

Modified forms were used in Soviet fighters, and in such aircraft as the BF 109, FOCKE-WULF 190, and P-39 AIRACOBRA, as well as in most Japanese interceptors.

Intruder

During World War II the name "Intruder" came to be applied to night fighter/bomber aircraft operating offensive missions deep over enemy territory. To some degree the use of such aircraft stemmed from the absence of Luftwaffe planes over Britain after May 1941, but it was a natural corollary of the "Rhubarb" daylight sweeps over Europe practiced by the day fighters. Until 1943 no aircraft carrying AI (airborne interception) radar were allowed over hostile territory, and the main intruder aircraft were the Douglas Havoc, a few Bristol Blenheims and Beaufighters, and de Havilland MOSQUITOES. Most intruder missions had the threefold objective of reconnaissance, harassment, and the destruction of enemy planes in the dark. Without radar, actual interceptions were difficult, but in 1943 radar was at last released for use over Europe, and

on D-Day (June 6, 1944) all radar was cleared for intruder missions. By this time the Luftwaffe had suffered heavy losses from Mosquito intruders, and almost as many losses from desperate measures taken to avoid them. The Luftwaffe, for its part, sent NJG (night fighter) wings with the twin-engined Ju 88G bomber and other aircraft on intruder missions over British airfields, where they could mingle with the returning heavy bombers and shoot them down over their lighted home bases.

The name "Intruder" has also been given to the GRUMMAN A-6, a long-range, low-level carrier-based attack aircraft of the U.S. Navy and Marine Corps, with two turbofans of 8,500 lb. (3,850 kg) thrust and up to 18,000 lb. (8,154 kg) of armament. Intruder missions are now called interdiction.

Invader

Twin-engined American tactical bomber which began active service in the last months of World War II. Intended as a replacement for the Douglas A-20 Havoc ground-attack bomber, the Douglas A-26 Invader made its maiden

flight on July 10, 1942. In general outline it resembled the slower A-20 but was distinguished by its square-cut wing tips and tail. The Invader was built in two principal models, the A-26B with a "solid" nose and six or eight fixed .50 machine guns, and the A-26C with a glassed-in nose, bombardier's compartment, and two fixed nose guns. Both models had a 4,000-lb. (1,814-kg) bomb-bay capacity and remote-controlled ventral and dorsal turrets each with two .50 guns. Pratt & Whitney R-2800 radials gave a top speed of 373 mph (602 km/h) on late models. Wingspan was 70 ft. (21.34 m) and length 50 ft. (15.24 m). The A-26 was given its first operational test over Europe in September 1944 and thereafter began to replace the A-20s and Martin B-26 Marauders of the Ninth and Twelfth Air Forces. The Invader was also used in the Pacific during 1945, and became the standard light bomber of the immediate postwar years. Redesignated the B-26 (the original Martin B-26 was no longer in service), the Invader saw extensive use during the Korean War as a night interdictor. Rebuilt Invaders (B-26Ks) were used for counterinsur-

Douglas Invader

gency operations in Southeast Asia during the Vietnam conflict. A total of 2,451 Invaders were built between 1943 and 1946.

J

Jabara, James (1923–   )

First American jet fighter ace and second highest scoring pilot of the Korean War. Jabara was a fighter pilot in World War II, serving with the 354th Fighter Squadron, 355th Fighter Group in England, with which he achieved six ground-strafing victories and one and a half in air combat.

A member of the 334th Fighter Squadron, 4th Fighter Interceptor Wing, which was sent to the Korean War zone in November 1950, Captain Jabara, flying an F-86 SABRE, had a number of successful engagements during the early months of 1951. On May 20 he shot down his fifth MiG-15 to become the first Allied air ace of the Korean conflict, and the first U.S. Air Force pilot to achieve ace status flying jet aircraft and fighting jet aircraft. At the end of his tour Jabara had a total of

six MiG victories.

In January 1953 Jabara returned to the 4th Wing for a second combat tour. During the next six months he destroyed another nine MiGs, claiming his 15th and last on July 15.

Jaguar

Produced as a joint project by the Anglo-French group Sepecat (made up of Breguet and BAC), the Jaguar was intended as a tactical strike fighter and advanced trainer and made its maiden flight in 1968. Versions include the Jaguar A (single-seat tactical support aircraft for the Armée de l'Air), Jaguar B (RAF T2 two-seat trainer), Jaguar E (French two-seat trainer) and Jaguar S (RAF GR1 tactical support). All are powered by two afterburning Adour 102 engines of 8,000 lb. (3,625 kg) thrust. Wingspan is 28 ft. 6 in. (8.68 m) and length (Jaguar A and S) is 50 ft. 11 in. (15.52 m). Types A, S, and E have twin 30-mm cannon, Type B a single gun; both trainers and single-seaters are equipped with under-wing and fuselage center-line mountings for missiles, bombs, or drop tanks. Export models have Adour 804 engines of greater power. Jaguars entered service with the

Armée de l'Air in 1972 and the RAF in 1973.

Japan Air Lines

Formed in 1951 as, initially, a domestic airline, JAL (Nihon Koku Kabushiki Kaisha) was already operating its own DC-4s and establishing a Tokyo–San Francisco service in 1952. Reorganized in 1953 with government shareholding as the Japanese national airline, it went on to introduce DC-7s on its transpacific service. A route across the Pole to Paris opened with leased Boeing 707s in 1959; DC-8s came into service the following year. In March 1970 JAL opened a Tokyo–Moscow–Paris route; it acquired Japan Domestic Airways in 1971. An extensive network of scheduled passenger and cargo services now links cities in Asia, Australia, North and Central America, the Middle East, and Europe. The present fleet includes 26 Boeing 747s, 45 DC-8s, 3 Boeing 727s, 3 Falcon 20s, and 3 Beech 18s. Over 9,000,000 passengers are carried annually.

Japanese military aviation

During the Russo-Japanese War in 1904, two Japanese balloons had made 14 ascents in support of the Imperial Army forces besieging Port Arthur, and five years later members of the Army, the Imperial Navy, and Tokyo Imperial University, formed the Provisional Military Balloon Research Society. In fact, this body spent most of its time studying heavier-than-air aviation, and in 1910 it sent two Army officers to Europe to buy two airplanes and train as pilots. They returned with a Farman and a Grade, making the first-ever flight in Japan on December 19, 1910. Over the next five years more than 1,000 Japanese officers were taught to fly, many of them in Japan, but only

RAF Jaguar S tactical support aircraft

a handful served as pilots in World War I (with the French Aviation Militaire). In December 1915 the Army pilots were finally organized into an Air Battalion, expanded into the Army Air Division in April 1919.

A French mission was sent to Japan with 63 flight instructors, and this cemented the Army Air Division's preference for French military aircraft (such as the NIEUPORT 24, SPAD XIII, and Salmson 2A-2). Gradually Japanese industrial giants such as KAWASAKI, MITSUBISHI, and NAKAJIMA, began to build foreign designs, almost all of them French in origin. On May 1, 1925, the Army Air Division became an autonomous branch of the Army as the Imperial Army Air Corps. It acquired combat experience in the increasing level of Japanese involvement in China, which eventually led to the outbreak of a full-scale Sino-Japanese War on July 7, 1937. One of the results of this conflict, which involved by far the biggest display of air power anywhere in the world since 1918, was the reorganization and expansion of the Air Corps into a number of Kokuguns (Air Armies), each comprising two to three Hikodishans (Air Divisions), which in turn each consisted of two to three Hikodans (Wings or Air Brigades). A Hikodan was made up of three Sentais (Groups), which were flexible units of three Chutais (Squadrons) each of 9 to 12 aircraft. This command structure survived to the end of World War II, and proved itself in China and in serious incidents with the Soviet Union (Changkufeng and Nomonhan), which apparently involved more than 3,900 aircraft and led to exaggerated Japanese claims of over 1,900 Russian machines destroyed (the true figure was about 600).

When Japan entered World War II on December 7, 1941, the 5th Hikodishan struck at the Philippines and the 3rd attacked Malaya and Burma, achieving overwhelming success until about July 1942. Thereafter these two great air armies gradually crumbled, while the Hikodishans in China likewise found they were losing the total superiority they had enjoyed since 1937. Not until October 1944 did newer combat aircraft begin to reach the front-line Chutais in quantity, but by this time the entire Japanese war machine was becoming outclassed and outnumbered. In its final last-ditch defense against B-29 SUPERFORTRESS raids, the Army fighter force was crippled by the inability of its interceptors to climb either fast enough or high enough.

Throughout World War II the main air forces facing the Western Allies were those of the Imperial Navy. Naval aviation began with an aeronautical research committee established in 1912; two officers were soon dispatched to France and the United States to learn to fly and to buy aircraft. In the early months of World War I four Imperial Navy seaplanes sank a German mine-layer at Tsingtao, but thereafter development was slow until the arrival of a British mission in 1921. The result was adoption of designs from such British manufacturers as Sopwith, VICKERS, BLACKBURN, and GLOSTER, as well as the swift build-up of a Japanese design and manufacturing capability. Naval air units were flung into the Sino-Japanese War in major strength in 1937; they were equipped with such new and successful aircraft as the ZERO, the result of the 9-shi plan of 1934.

At the time of Pearl Harbor the Imperial Navy had over 3,000 front-line combat aircraft, organized into Koku Kantais (Air Fleets) made up of Koku Sentais (Air Flotillas) subdivided into Kokutais (Naval Air Corps) of 30 to 150 aircraft, although the command structure was modified to suit deployment aboard carriers and surface ships.

For the first seven months of the war the Imperial Navy swept all before it—especially effective were the big carrier forces of the 1st Koku Kantai—but the offensive gradually ground to a halt, and the carrier force was totally annihilated by late 1944 in such great sea battles as Leyte Gulf and the Philippine Sea. During the final six months of the war the only effective operations flown by the Imperial Navy were KAMIKAZE (suicide) attacks, despite the availability of such new combat aircraft as the RAIDEN and SHIDEN.

Present-day Japanese military aviation is divided into air and naval forces together with an air component of the army. The Koku Jiei Tai, created on July 1, 1954, is the Air Self-Defense Force. Originally equipped with American aircraft, it is today receiving Japanese-built F-4EJ PHANTOMS and the entirely Japanese Mitsubishi FST-2 and T-2A supersonic attack and training machines (strongly resembling the JAGUAR). The Kaijyoe Jiei Tai (Maritime Self-Defense Force) was formed on April 26, 1952, and deploys large numbers of patrol and antisubmarine aircraft, including the Japanese-developed twin-turboprop P-2J derivative of the Neptune, the Grumman S-2 Tracker, and the all-Japanese SHIN MEIWA STOL flying boat and amphibian. The Kikujyo Jiei Tai (Ground Self-Defense Force, or Army) has units equipped with helicopters and liaison and reconnaissance machines.

Jenny

The JN two-seat biplane was one of the first landplanes built by Curtiss and one of the most widely used trainers of World War I, in Britain as well as America. The Jenny combined features of earlier Curtiss J and N machines, Model J having been designed in England ·by a Sopwith engineer, B. D.

Curtiss Jenny

Thomas, whom Glenn Curtiss hired for the purpose. The JN was evaluated by the U.S. Army in 1914 and built in large numbers throughout the war by Curtiss and by the Canadian Aeroplane Corporation. The name "Jenny" was entirely unofficial; Canadian machines were known as "Canucks." Production of the Jenny ran from the JN-2 version of 1915 through the JN-6H; the most numerous single model was the JN-4D. Wingspan was 43 ft. 7 in. (13.29 m) and length 27 ft. 4 in. (8.33 m); the engine most frequently employed was 90-hp Curtiss OX-5. The Jenny was used by American, Canadian, and British forces and performed observation duties with General Pershing's expedition against Pancho Villa in 1916. War-surplus Jennies were bought by private owners, flying schools, barnstormers, and stuntmen. One aircraft flew the first experimental Washington Philadelphia New York airmail service in May 1918. A number of Jennies remained in Army service until 1927; many of these were fitted with 150-hp Wright-Hispano engines.

Jet engine

A jet engine draws in air at the front, compresses it, and mixes it with fuel in the combustion chamber. The resulting hot gases stream out of the jet nozzle, producing forward thrust by reaction in accordance with Newton's third law ("For every action there is an equal and opposite reaction"). While a propeller provides thrust by forcing a large volume of air back at a relatively slow speed, a jet engine ejects a comparatively small volume of gas at a very high speed.

The greatest speed attained by a propeller-driven plane is about 550 mph (885 km/h), while the fastest jets reach well over 2,000 mph (3,218 km/h).

There are four basic elements in a jet engine: the compressor, which sucks air in at the front; the combustion chamber; the turbine, which is turned by the exhaust gases and drives the compressor (and, in the case of a TURBOPROP, the propeller); and the exhaust nozzle.

Compressors can be either centrifugal or axial. The centrifugal type distributes air to the combustion chambers from the perimeter of the impeller disk and can compress air to a ratio of about 5:1 in a single stage. In the more common axial compressors the air passes along a series of perhaps nine rings of compressor blades that become progressively smaller and increase the degree of compression. A second spool, comprising another seven or so rings of blades, may be placed behind the first one to give compression ratios of 25:1 or more. The compressed air passes into the combustion chambers through a diffuser, which ensures that the air is at the optimum pressure and velocity for the best fuel combustion. Combustion chambers may be of the individual *can* type, the *annular* type (basically a single chamber that completely surrounds the engine), or the *cannular* (or turbo-annular) variant, which is a series of small flame tubes inside a common air casing that encircles the engine.

Jet engines usually have only two ignition sources, located in flame tubes on opposite sides of the power unit; at start-up the flame travels from these tubes to every other tube in turn via cross-ignition tube connections.

In the combustion chambers the compressed air mixes with fuel sprayed in from a nozzle and this mixture is ignited. Exhaust gases pass backward and drive turbines, the first (high-pressure) turbine powering the high-pressure compressor while the shaft from the second (low-pressure) turbine runs forward inside the high-pressure shaft to drive the low-pressure compressor.

Finally the exhaust passes out of the tail pipe, where an AFTERBURNER may be located. Power can also be boosted by 25% if water, or a water and alcohol mixture, is injected either into the compressor inlet or into the diffuser. Reverse thrust is achieved by fitting clamshell vanes to the exhaust nozzle. These can be used to redirect the gas flow forward, while swiveling nozzles are a feature of vertical takeoff and landing aircraft (see HARRIER; VTOL).

In bypass engines, some of the air from the first stage of an axial compressor is led down ducts that bypass the combustion chambers and discharge into the jet nozzle. This increases the mass of the air being accelerated out of the engine and hence increases the thrust. Turbofans have the first rows of the low-pressure compressor designed as fans to accelerate a cold air stream and thus provide a low-velocity jet efflux with accompanying high propulsive efficiency (in some engines this fan is at the rear and forms part of the turbine assembly).

Early development of the jet engine took place in Britain during the 1930s under the direction of Frank WHITTLE, who took out a patent on the principle in 1930. Parallel experimentation in

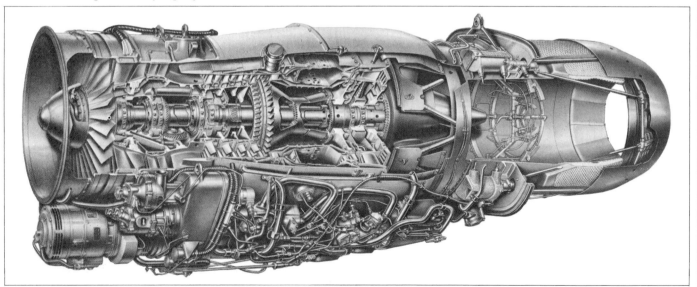

Cutaway view of a jet engine: the Rolls-Royce/Turboméca Adour two-shaft turbofan specially designed to power the Sepecat Jaguar advanced trainer/strike aircraft. The rings of compressor blades can be seen at left; the variable afterburner (far right) boosts thrust by over 40 percent

Germany was initiated by Ernst Heinkel and Hans von OHAIN in 1936, and the world's first jet-propelled flight was made by the HEINKEL He 178 in August 1939.

The first two Heinkel jet engines had been unsuccessful because a small combustion chamber size had been dictated by the lack of ram effect at the air intake. The He 178's engine, the centrifugal S2, developed just under 1,000 lb. (454 kg) thrust and overcame this difficulty by spraying vaporized fuel into its 16 combustion chambers.

The Whittle-designed W.1 engine of 1,000 lb. (454 kg) thrust powered the Gloster E.28/39 (see METEOR) research aircraft when it made Britain's first jet flight at Cranwell in May 1941, and in America the Bell P-59 AIRACOMET used two General Electric copies of the Whittle W.2B power unit when it flew in October 1942.

During the latter stages of World War II the Germans developed the 1,980 lb. (898 kg) thrust Junkers Jumo 004 axial-type engine, which powered both the Messerschmitt ME 262 fighter and the ARADO 234 bomber. British jet engine development tended initially to concentrate on centrifugals (Rolls-Royce Derwent and Nene) and the sale of 55 of these engines to the Soviet Union in 1948 had an influential effect on early Russian jet engine development. In America the 4,600 lb. (2,086 kg) thrust J-33 centrifugal that powered early military jets was followed by the axial J-47 developing 7,500 lb. (3,401 kg) thrust, and the axial has now virtually supplanted the centrifugal. Afterburning had been applied to the J-33 by 1952, and bypass engines were coming into service in 1959, with turbofans making their appearance by the early 1960s. Current research concentrates on increasing efficiency and power, and on reducing noise (by higher bypass ratios, for example). (See also PULSE JET; RAMJET.)

Jet stream

A strong, narrow current of air flowing at speeds of from 70 to over 300 mph (110 to over 480 km/h), at heights of between 30,000 and 40,000 ft. (9,000–12,000 m). These currents are generated by energy released during the massive heat transfer occurring between cold air from the poles and warm air from the tropics. They flow eastward and are usually a few miles deep, up to about 100 mi. (160 km) wide, and well over 1,000 mi. (1,600 km) in length. The strongest jet streams occur over Japan.

Discovered by World War II pilots flying over Japan and the Mediterranean, jet streams have become of major significance with the advent of airliners capable of cruising at over 30,000 ft. Aircraft can of course take advantage of a jet stream's existence to increase their ground speed, but flying in the opposite direction in a jet stream requires additional fuel, while the turbulence commonly found at the margins represents a considerable hazard.

Johnson, Amy (1903–1941)

One of the best-known long-distance women aviators of the 1930s, Amy Johnson received her pilot's license in July 1929, becoming the first woman in Britain to qualify as a licensed aircraft engineer a few months later. The following year she became world-famous when she made the first solo flight by a woman from England to Australia (May 5–24, 1930), flying the Gipsy MOTH *Jason*. On her return she was presented with a new Gipsy Moth *Jason III* by newspaper readers, and a de Havilland Puss Moth *Jason II* by the manufacturers (which she flew from London to Tokyo and back with C. G. Humphreys in 1931).

In 1932 Amy Johnson married Australian pioneer aviator James Mollison and made a record solo flight from London to Cape Town in 4 days 6 hours, flying the Puss Moth *Desert Cloud*, and returning in 7 days 7 hours. An east-west Atlantic crossing was achieved in 39 hours with her husband (from Pendine Sands, Wales, to Bridgeport, Conn.) in the twin-engined de Havilland Dragon *Seafarer* (July 22–23, 1933), and the couple competed in the 1934 England–Australia Race with the de Havilland DH88 COMET *Black Magic*. Amy Johnson later set a new out-and-home record between London and Cape Town of 7 days 22 hours in a Percival Gull Six (May 4–12, 1936), and at the beginning of World War II became a ferry pilot with the Air Transport Auxiliary. She was killed when her aircraft crashed into the Thames Estuary on January 5, 1941.

Johnson, James Edgar (1915–)

Second-ranking World War II RAF ace, with 38 victories; in 4¼ years of operational flying, his aircraft was only once damaged by an enemy fighter during combat. A prewar RAF volunteer reserve sergeant, he began active service in 1939 and by December 1940 had become a Spitfire pilot in 616 Squadron. Given command of 610 Squadron in the summer of 1942, he went on to lead a wing in 1943. In 1944 he assumed command of another wing (in the 2nd Tactical Air Force), which operated from improvised landing grounds behind the Normandy beachheads. A group captain at the end of the war, he continued his air force career, flying Sabres with the U.S. Air Force in Korea, and retired as an air vice-marshal in 1966.

Johnson, Robert S. (1920–)

Second-ranking U.S. fighter ace in Europe during World War II, Johnson enlisted in November 1941 and was commissioned in July 1942. He was assigned to the first P-47 THUNDERBOLT unit, the 56th Fighter Group, and went with it to Britain in January 1943. Flying with the 61st Fighter Squadron he destroyed his first enemy aircraft, an Fw 190, over Belgium on June 13, 1943. All Johnson's 27 victories, obtained in a year of combat while flying 91 missions, were fighters. His last successes were over north Germany on May 8, 1944. Johnson had reached the rank of major when he returned to the United States in May 1944. After the war he joined Republic Aviation.

Ju 87. See STUKA

Ju 88

Originally designed as a fast twin-engined medium bomber, the Junkers Ju 88 was one of the most versatile and successful aircraft to serve with the Luftwaffe during World War II. The prototype first flew on December 21, 1936, and initial production models (the Ju 88A, with 1,200-hp engines) began reaching combat units in 1939, taking part in the Battle of Britain the following year. Wingspan of the original Ju 88A was 62 ft. (18.90 m) and length 51 ft. (15.54 m). As a result of combat experience various design improvements were made. These included the provision of additional guns and more armor protection for the crew—the Ju 88's maximum speed of 286 mph (305 km/h) had proved insufficient to evade defending fighters.

The Ju 88 was modified for a number of different roles, and variants were produced for dive-bombing, anti-shipping strikes, strategic reconnaissance missions (Ju 88D), close support work (Ju 88P), and torpedo-bombing, as well as for night fighting (Ju 88C, Ju 88R, Ju 88G).

By mid-1943 the standard Ju 88A bomber was too slow to operate unescorted in daylight, and the Ju 88S with 1,700-hp engines was evolved. The later Ju 188 series had a redesigned

Ju 88A

nose section and could attain over 300 mph (480 km/h); a special high-altitude reconnaissance variant with a pressurized cabin (the Ju 188S-1) was capable of 429 mph (690 km/h).

About 1,000 Ju 188s were built before the Ju 388 replaced them on the production lines. These aircraft were a development of the Ju 188S and featured pressurized cabins for high-altitude flying. Both reconnaissance (Ju 388L) and bomber (Ju 388K) versions were built. The prototypes of the Ju 488 were destroyed by Allied bombing and never flew, but they had been designed for a cruising speed of 385 mph (619 km/h) at 42,600 ft. (13,000 m), giving a range of 1,280 mi. (2,060 km); the engines were 2,500-hp Jumo radials.

During the last year of the war a number of Ju 88s were used as pilotless flying bombs, carrying a Bf 109 or Fw 190 control aircraft which left the Ju 88 as it approached its target (so-called Mistel combinations).

Total production of the Ju 88 series amounted to over 15,000 aircraft.

Junkers

Hugo Junkers (1859–1935) established the company that bore his name in 1895, and during 1915 produced the world's first all-metal aircraft, the Mercedes-engined J1 single-seat monoplane —dubbed the "Tin Donkey." An iron tube framework and sheet-iron skin was also employed for the heavily armored J4 low-level patrol biplane (1917), but a corrugated Duralumin skin was chosen instead for the J9 cantilever monoplane fighter and the

J10 two-seat attack monoplane, both of 1918.

After World War I, Junkers concentrated on civil designs, beginning with the F13 four-passenger cabin monoplane with a single 185-hp BMW engine (1919), which was operated by the U.S. Post Office as well as by Junkers-Luftverkehr (which had 60) and airlines throughout Europe and South America. The W33 mail/freight plane (the first to cross the Atlantic from east to west nonstop) and W34 passenger monoplane of 1926 were direct developments. The trimotor G23 (1924) led to the G24 (1925), the first all-metal low-winged trimotor passenger airplane, of which over 50 were built for commercial use. The larger G31 airliner, accommodating 15 passengers (1926), flew with Luft Hansa, Österreichische Luftverkehrs, and two operators in New Guinea.

Only two of the enormous four-engined 34-passenger G38 "flying wings" were built (1929), one of which (D-2500) survived in Lufthansa and Luftwaffe service until destroyed by RAF bombing at Athens in 1941.

The original Ju 52 (1930) had only a single engine, and five examples were built. The rugged trimotor Ju 52/3m (1932) seated 17 passengers and became one of the most successful airliners and military transports ever produced, nearly 5,000 of them serving all over the world.

The Ju 46 (1932) was in effect a strengthened W34 originally designed as a mail-carrying seaplane for catapulting off ships. A smooth alloy skin was used on the Ju 60 high-speed single-

engined six-passenger airliner (1932), only four of which were built; it was replaced by the Ju 160 (1934) which flew with Lufthansa until World War II.

Produced after the company became state-owned in 1934, the sleek Ju 86 was intended to be both an airliner and a bomber, about 15 entering Lufthansa service on domestic routes in 1936. The bomber version served with the Luftwaffe from 1936 onward, but proved too slow for bombing operations during World War II, although high-altitude reconnaissance versions were quite widely used.

The STUKA dive-bomber and the JU 88 bomber were the company's best-known World War II aircraft, but the Ju 89 four-engined bomber prototype (1936) led to the Ju 90 Lufthansa airliners, the series being continued as the Ju 290 transport or long-range reconnaissance plane and the six-engined Ju 390 (1943).

Though intended as Ju 52 replacements, neither the Ju 252 nor the Ju 352 served in any numbers, and only two examples of the Ju 287 four-jet bomber with swept-forward wings and fixed undercarriage were built.

Junkers aircraft engines included the six-cylinder 400-hp L5 (used in the A35, W33, G23 and G24, and Heinkel 42), and a series of V-12s that included the 800-hp L88 (Ju 49), the Jumo 205 and 207 (Ju 86), the Jumo 210 (Ju 87), and the Jumo 211 (Ju 88, Heinkel 111). The Jumo 004 was an eight-stage axial turbojet producing 1,980 lb. (897 kg) thrust. It was installed in the ME 262 and ARADO Ar 234.

The Junkers factories were nearly all taken over by the Russians in 1945.

The Junkers Ju 52/3m, most successful German airliner of the 1930s and a highly valued Luftwaffe transport

Junkers Flugzeug-und-Motorenwerke near Munich was absorbed by MESSER-SCHMITT in 1965.

K

Kamikaze

Known in Japanese as Taiatari, suicide attacks were a major feature of all Japanese military air operations in the final nine months of World War II. Previously there had been instances of fighter pilots—notably in the Soviet Union in 1941–42—deliberately ramming hostile aircraft, but this was an exceptional act often undertaken either when ammunition was exhausted or in desperation in a crisis. The Japanese made it routine, and not for defense but for offense.

Kamikaze means Divine Wind, a reference to a Heaven-sent gale which scattered the ships of a Mongol invasion fleet in 1281. When Japan's situation in the Pacific war began to deteriorate there was no shortage of volunteers who considered the best way they could serve their country would be to dive their aircraft, laden with bombs, straight into an Allied warship. Such attacks were first mounted by the Imperial Navy on October 25, 1944, during the battle of Leyte Gulf. It took some time for the deliberate intent to be recognized, but in time the incessant attacks took a severe toll of Allied shipping, and in particular had a demoralizing effect on ship crews because they were very difficult to stop and had devastating

effect. Kamikaze pilots often indulged in ritual before takeoff, and wore a special scarf, but almost all attacks were made by regular squadron aircraft (though sometimes overloaded with bombs or other high-explosive charges). More than half the total number of 2,900 Kamikaze sorties were flown during the defense of Okinawa in April 1945 when about 1,900 suicide missions were dispatched (compared with some 5,000 conventional attacks).

The most common Kamikaze attacker was the A6M2 ZERO fighter, because this was available in the largest numbers. Other important types were the Aichi D3A, Nakajima B6N, Yokosuka D4Y, and a number of Army aircraft. One conventional aircraft, the

Nakajima Ki.115 of the Army, was explicitly designed and put into production as a Kamikaze vehicle. A crude fighter-like machine made by semiskilled labor from low-quality parts, it carried a bomb of up to 1,764 lb. (800 kg) recessed into the belly, and after jettisoning its steel-tube undercarriage could fly at 342 mph (573 km/h). Though 105 were built, none was used.

A more radical Kamikaze attacker was the Yokosuka MXY7 Ohka (Cherry Blossom), called Baka (Japanese for fool) by the Allies. Essentially a manned missile, this was a tiny rocket aircraft, carried aloft by a G4M (BETTY) bomber and released about 20 mi. (32 km) from the target. In the final dive the three rockets boosted its speed to 576 mph

A rare example of the Yokosuka MXY7 Ohka (Cherry Blossom), a rocket-powered Kamikaze aircraft capable of reaching 576 mph (922 km/h) after release from its mother plane

Kamov Ka-25 "Hormone" antisubmarine helicopter

(922 km/h), making it almost impossible to shoot down with contemporary weapons. The nose was occupied by a 2,645-lb. (1,200-kg) warhead.

Though over 740 were built, only a few were released near their targets, making only a modest contribution to the 300-odd Allied vessels hit by Kamikaze attacks. Ships of all kinds up to fleet carriers were sunk, but British carriers had armored decks, and were much less severely damaged.

Kamov, Nikolai Ilich (1902–1973)

Soviet AUTOGIRO and helicopter designer. Kamov first became associated with rotary-winged aircraft in 1929, when he helped design the Kaskr-I and Kaskr-II autogiros. From 1931 to 1940 he was in charge of an autogiro design team and was responsible for the A-7, which was produced in small numbers.

The design team was disbanded in 1943, and shortly after World War II Kamov began to develop a series of helicopters using a twin contra-rotating coaxial rotor layout. The Ka-15 was a two-seat cabin helicopter developed in 1952 from earlier single-seat open-frame models; it was used in a variety of roles by AEROFLOT and by the Soviet Navy as a spotter aircraft.

The Ka-26, which first appeared in 1964, is a twin-engined coaxial helicopter employed mainly for agricultural work. The Ka-25 is a larger naval helicopter of 15,562 lb. (7,100 kg) gross weight with sonar equipment and missile armament.

The experimental Ka-22 heavy-lift helicopter (1961) had two horizontally-opposed rotors; it was the only Kamov helicopter not to have a coaxial layout. Kamov also designed several successful propeller-driven sleighs.

Kartveli, Alexander (1896–1975)

A Russian-born American aircraft designer, Kartveli had served as an officer in the Russian Imperial Army during World War I, and in 1918 was sent to Paris to study military tactics. He never returned to Russia, and instead studied aeronautical engineering, finding employment with French companies before moving to the United States in 1927. Kartveli worked for Columbia and Fokker, but in 1931 a fellow Russian emigré, A. P. Seversky, invited him to become assistant chief engineer with the newly founded Seversky Aircraft Corporation. The Seversky designs, which concentrated on all-metal low-wing monoplanes with radial engines, advanced through a number of amphibian and sporting designs to a single-seat fighter, eventually produced as the Seversky P-35, which was accepted by the Army Air Corps in 1937. In 1939 Seversky himself left the company, which was renamed Republic Aviation (see FAIRCHILD-REPUBLIC). Kartveli became Republic's chief designer, and produced the P-43 Lancer, followed in 1941 by the outstanding P-47 THUNDERBOLT. Kartveli was also responsible for the F-84 THUNDERJET and F-105 Thunderchief, aircraft that proved valuable tactical fighters in Korea and Vietnam, respectively.

Kawanishi

Since its foundation in 1928 Kawanishi Kokuki KK (Kawanishi Aircraft Co.) has been the principal Japanese producer of flying boats and seaplanes. It took over the assets of Kawanishi Engineering at Kobe, which since 1921 had manufactured a number of seaplanes and land biplanes of foreign design. Although it continued to produce several foreign types, mainly of SHORT BROTHERS origin, Kawanishi's most important work began in 1933 with the design of what became the Experimental 9-Shi Large Flying Boat, first flown on July 14, 1936. Bearing a strong resemblance to the Sikorsky S-42, this was developed into the H6K series of long-range reconnaissance and transport flying boats (code-named "Mavis" by the Allies), of which nearly 220 were delivered in five main variants. They were outstanding aircraft, but were eclipsed by the H8K, first flown in January 1941. At first thought to be merely a copy of the British Sunderland, the EMILY proved to be probably the best flying boat of World War II.

Only a few E15K Shiun high-speed seaplanes were built, despite their extremely advanced design and outstanding performance with a contra-rotating propeller powered by a 1,500-hp engine, but the Kawanishi Kyofu ("Mighty Wind") seaplane unexpectedly led to the successful SHIDEN navy fighter. The Kyofu was a single-seat fighter floatplane intended to gain air supremacy until amphibious landing forces could capture an airfield. First flown on May 6, 1942, the Kyofu was a fighter with a single central float, auxiliary wingtip floats, and armament of two cannon and two machine guns. It was a formidable aircraft, but Japanese forces were soon no longer on the offensive, and the role for which it was designed had vanished. The Shiden, on the other hand, proved to be one of the finest fighters of the war, and over 1,400 were delivered.

Kawanishi has been reorganized as SHIN MEIWA, whose main products are the PS-1 (SS-2) and PS-1 Mod (SS-2A) ocean-patrol and antisubmarine flying boats and amphibians.

Kawasaki

Kawasaki Kokuki Kogyo KK was formed in 1918 as the aircraft subsidiary of Kawasaki Heavy Industries of Japan. From 1930 it built all its aircraft (but not its engines) exclusively for the Imperial Army, though in its early years it had also manufactured marine aircraft. Among these were versions of the German Dornier WAL flying boat, and several machines designed by Dr. Richard Vogt, later technical director of BLOHM UND VOSS. Under Vogt's direction Kawasaki produced the Type 88 two-seat reconnaissance biplane in 1927, the Type 92 single-seat fighter in 1930, and Ki.3 biplane bomber in 1933 (some of the 243 built were delivered by Tachikawa Arsenal), and no fewer than 644 of the excellent Ki.10 fighter in 1935–38 (75 were rebuilt as dual trainers). The Type 98 (Ki.32, codenamed "Mary" by the Allies) was an outstanding light bomber, with spatted landing gear and in-line engine, and 854 were delivered by 1940. The Ki.45 Toryu or Dragon-slayer ("Nick") was a twin-engined fighter; 1,698 were produced between 1942 and 1944 (some equipped with radar as night fighters). Nearly 2,000 Ki.48 ("Lily") twin-engined bombers were built, and in 1942 this machine was the Army's most important light bomber serving outside China. The Ki.56 was a transport based on the American Lockheed Model 14, 121 being delivered.

In 1941 Kawasaki built three prototypes of the Ki.60 fighter, almost unique among modern Japanese aircraft in having a liquid-cooled engine (the 1,100-hp Daimler-Benz DB 601). Although armed with two of the very deadly new Mauser MG 151 cannon, of which 300 had been imported by submarine from Germany, the Ki.60 soon gave way to the Ki.61 HIEN ("Swallow"), with lower drag, a larger wing area and better maneuverability. The Ki.61, codenamed "Tony" by the Allies, was developed with a new and more powerful radial engine as the Ki.100, one of the few fighters capable of intercepting the B-29 Superfortress. Kawasaki's other wartime products included the Ki.102 ("Randy") twin-engined heavy fighter, built in special night or high-altitude versions.

Kawasaki Aircraft was resurrected in the 1950s and is now one of the largest Japanese aerospace companies. Its products include the P-2J, a stretched version of the Lockheed Neptune maritime patrol aircraft with turboprop engines, the company's own C-1 twin-jet military freighter, a range of license-built Bell, Boeing-Vertol, and Hughes helicopters, and components for the Lockheed TriStar, YS-11 airliner, and Japanese-built F-4EJ Phantom.

Kazakov, Alexander Alexandrovich (1889–1919)

Russia's top-scoring air ace of World War I, Kazakov achieved 17 confirmed victories, and was unofficially credited with 32. At the time of his death on

Twin-engined Kawasaki Ki.45 Toryu (Dragon-slayer)

September 1, 1919, he held 16 decorations, including the highest Russian decoration, the Gold Sword of the Order of St. George. Kazakov graduated as a pilot in September 1914, and achieved his first victory using a grapnel and then ramming a German Albatros two-seater. In September 1915 Kazakov became commander of the 19th Corps Aviation Otryad, then equipped with Morane-Saulnier scouts, and in February 1917 was appointed commander of the 1st Military Aviation Group, flying Nieuport 17s and Spad VIIs.

After the Bolshevik Revolution, Kazakov joined a British force in North Russia, and as a major in the RAF he commanded the Slavo-British No. 1 Squadron at Bereznik, winning the DSO in October 1918. Kazakov died in a mysterious flying accident soon after the British announced their withdrawal from North Russia, when his Sopwith Snipe dived to the ground under full power from 300 ft. (480 m).

Kfir

A development of the French-designed delta-wing MIRAGE 5J, the Kfir is a jet fighter produced in Israel and powered by a General Electric J79 engine. Design and construction of the Kfir were the result of an increasingly severe French arms embargo against Israel, first imposed in 1967. It soon became impossible for the Heyl Ha'Avir (Israeli air force) to maintain the 72 Mirage IIICJ air-superiority fighters it already possessed, and delivery of the more advanced Mirage 5J was canceled,

although Israeli engineers had co-operated with the French designers to evolve a machine specially tailored to Heyl Ha'Avir requirements. The Israelis decided therefore to build the new Mirage on their own, with the aid of drawings obtained clandestinely from France. SNECMA Atar 9C turbojets were at first installed in both the old Mirage IIICJ fighter and the new machine, which was given the name Nesher ("Eagle"). When J79 power plants were obtained from the United States, the new name Kfir ("young lion") was adopted.

The Nesher prototype flew in September 1969, and about 40 aircraft were available during the Yom Kippur War (1973), operating as air-superiority fighters armed with twin 30-mm cannon and two Rafael Shafrir ("Dragonfly") air-to-air missiles. The Kfir officially joined the Heyl Ha'Avir in the spring of 1975 in two versions with different radar fits: one for the aircraft's air-superiority role, the other primarily for ground-attack use. The following year the Kfir-C2 appeared; it possessed aerodynamic refinements that included canard surfaces projecting from the engine air-intake cowlings, dog-tooth extensions to the outboard wing leading edges, and small horizontal strakes immediately behind the nose radome. Earlier Kfir fighters are to be modified to C2 standard. Israeli plans to sell an export version of the Kfir to Ecuador were vetoed by the United States in 1977.

With twin 30-mm cannon and provision for up to four Rafael Shafrir missiles, the Kfir-C2 is capable of Mach 2.3 above 36,000 ft. (11,000 m) and has a service ceiling of 52,500 ft. (16,000 m). Its wingspan is 27 ft. (8.23 m) and length 51 ft. (15.54 m).

Kingsford Smith, Charles Edward
(1897–1935)

After serving in the Royal Flying Corps and RAF during World War I, the Australian aviator Charles Kingsford Smith barnstormed in England, flying DH 6s (1919), and in California, piloting Avro 504Ks (1920), and eventually became chief pilot for West Australian Airways in 1924. The first of his record-breaking long-distance flights was a 7,500-mi. (12,065-km) round-Australia trip in a Bristol Tourer, accompanied by Charles T. P. Ulm. Following this achievement, and now flying the Fokker F.VIIb-3m *Southern Cross*, he made the first transpacific flight (Oakland–Hawaii–Fiji–Brisbane), with Ulm (copilot), Harry Lyon (navigator), and James Warner (radio operator), May 31–June 9, 1928, and went on to pilot *Southern Cross* on numerous record flights. These included 2,090 mi. (3,363 km) nonstop from Melbourne to Perth, August 8–9, 1928; the first of three 1,624-mi. (2,613-km) crossings of the Tasman Sea, from Sydney to Christchurch, New Zealand, September 10–11, 1928; a new Sydney–London record of 12 days 18 hours (June 1929); and a 1,900-mi. (3,060-km) crossing of the Atlantic from Portmarnock Strand, Ireland, to Harbour Grace, Newfoundland, June 24–25, 1930.

Kingsford Smith later lowered the England–Australia solo flight time, first to 9 days 21 hours in the Avro Avian *Southern Cross Junior* (October 1930), and then to 7 days 4 hours in the Percival Gull Four *Miss Southern Cross* (December 1933). He had already founded Australian National Airways in 1930 and in 1934 went on to make the first eastbound transpacific flight (Brisbane–San Francisco), in the Lockheed Altair *Lady Southern Cross* (October 22–November 4). He left London bound for Sydney on November 6, 1935, and disappeared over the Bay of Bengal on November 8.

KLM

KLM, Royal Dutch Airlines (Koninklijke Luchtvaart Maatschappij NV), was formed in 1919 and inaugurated an Amsterdam–London service on May 17, 1920, with a chartered DH 16. A fleet of two DH 9s and two Fokker F.IIIs was then purchased, and European routes were established with Fokker monoplanes during the 1920s. The first Amsterdam–Djakarta service, in a Fokker F.VIIb-3m flown by Capt. I. W. Smirnoff, departed on September 12, 1929. KLM was the first European airline to purchase DC-2s and DC-3s; the KLM DC-2 *Uiver* came first in the handicap section of the England–Australia Air Race in 1934. The first KLM transatlantic flight was flown to Curaçao in December 1934.

German bombing destroyed 18 KLM aircraft at Amsterdam on May 10, 1940, and 11 more were seized by the Nazis. DC-3s that avoided capture or destruction were leased to BOAC for its wartime Bristol–Lisbon route. Full services were restored in 1945 with DC-4s, which were progressively replaced by DC-6, DC-7, and Electra aircraft. In 1977 scheduled passenger and cargo services served Europe, the Western Hemisphere, Africa, the Middle and Far East, and Australia. Subsidiaries include KLM Noordzee Helicopters and KLM Air Charter. The present fleet consists of 9 Boeing 747s, 7 DC-10s, 19 DC-8s, and 19 DC-9s. Some 3,500,000 passengers are carried annually.

Komet

The only operational World War II fighter that relied exclusively on rocket propulsion was the Messerschmitt Me 163 Komet, powered by an HWK 509 liquid-fuel motor of 3,748 lb. (1,700 kg) thrust. Developed from the DFS 194 rocket-powered glider of 1940, designed by Alexander LIPPISCH, the tailless Me 163 was intended as a high-speed research aircraft, but after Heini Dittmar attained 623.85 mph (1,003.77 km/h) in a test on October 2, 1941, it was developed as an interceptor, armed with

Me 163B Komet

two 30-mm cannon and rocket projectiles. The Me 163A was only a training glider, and the first operational aircraft was the Me 163B, with a maximum speed of 597 mph (960 km/h). Wingspan was 30 ft. 7 in. (9.32 m) and length 18 ft. 8 in. (5.69 m). The first Komets were sent to units of *Jagdgeschwader* 400 in mid-1944, and they first saw action intercepting a B-17 raid on August 16, 1944, from a base at Brandis, near Leipzig.

Some success was achieved by the Komet (about a dozen Allied aircraft were shot down), but it was dangerous to fly and accidents were frequent. More aircraft, in fact, exploded than were brought down by opponents. The Me 163C had a powered-flight endurance of 12 minutes (4 minutes longer than that of the Me 163B) and a pressurized cabin, but it did not see operational use, and the larger, faster (620 mph, 997 km/h) Me 263 never entered production.

Kondor

Designed as a 26-seat airliner, the four-engined Focke-Wulf 200 Kondor was developed as a maritime reconnaissance bomber during World War II, and despite structural weaknesses (leading to failure of the rear fuselage and the rear wing-spar) proved relatively successful.

First flown in July 1937, the early Kondors made a number of long-distance publicity flights, notably Berlin–New York in 24 hours 55 minutes, returning in 19 hours 47 minutes (August 1938). The D-2600 (*Immelmann III*) became Hitler's personal aircraft.

When war broke out in 1939 the Kondor was the only long-range German aircraft that could readily be adapted for patrolling the Atlantic. Defensive armament was installed, together with radar equipment, and bombs were carried both in a ventral gondola and externally under the wings. Maximum speed with 1,200-hp Fafnir engines was 224 mph (360 km/h), and the service ceiling 19,000 ft. (5,800 m). Wingspan was 107 ft. 9½ in. (32.85 m) and length 76 ft. 11½ in. (23.45 m).

Once Allied shipborne fighters and long-range patrol planes became available, the Kondor proved highly vulnerable, and by 1943 it was being replaced by the He 177 and the Ju 290. In the closing stages of the war it was used principally as a transport.

Korean War

The Korean War was the first in which jet warplanes played a major role.

Tactically, however, their operations differed little from those of World War II, except that in the context of the peculiar ground rules for "limited warfare" bombing missions were not allowed against enemy supply lines or airfields beyond the Yalu River inside China. Thus the air war was not a true test of strategic capability. Another significant innovation of the Korean War was the deployment of large numbers of helicopters. These immediately proved their value as troop transports (moving infantry battalions from one section of the front to another, for example, or evacuating the wounded); as aerial taxis in which brigade commanders could visit and control combat units separated by many miles of difficult terrain; and as transports for supplies and ammunition. U.S. air superiority was achieved largely by the F-86 SABRE and is demonstrated by the fact that toward the end of the conflict 13 MiG-15s were being shot down for every F-86. Around 2,300 Communist aircraft were destroyed or damaged in the air against a UN Forces (overwhelmingly U.S.) loss of 114.

The North Korean invasion of South Korea on June 25, 1950, was supported by some 120 piston-engined combat aircraft of Russian manufacture. A United Nations armed force was established to aid South Korea and air superiority was quickly established by the U.S. Air Force. The 5th Air Force, based in Japan, had three groups equipped with F-80 SHOOTING STAR jet fighters and four squadrons with long range F-82 Twin-Mustangs (one of these scored the first U.S. air kill by shooting down a Yak-7 near Seoul on June 27). The North Korean advance was only checked when the UN forces had been driven into a small area around Pusan in the southeast corner of the peninsula. Air Force tactical fighters used forward airfields within this perimeter for daily ground-support operations, and their constant rocket, napalm, and strafing attacks on enemy positions played a major part in relieving the situation. Owing to range limitations of the F-80 jets, a large number of F-51 MUSTANGS were employed for ground attack, assisted by Marine Corps F4U CORSAIRS.

By October 1 a UN offensive had driven the North Korean forces back across the 38th parallel, following strategic bombing of depots and communications in North Korea by a small force of B-29 SUPERFORTRESSES operating from Japan. At the end of October 1950 Chinese forces began to make an appearance, and the following month Allied aircraft began to meet MiG-15 jet fighters flown, initially, by Russian

pilots. To counter these new aircraft, a group of F-86 Sabres was hurried from the United States. Although the MiG-15 had a superior performance on many counts, the Sabre units were able to defeat their adversaries on most occasions by better tactics and airmanship. The appearance of the MiGs eventually forced the Superfortresses and INVADER medium bombers to turn to night operations.

Chinese intervention pushed the front back toward the 38th parallel, and this was followed by a UN counteroffensive in the spring of 1951. For the rest of the war the ground situation was basically a stalemate, with air operations playing a major part in containing the Communist forces.

F-84 THUNDERJETS became the principal ground-attack aircraft, gradually replacing the F-80s and F-51s. Further F-86 units were established to counter increasing numbers of MiG-15s during enemy attempts to regain air superiority during 1952 and 1953. Sabre pilots had claimed more than 800 of the Russian jets by the date of the truce (July 27, 1953), but about 90 Sabres had been lost in air combat—most of them before the arrival of the F-86F which performed much better than earlier models.

Air units from other nations with the UN included an Australian squadron that flew Mustangs and (later) METEORS, and a South African fighter squadron flying Mustangs and Sabres. The RAF had Sunderland flying boats in the war zone, but the principal British air contribution was Royal Navy carrier squadrons operating Sea FURY, Firefly, and Seafire types. U.S. Navy carriers gave extensive support using Banshee, Panther, Corsair, and SKYRAIDER aircraft in strikes against targets in North Korea.

Kozhedub, Ivan Nikitovich (1920–   )

Top-scoring Soviet air ace of World War II, Kozhedub was credited with 62 victories in a total of 520 sorties, which places him as the highest-scoring Allied pilot of the war. He was also the only Soviet airman to shoot down a German jet aircraft.

Kozhedub scored his first victory on March 26, 1943, flying an La-5 in the Kharkov area. He received his first Hero of the Soviet Union gold star on February 4, 1944, as a senior lieutenant commanding a squadron in the 240th Fighter Aviation Regiment. On May 2, 1944, he was given an La-5FN paid for by public subscription, with which he shot down eight enemy aircraft in seven days, including five FOCKE-WULF 190s. He was promoted to captain and received an LA-7 in July 1944 on transfer to the 176th Guards Fighter Regiment as deputy commander. Kozhedub was awarded his second gold star on August 19, 1944, and received the third at the end of the war. The Soviet ace's jet victim was an ME 262, shot down while he was flying his La-7 over Berlin on February 24, 1945. His last two victories, both Fw 190s, were scored on April 19, 1945.

Kozhedub ended the war as a major and eventually became a colonel general. His La-7 is preserved in the museum at the Gagarin Air Force Academy.

L

La-7

The LAVOCHKIN series of single-seat fighters made up, together with the YAKOVLEV series, almost all of Russia's fighter strength in World War II. Nearly 23,000 Lavochkin fighters were built; 5,733 of them were La-7s. This aircraft was developed during the winter of 1943/4 and had a single 1,850-hp radial engine that gave a maximum speed of 404 mph (648 km/h). Wingspan was 31 ft. 9 in. (9.68 m) and length 28 ft. 7 in. (8.71 m). Construction was of wood with a bakelite ply covering, and armament was generally two or three 20-mm cannon. Apart from certain aerodynamic refinements, the main difference between the La-7 and the earlier La-5 was the replacement of the wooden box mainspar by a metal I-section spar, which saved both weight and space, and provided greater fuel capacity. An La-7

was used to test the laminar-flow wing for the all-metal La-9. It was also used in experiments with auxiliary rocket engines.

Lafayette Escadrille

An all-American volunteer squadron that fought in support of the French before formal American entry into World War I. The *Escadrille Americaine* came into official existence on April 18, 1916. Equipped with SPADS and NIEUPORT Scouts, the "American Squadron" flew its first patrol on May 13, 1916. Five days later Kiffin Rockwell scored the unit's first aerial victory, and for the rest of the year its pilots steadily increased their tally. More Americans joined the unit, the most outstanding of these, Raoul LUFBERY, arriving on May 24. The squadron's first fatal casualty was Victor Chapman, who was killed in combat on June 23, 1916. German diplomatic protests over the squadron's title, which blatantly advertised the fact that pilots from a neutral nation were fighting for France, were finally heeded, and the unit name was changed to *Escadrille Lafayette* after December 6, 1916 (it appeared in French official records as *Escadrille Spa 124*). About this time the unit adopted its own insignia—an Indian warrior chief's head painted on both sides of their aircraft. After April 1917, when the United States entered the war against Germany, a number of American pilots were formed into the Lafayette Flying Corps, though this did not include the original Lafayette unit. Gradual pressure was brought to have the Lafayette pilots transferred to

American service, and finally, on February 18, 1918, the Lafayette Escadrille was disbanded, and most of its pilots became the nucleus of the 103rd Aero Squadron of the U.S. Air Service. In its career, the Lafayette unit had scored 57 confirmed victories and lost nine pilots killed.

Laminar flow

Streamline or viscous flow, in which the particles are in continuous, steady motion in a regular path. A state of laminar flow will exist in the airstream around a wing up to a certain velocity, beyond which the flow pattern becomes random and haphazard (turbulent). (For laminar-flow airfoils see SUPERSONIC FLIGHT.)

Lancaster

The largest and most successful RAF heavy bomber of World War II, the AVRO 683 Lancaster was derived from the 1939 Avro Manchester Mark III project—a four-engined variant of the twin-engined Manchester bomber that was under consideration at least a year before the ill-fated Manchester saw RAF service. (Due to the unreliability of its Vulture engines the Manchester served only briefly during 1941–42.)

The Lancaster I (1,280–1,640 hp Merlins) flew its first operational missions in March 1942; Mark II aircraft (only 300 built) had Hercules engines. The Mark III, which was produced in quantity, was powered by Packard-Merlin engines, giving a maximum speed of 270 mph (432 km/h) and a ceiling of 21,500 ft. (6,550 m). Arma-

Lavochkin La-7

Lancaster heavy bomber of World War II

ment consisted of .303 machine guns in nose, tail, and dorsal turrets. Wingspan was 102 ft. (31.09 m) and length 68 ft. 11 in. (21.00 m).

By May 1945 a total of 61 RAF squadrons were equipped with Lancasters. Normally capable of lifting an 8,000-lb. bomb load, special versions equipping 617 Squadron carried a single 22,000-lb. (10,000-kg) "Grand Slam" bomb for specific targets. Total production of Lancasters amounted to 7,366, and during World War II they flew a total of 156,000 individual sorties, dropping 608,612 tons of bombs. After the war a number of Lancasters were converted for use as civil airliners (Lancastrians) and saw wide service.

Landing

Landing is generally considered the most demanding aspect of basic flying. To land a conventional fixed-wing airplane it is necessary gradually to destroy the lift provided by the wings. The speed is therefore reduced until a controlled rate of descent at the chosen air speed

can be maintained on the approach by using the throttle(s) and adjusting the attitude of the airplane (a glide approach is more difficult since no throttle control is possible). As the machine's wheels near the runway surface, the pilot reduces power and rounds out so that the plane is flying just above the runway. He then holds off while the speed decays and the aircraft settles to the ground. Touchdown is initially achieved on only the main wheels of a tricycle undercarriage; with a tailwheel undercarriage all the wheels can be set down together in a three-point landing. Landings are normally made into wind to reduce the ground speed, but a cross-wind approach is sometimes necessary. The normal procedure in a cross wind is to hold the aircraft's nose into the wind with the rudder on the approach, only straightening up just before the wheels touch. Alternatively a pilot can SIDE-SLIP into the wind as he comes down. In a strong wind or with large, heavy craft, substantial power may need to be maintained up until touchdown (propeller slipstream will maintain a strong

airflow over control surfaces and improve control).

Langley, Samuel Pierpont (1834–1906)

The first American to make a serious attempt at sustained, powered flight, Langley was an eminent astronomer and secretary of the Smithsonian Institution. He began to study the problem of powered flight as early as 1886, and soon decided the best answer was the tandem-wing layout as flown in model form by D. S. Brown in England in the 1870s. Eventually, in 1896, he managed to fly a successful steam-driven model. He called his flying machines *Aerodromes*, and built further models which made flights of up to 4,200 ft. (1,280 m). Langley intended to leave further development to others, but in 1898 the federal government asked him to explore the possibility of building a man-carrying aircraft, which might be of use in war. By 1901 he had successfully flown a model powered by a gasoline engine, and his assistant, Charles Manly, was developing a full-sized air-cooled 52-hp

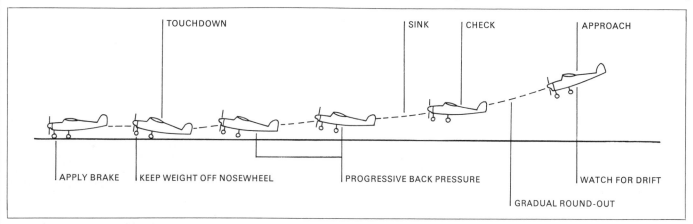

Procedure in a typical landing sequence for an aircraft with tricycle undercarriage

engine considerably ahead of its time in its light weight and efficiency. Unfortunately, Langley chose to launch his *Aerodrome* by catapult from a houseboat, and on both flight attempts, on October 7 and December 8, 1903, the aircraft fouled the launching mechanism and fell into the Potomac River. Langley became a national laughingstock. The second failure occurred nine days before the epoch-making achievement of the Wright Brothers at Kitty Hawk.

In 1914 Glenn CURTISS flew the *Aerodrome* at Lake Keuka, N.Y., but only after surreptitious alterations and installation of a more powerful engine. Whether it could ever have flown in its unaltered state remains an open question to this day.

Lavochkin, Semen Alekseevich
(1900–1960)

S. A. Lavochkin was a Soviet aircraft designer best known for a series of fighters produced in large numbers during World War II. Between 1931 and 1938 Lavochkin worked in various design teams on a number of fighter projects, and in 1936 he received a position in the chief administration of the aviation industry. In 1938 he joined V. P. Gorbunov and M. I. Gudkov in creating a new design bureau to develop the wood-and-canvas LaGG-1 single-seat fighter. The production version, the LaGG-3, was the first in a series of Lavochkin piston-engined fighters of which a total of some 23,000 were eventually built. The La-5 was essentially a LaGG-3 with a 1,330-hp M-82 14-cylinder air-cooled radial engine—in 1942 the most powerful production engine available in the Soviet Union—which gave the prototype La-5 an increase of 25 mph (40 km/h) in speed over the in-line engined LaGG-3; altogether 10,000 La-5s were produced. Next in the series was the La-7, the last to see combat in the war, which was followed in 1946 by the all-metal La-9 and the La-11 escort fighter. The La-11 was the last piston-engined fighter in service with the Soviet air force; it had a maximum speed of 429 mph (690 km/h) at 20,340 ft. (6,200 m).

Between 1947 and Lavochkin's death in 1960 the design team produced over a dozen prototypes of jet-engined fighters, but only one was manufactured in quantity. This was the La-15 (1948), which was powered by an RD-500 Soviet copy of the Rolls-Royce Derwent and had a maximum speed of 637 mph (1,026 km/h) at 9,840 ft. (3,000 m).

In 1956 Lavochkin's design team produced the delta-wing La-250 long-range all-weather fighter, but problems were encountered during the test program and the bureau was closed on Lavochkin's death.

Liberator

The Consolidated B-24 Liberator heavy bomber was built in greater numbers—19,256—than any other U.S. military aircraft of World War II. First flown in 1939, the twin-finned B-24 was roughly similar in size and equipment to the B-17 FLYING FORTRESS, but had greater speed, range, and payload. Powered by four Pratt & Whitney R-1830 radial engines, early models could attain speeds approaching 300 mph (480 km/h). The first major model to see combat was the B-24D, which equipped heavy-bomber units in all theaters of the war. The B-24H and B-24J introduced models during 1943–44, when five factories—including the giant Ford Willow Run plant—were all producing Liberators. Wingspan of the B-24J was 110 ft. (33.53 m) and length 67 ft. 2 in. (20.47 m). Bomb load was 8,800 lb. (4,000 kg) and armament consisted of ten .50 machine guns.

The Liberator was credited with dropping some 635,000 tons of explosives on its targets, and its gunners shot down over 4,000 enemy aircraft. "Civilianized" Liberators operated an important transatlantic passenger service during the war, and were also employed on maritime reconnaissance and antisubmarine duties by the U.S. Navy (PB4Y) and by the RAF. A model with a single fin, the PB4Y-2 Privateer, had begun to reach American naval units by the end of the war.

Liberty engine

The 1,650-cu. in. (27-liter) V-12 Liberty engine, which had an eventual output of 420 hp, was the work of a commission led by engineers from the Packard and Hall-Scott automobile companies. It was reputedly designed and built in only three months after the United States entered World War I in 1917.

The Liberty engine powered a large number of aircraft, including the Martin MB-1 and 2, license-built DH4, Douglas DT-1, and the Curtiss flying boat series. In 1923 a 400-hp inverted Liberty engine was developed by Allison and installed in the Loening single-float amphibians.

Lift

The principal function of an airplane's WINGS (or helicopter's rotors) is to provide lift. This is achieved by accelerating the airflow around an airfoil of cambered section. As the leading edge of a moving wing separates the air, the flow must speed up to pass around the curve of the camber. The curvature is normally greater on the upper surface than the lower, so that the molecules of air travel much faster around the upper surface relative to the molecules passing below the wing.

Increasing the speed of any flowing gas (or liquid) results in a pressure reduction (Bernoulli's Principle). Accelerating the airflow around the cambered upper surface therefore reduces the pressure above the wing and consequently creates lift by "negative air pressure," or suction. Depending on the wing's angle of attack, this suction

U.S. Navy Liberator long-range search bomber (PB4Y-1) of World War II, a close counterpart of the B-24 bomber

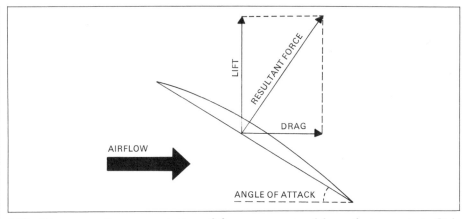

Lift: Diagrammatic representation of the components of force that operate in flight

above the wing may be supplemented by increased pressure from air striking the lower surface. During a landing approach the angle of attack is increased to augment lift. This is necessary to compensate for loss of suction lift caused by the lower airspeed. However, beyond an angle of attack of about 15–20° the airflow over the top of the wing breaks down, causing STALLING. (See also AERODYNAMICS.)

Light aircraft

The term "light aircraft" generally includes all planes with an overall weight of less than 6,000 lb. (2,722 kg).

The earliest example of a machine designed for cheapness and simplicity of operation by a private pilot was probably SANTOS-DUMONT's Demoiselle of 1909. With a bamboo and piano-wire airframe partly covered by fabric, and a 30-hp engine, it could carry only pilots weighing less than about 120 lb.

(54 kg).

After World War I, huge military surplus stocks coming onto the market set the lightplane industry back. But in France the Farman David biplane, with a 50-hp Gnôme engine, entered production with some success, to be followed by the Farman Sport of 1923. Under the stimulus of competitions, the German aircraft industry developed a number of novel designs during the mid-1920s, including the first all-metal lightplane to be series-built, the Dornier Libelle flying boat. In 1925 the British de Havilland MOTH made its appearance. Over 1,900 examples of the Moth were constructed; production of the two-seat Avro Avian of 1926 ran to about 740, and over 1,000 Klemm Kl 25s were built in the late 1920s.

Light racing planes were numerous in the early 1930s, particularly in America. The first of over 8,000 de Havilland Tiger Moth biplanes appeared in 1931, the year that also saw

the introduction of the highly successful 35-hp Taylor J-2 CUB and the AERONCA C-3 in the United States. Experiments in Europe with ultra-light aircraft such as the Flying Flea were largely failures. New and more powerful versions of the ubiquitous Cub were marketed in America during the late 1930s, and several outstanding light aircraft, such as the Fieseler Storch in Germany, and the Taylorcraft Auster, were developed during World War II.

The CESSNA 120/140 series appeared in 1948, and the later 150 became one of the most widely used two-seat lightplanes ever built, with some 25,000 coming off the production lines by the mid-1970s.

American aircraft now predominate in the lightplane market. Cessnas (four and six-seaters) and Pipers (Comanches, Cherokees, and Super Cubs) have been particularly successful and over 10,000 BEECHCRAFT Bonanzas have been built. The Pitts Special biplane and the ZLIN have been outstanding aerobatic aircraft.

In some lightplanes (for example the French Rallye), a system of slats makes it almost impossible to spin, and difficult to stall. Because of the efficiency of propellers at low airspeeds, jets have not replaced air-cooled piston engines in light aircraft.

Lightning (BAC)

For 15 years the RAF's principal interceptor, and its first Mach-2 combat aircraft, the twin-engined BAC (originally English Electric) Lightning first flew (as the P.1B) in 1957.

Mark 1 aircraft with Avon engines of

Piper PA-18 Super Cub, postwar version of a classic light airplane

117

BAC Lightning Mark 2A

14,430 lb. (6,500 kg) began to reach service squadrons in 1960, followed by Mark 1A machines with an in-flight refueling probe. The Mark 2 (1961) was an interim version later modified (as the 2A) to take a Mark 6 ventral fuel tank or combined fuel tank/twin-Aden gun pack.

The Mark 3 (1962), with afterburning Avon 301s (12,690 lb., 5,650 kg, thrust) relied entirely on a missile armament. It was followed by the Mark 6 (1964), which had a greater range (800 mi., 1,287 km, with ventral tank), and an optional twin-Aden gun pack below the fuselage. With a length, including probe, of 53 ft. 3 in. (16.23 m) and a wingspan of 34 ft. 10 in. (10.61 m), the Lightning has a top speed of about 1,450 mph (2,335 km/h).

The two-seat trainers T4 and T5 were versions of the Mark 1A and Mark 3 respectively. Export multirole versions of the Lightning were sold to Kuwait and Saudi Arabia.

Lightning (Lockheed)

One of the most famous American fighters of World War II, the twin en-gined P-38 single-seat Lightning had an unusual design, with a fuselage pod for the pilot centered in the wing between two Allison engines, nacelles extending into tail booms, and a tricycle under-carriage—the first used on a fighter. The prototype made its maiden flight in January 1939, but production de-liveries were slow and few P-38s were in service when the United States entered World War II, though more than 9,900 were eventually built. The Lightning's wingspan was 52 ft. (15.85 m) and its length 37 ft. 10 in.

(11.53 m).

Although Lightnings were among the best aircraft used in the Pacific, initially the cold wet weather of Europe seriously affected the Allison engines. The later P-38J and P-38L (414 mph, 666 km/h) incorporated modifications that largely overcame these and other troubles exhibited by earlier models. The two top-scoring American aces, Richard BONG and Thomas MCGUIRE, both flew P-38s in the Pacific, where Lightnings accounted for more Japanese aircraft than any other U.S. fighter.

Photo-reconnaissance versions of the Lightning flew under the designations F-4 and F-5.

Lilienthal, Otto (1848–1896)

German aviator, known for his pio-neering experiments with manned, con-trollable gliders, which helped pave the way for the powered airplane. While still boys he and his brother Gustav (1849–1933) tried to make themselves gliding wings, and later constructed ORNITHOPTERS, models with flapping wings. After many years of experiments he made his first successful flight in 1891, using a delicate silk-covered bat-wing glider to cover a distance of 100 ft. (30 m).

Lilienthal's machines, usually mono-planes, bore a broad resemblance to today's hang gliders. He launched him-

Lockheed P-38 Lightning

self from a hill and hung onto the glider by his shoulders and arms and steered by swinging his body and legs. He went on to make more than 2,000 flights over a period of five years, covering distances of 650–980 ft. (200–300 m). On August 9, 1896, his glider went into a sudden dive and he broke his back. He died the next day. Lilienthal's book *The Flight of Birds as a Basis for the Art of Flying* (1889) was a pioneering work in the field of aeronautics, and it was descriptions of his achievements and news of his tragic death that were the direct inspiration of the Wright Brothers.

Lindbergh, Charles Augustus
(1902–1974)

Pioneer American aviator who made the first solo crossing of the Atlantic and excited the imagination of millions about the potential of airplanes as a means of long-range transportation. After learning to fly in Lincoln, Nebraska, he bought a World War I Curtiss JENNY and became a popular stuntman, touring the South and Midwest and earning himself the nickname "Daredevil." He then attended army flying schools in Texas, and in 1926 became an airmail pilot before deciding to try for the $25,000 prize offered for the first New York–Paris nonstop flight.

He persuaded a group of St. Louis businessmen to back his venture, and made the flight on May 20–21, 1927, in a single-engined Ryan monoplane SPIRIT OF SAINT LOUIS. His flight covered a distance of 3,609 mi. (5,809 km), lasted $33\frac{1}{2}$ hours, and caused an immediate sensation. Lindbergh became a popular hero overnight.

With his wife Anne Morrow, Lindbergh made many pioneering flights in the 1930s, and he was technical adviser to Transcontinental Air Transport and Pan American Airways. After the publicity that followed the kidnapping and murder of his infant son in 1932, and the conviction and execution of Bruno Hauptmann in 1936, Lindbergh retired for some years to Europe.

There he visited Germany, and issued many warnings about the growth of German air power. However, his acceptance of a decoration from the Nazi government, and his advocacy of American neutrality during the early stages of World War II caused much resentment. But once the United States had entered the war Lindbergh devoted himself to the war effort—advising and flying missions in the Pacific for the United Aircraft Corporation.

After the war he continued as consultant to Pan Am, the Defense Department and the National Advisory Committee for Aeronautics.

Linebacker II

Code name for the strategic bombing campaign against North Vietnam in December 1972. When negotiations for a peace settlement in Vietnam broke down, the U.S. Air Force commenced bombing the heavily defended Hanoi–Haiphong area, using Strategic Air Command B-52 STRATOFORTRESSES for the first time. There were approximately 240 B-52D and G models based in Thailand and Guam, supported by KC-135 tankers for in-flight refueling. Carrying 500-lb. (227-kg) high-explosive bombs, operations began in darkness on December 18 and continued with only one break for 11 days. During the early operations the B-52s encountered intense surface-to-air missile firings with over one hundred reported on the first night. Toward the end of the campaign few SAMs were observed and there were signs that the defenses had collapsed. A total of 729 B-52 sorties were flown during the 10 nights of the campaign, and 15 bombers were lost to SAMs. Radar-directed bombing caused an estimated 85 percent destruction of assigned targets and is considered to have been the reason for the prompt return of the North Vietnamese to the negotiating table.

Lippisch, Alexander

The German aircraft designer Alexander Lippisch was one of the pioneers in the design of "tailless" aircraft (airplanes without horizontal tail surfaces), and later, while living in the United States, supervised the design of the first delta-wing jet aircraft. In 1929 Lippisch developed his first delta-wing design, Delta I, and in the early 1930s began a long association with the DFS (German Glider Institute), for which he built a series of tailless research machines. These culminated in the DFS 194 rocket-powered glider, successfully tested by Heini Dittmar in 1940, which ranks as the first liquid-fuel rocket aircraft. It led directly to the Me 163B KOMET, the world's first (and only) operational rocket-powered interceptor, used by the Luftwaffe in the final months of World War II.

In 1942 Lippisch left the Komet program and began a series of research projects that were intended to lead to a supersonic interceptor, the LP-13a. To explore low-speed handling, the Flugtechnische Fachgruppe at Darmstadt and Munich universities built the DM-1 towed glider. This was the first pure delta-wing aircraft to fly, but it did so in

American hands in 1945, and eventually found a home at the Smithsonian Institution. Work had reached the hardware stage on rocket developments of the design (DM-2, DM-3, DM-4), although the LP-13a itself would have been propelled by an integral ramjet. During these same years of the war Lippisch was also working on several projects for high-speed tailless aircraft, designs derived from the Me 163 and the unbuilt Me 265 turbojet/rocket bomber.

In 1946 Lippisch went to the United States. He played a major role in supervising the aerodynamic design of CONVAIR's first delta-winged aircraft, the Model 7002, which was originally intended to be developed into a turbojet/rocket interceptor. Eventually this project was recast as a pure research machine, the XF-92A, which first flew in February 1949, and was the first delta-wing jet aircraft. From it, again with Lippisch acting as consultant, were derived Convair's F-102 DELTA DAGGER interceptor and B-58 Hustler supersonic bomber.

Lockheed

One of the major U.S. aircraft manufacturers, Lockheed Aircraft came into prominence at the end of the 1920s with a series of fast monoplanes of exceptionally clean design. These included the VEGA, the ORION, the parasol-wing Air Express four-passenger mailplane (with a 425-hp Pratt & Whitney Wasp engine), and the two-seat Model 8 Sirius, the first of which was named *Tingmissartoq* and set a U.S. transcontinental record in 1930, flown by Charles LINDBERGH.

Lockheed established a considerable reputation with its twin-engined airliners of the 1930s, the first of which was the 10-passenger Model 10 ELECTRA (1934). The Model 12 (1936) was a somewhat smaller aircraft with an enhanced performance (maximum speed 225 mph, 360 km/h); the Model 14 Super-Electra (1937) was claimed to be the fastest airliner of its day (257 mph, 413 km/h), its deep fuselage offering greater comfort than the earlier Lockheed "twins." In July 1938 Howard HUGHES flew a special Model 14 with 1,100-hp engines in a 15,000-mi. (24,000-km) round-the-world trip from New York through Paris, Moscow, Omsk, Yakutsk, and Fairbanks in a record-breaking 3 days $19\frac{1}{4}$ hours. The Model 14 was followed by the Model 18 Lodestar (1939), which seated 17 passengers. Lodestars were operated by commercial airlines throughout the world and flew on military services during World War II.

The Hudson (1938) was developed from the Model 14, primarily as a coastal reconnaissance bomber for British use, though it also served with U.S. and Australian forces. So successful was the Hudson in RAF service that a heavier and more powerful development of the Model 18, the Ventura, was produced in 1941, but it proved vulnerable in combat as a medium bomber, and many were used instead by the U.S. Navy for maritime reconnaissance under the designation PV-1. The PV-2 Harpoon, with increased wingspan, larger fins and rudders, and better armament, served with the Navy in the Pacific.

A postwar maritime reconnaissance bomber, the P2V Neptune (1945), set a new world distance record in 1946 when Commander Thomas D. Davies' *Truculent Turtle* flew 11,250 mi. (18,000 km) from Perth, Western Australia, to Columbus, Ohio. The Neptune remained in U.S. Navy service until 1962. It also flew with the RAF (1952–1957), and was built under license in Japan. Its replacement was the four-engined P-3 Orion, a development of the ELECTRA turboprop airliner with antisubmarine radar. The CONSTELLATION, which first saw service as a military transport, went on to play a major role in postwar civil aviation. A number served with the U.S. Air Force as the C-69 (L-49) and C-121 (L-749, L-1049), some being converted for early warning operations (EC-121). The Navy also operated Constellations (R70-1, R7V-1), including WV-1 and WV-2 early warning aircraft and the WV-3 for weather reconnaissance.

The Electra turboprop airliner began operating with Eastern Air Lines and American Airlines early in 1959. Within a year it was in widespread use throughout the world, but a series of crashes resulted in an FAA restriction and ultimately required extensive modifications to rectify certain defects (notably in the wing structure). Lockheed's wide-bodied airliner was the 400-seat three-engined TriStar (1970), of which some 200 had been ordered by the mid-1970s.

Among Lockheed's fighters, the successful P-38 LIGHTNING (1939) remained in production until the end of World War II, and the SHOOTING STAR became the Air Force's first operational jet fighter, with nearly 6,000 of the T-33 two-seat trainer derivative being manufactured in the United States, together with a number of license-built Canadian (Silver Star) and Japanese examples. The F-94 Starfire all-weather fighter (1949) was basically a T-33 airframe modified to accommodate radar in the nose and an afterburner for the J33

Lockheed Constellation

engine. Lockheed produced the STAR-FIGHTER in 1954 and the U-2 high-altitude reconnaissance aircraft in 1955, at the same time developing the very successful C-130 Hercules, a four-engined turboprop military transport (1955), which entered service with the U.S. Air Force in December 1956 and eventually flew with some 30 other air forces. Civilian versions of the Hercules (L.100 series) found extensive commercial employment. The C-130E (1961), with additional fuel tankage, had a range of 2,420 mi. (3,900 km) fully loaded.

The unorthodox SR-71 strategic reconnaissance aircraft had a Mach-3 performance, unapproached by any comparable machine when it entered service in 1966. The elegant four-engined JetStar business jet and military VIP transport (1957) offered a 570-mph (915 km/h) cruising speed and sold well, over 200 going to commercial and private operators as well as 16 to the U.S. Air Force (as the C-140) and 3 to the Luftwaffe.

The S-3 Viking four-seat carrier-borne antisubmarine aircraft was developed for U.S. Navy use, with deliveries to operational units beginning in 1974.

Regarded as a replacement for the ageing C-97s, C-124s, and C-135s of the U.S. Air Force, the four-jet C-141 StarLifter (1963), had a wingspan of 160 ft. (48.7 m) and range of 4,000 mi. (6,400 km). Fourteen squadrons were equipped with this long-range strategic transport by mid-1968. The even larger C-5A GALAXY began to reach service units 18 months later.

Loran

This very accurate Long Range Aid to Navigation measures the different times taken by pulses from separate (master and slave) transmitting stations to reach the receiver on an aircraft. A video representation of the pulses is superimposed by the operator and a digital readout gives the time separation (in microseconds) between them. Special charts are then used to establish the position. Master and slave stations are usually about 300 mi. (480 km) apart, operate at frequencies of 1,700 to 2,000 kilohertz, and have a useful range of some 700 mi. (1,126 km) by day (and double this distance at night). See NAVIGATION.

Lufbery, Gervais Raoul (1885–1918)

Third-ranking U.S. World War I air ace, born of French parents. At the age of 17, Lufbery left his home in Connecticut and traveled through Europe, North Africa, and the Balkans. In 1908 he enlisted in the U.S. Army and served in Cuba, later he worked as a customs official in China, and while in India met a French stunt pilot, Marc Pourpe. Lufbery became Pourpe's mechanic, joined

Gervais Raoul Lufbery, American World War I ace

the French Foreign Legion in 1914 when the Frenchman enlisted in the French Air Service, and obtained a transfer to Pourpe's squadron (still as a mechanic).

After Pourpe's death in combat in December 1914, Lufbery learned to fly, and on October 7, 1915, was assigned to a bomber squadron. By May 1916 he had transferred to the LAFAYETTE ESCADRILLE, becoming a major in the U.S. Air Service in November 1917. He had been credited with 17 victories before he jumped to his death when his Nieuport 28 burst into flames during a dogfight on May 19, 1918.

Lufthansa

In January 1926, Deutscher Aero Lloyd AG (which incorporated Deutsche Luft-Reederei) and Junkers Luftverkehr amalgamated to form Luft Hansa. Competition between the two companies had been proving financially ruinous, and government subsidies were considerable. The new airline inherited some 100 aircraft of 21 different types, and it immediately began to rationalize its fleet. Junkers types, such as the G 23, were extensively employed, together with Rohrbach Rolands and Dornier flying boats.

On May 1, 1926, a night service from Berlin to Koenigsberg was inaugurated with G 24 aircraft, and in the same year two G 24s made an experimental flight to Peking and a Dornier Wal explored a route to West Africa. A postal service was established to Las Palmas in the Canary Islands, so that mail could then be placed on ships to cross the South Atlantic and be collected by another Dornier from Rio de Janeiro. A similar service, using an airship to fly the stage between Lake Constance and Recife, opened in 1930. The previous year, a Heinkel 12 floatplane carrying mail for America was catapulted off the liner *Bremen* 280 mi. (450 km) from New York, and Junkers 46 aircraft were subsequently launched as much as 1,250 mi. (2,000 km) out at sea.

Renamed Lufthansa in 1930, the airline stationed seaplane tenders in the Atlantic to refuel its mail-carrying flying boats, and in 1931 began operating over the Alps from Munich to Milan and on to Rome. The route network by this time extended to Oslo, Stockholm, Paris, Budapest, and London.

Ju 52/3m aircraft entered service in 1932 and quickly became the most numerous of the company's airliners (78 at the end of 1940), while single-engined He 70 and Ju 160 high-speed mailplanes came into use in 1934 on domestic routes, followed by the twin-

Junkers Ju 52/3m, mainstay of Lufthansa services in the 1930s

engined Ju 86. One of the only two massive Junkers G38s ever built (D-2500) was a familiar sight over all Europe from September 1931 onward, and in the last few years of peace the Kondor pioneered services from Berlin to New York and Tokyo (1938). Eleven fast, four-engined Ju 90s began operating in 1938 (notably between Berlin and Vienna), and Ju 52s inaugurated a short-lived route to Bangkok and Tokyo in the summer of 1939.

The airline was reestablished (as Luftag) on January 6, 1953, and on April 1, 1955, began to operate DC-3s on domestic routes, with Convair 340s flying European services. Lockheed 1049G Constellations were acquired for international operation, reaching out to Bangkok in 1959 and then initiating a first-class-only North Atlantic service with L.1649As. Viscounts were used in Europe and the Middle East, and Boeing 707 jets arrived in 1960 for services to New York and Chicago, 727s starting work on European routes in 1967. Lufthansa was the first airline to place an order for the Boeing 737 and acquired Boeing 747s in 1971, DC-10s in 1973, and the A.300 Airbus in 1976. The first 747 freighter to enter service with any airline joined Lufthansa in 1972. Lufthansa carries some 9,600,000 passengers and 290,000 tons of freight annually (including the business of its associated charter company, Condor). The fleet includes 10 DC-10s, 4 Boeing 747s, 1 747F, 18 707s, 30 727s, and 28 737s.

Luftwaffe

The German High Command established a Military Aviation Service in 1912. By 1914 it possessed 246 airplanes of a reconnaissance type. Six months after World War I had broken out, the requirement for proper warplanes resulted in C-types (armed two-seater biplanes of over 150 hp), followed by J-types (armored C-types), G-types (twin-engined bombers), N-types (night bombers), and D-types (scouts). Initial success with Fokker EINDECKER scouts mounting twin machine guns and INTERRUPTER GEAR was followed by phases of relative dominance over the western front achieved in turn by the ALBATROS D.III, Fokker TRIPLANE and Fokker D.VII. Tactics led to the use first of *Jagdstaffeln* and then of *Jagdgeschwader* ("aerial circuses") consisting of four Jastas. Bombers operated included GOTHAS, Friedrichshafens, and the massive R-PLANES, used against both tactical and strategic targets (including attacks on England). By March 1918 the German air force had approximately 1,500 planes on the western front, with smaller forces deployed against Russia and Italy and in use in the Middle East. The Treaty of Versailles (1919) prohibited the maintenance of a German military air arm and required the surrender or destruction of all existing warplanes. Even the manufacture or import of civil aircraft was forbidden for a period. Eventually 516 airplanes were given up to the Allies and much larger numbers destroyed. A small German military establishment was, however, permitted, and an aviation department contained a nucleus of ex-wartime flyers. Gliding was actively encouraged by the military authorities, but a request from Russia for assistance in forming a new Soviet air force provided the opportunity to train pilots and manufacture warplanes in the Soviet Union (which had not been a signatory to the Versailles treaty).

From April 1922, trainers and light sporting planes were allowed to be constructed in Germany itself, and amateur flying centers were opened with the aid of Reichswehr funds. Restrictions on the construction of civil aircraft were eventually lifted and LUFTHANSA was established in 1926. The new airline became an ideal training medium for future bomber crews. Embryo fighter pilots went secretly to Lipetsk, in the Soviet Union, for training; new German warplanes were also sent there for testing. In 1933 Hitler came to power. The same year the Germans left Lipetsk, Goering became Air Commissioner, and Erhard Milch was made State Secretary for Air. On October 14 Germany withdrew from the League of Nations, and in March 1935 disclosed the existence of the 20,000-strong Luftwaffe, equipped with 1,880 aircraft (including the Do 23, Ju 52, He 45 and 46, Ar 64 and 65, He 51, and Dornier Wal). By 1936 new aircraft were being tested at the Rechlin experimental base (BF 109, ZERSTÖRER, STUKA, HE 111, DORNIER, FLYING PENCIL) and in November of that year the Condor Legion was formed to fight in the Spanish Civil War. Luftwaffe strength by the summer of 1939 included 1,270 bombers, 1,850 Bf 109s, 195 Zerstörers, and 335 Stukas, with 1½ million men.

Hitler launched his invasion of Poland on September 1, 1939, and the Polish Air Force was destroyed in only four weeks. Eight months later similar Blitzkrieg tactics placed the Nazis on the Channel coast. The Luftwaffe concentrated its maximum effort on one specific task after another, systematically eliminating each obstacle in its path; it was not designed for a long campaign and had no strategic long-range bombers—Gen. Wever, the chief German protagonist of strategic bombing, had been killed in an air crash on June 3, 1936, and the Do 19 and Ju 89 bombers were abandoned.

The Luftwaffe met its first setbacks in the Battle of BRITAIN and the BLITZ and in 1941 most of its forces were transferred to the newly-opened eastern front. Some 2,770 aircraft were committed to the invasion of Russia, but by the fall the German advance was stalled. At the same time 650 Luftwaffe aircraft were required in the Mediterranean; a fighter force had to be retained in northern France (including the latest FOCKE WULF 190s), and more than 150 aircraft (mostly KONDORS and JU 88s) were assigned to a maritime strike role over the Atlantic.

In 1942 the Luftwaffe was required to provide more fighters (including night fighters) for northwest Europe in the face of increasing Anglo-American night and day raids, while the situation in Russia and the Mediterranean deteriorated. Some 2,500 German planes were based on the eastern front, but a shortage of first-line aircraft led to older types such as the He 46 being pressed into service. Some night raids were made against England (by He 111s, Ju 88s, and Do 217s) as well as daylight fighter-bomber attacks (Bf 109s, Fw 190s).

Pressure on the Luftwaffe increased steadily during 1943. The German air force had been equipped to fight only a short Blitzkrieg-type conflict, and after over three years of war now found itself with only 4,000 aircraft, no reserves, and few new designs approaching availability. Erhard Milch wanted fighter production increased from 500 a month (spring 1942) to 3,000 a month by the summer of 1944, but Hitler and Goering continued to insist on a bomber production program. Massive Allied air raids early in 1944 followed by the D-day landing eventually earned priority for the German fighter arm, but deployment of the jet-propelled ME 262 had been seriously delayed by Hitler's insistence on its development as a bomber. Only in the last few months of the war were Me 262s, ARADO 234 jet bombers, and Me 163 KOMET rocket fighters available in even small numbers. Although they were excellent aircraft technically, they appeared too late to prevent the Allies from maintaining their air superiority. Near the end, lack of fuel grounded many Luftwaffe aircraft, and few really experienced pilots were still alive. A new Luftwaffe was established in 1956, equipped primarily with F-84F Thunderstreaks, RF-84Fs, and F-86 Sabres. Today its aircraft include PHANTOMS, STARFIGHTERS, ALPHA JETS, and G-91s; personnel number 111,000.

Lycoming

One of the world's largest designers and producers of engines for light aircraft, Lycoming was originally a manufacturer of automobile engines. It produced a 185-hp nine-cylinder radial aircraft engine as early as 1928. This eventually led to the 225–300-hp R-680 series (used by the Reliant, Martin B-10, and Grumman F2F). A seven-cylinder radial (200-hp R-530) appeared in 1937, and the following year a series of air-cooled, horizontally opposed four-cylinder units of 50–57 hp was begun (O-145, GO-145). Later flat-4s included the O-235 (104 hp) and O-290 (125 hp), while flat-6s (O-350 of 150 hp, O-435 of 190 hp) joined the range.

After World War II a flat-8 (GSO of 350 hp) was produced, and the company also developed a series of small gas turbines (T53, T55) before becoming part of Avco Corporation in the mid-1960s. Current production by Avco Lycoming includes horizontally-opposed piston engines for an enormous range of light aircraft, and numerous small turbofans and turboprops.

M

Macchi

The Italian aircraft manufacturer Macchi (now Aermacchi) came into being in 1912 as Nieuport-Macchi, established principally to build French NIEUPORT aircraft under license. During World War I it produced hundreds of Nieuport 11s and 17s, and also began large-scale manufacture of aircraft of its own design. Among these were flying boats, which were introduced after the capture of an Austrian Lohner L.40 flying boat in 1915, and the request that Macchi produce a copy. The resulting Macchi L.1 started the company's long association with the field of marine aviation. Large numbers of L.1 and L.2 flying boats were built, followed by several hundred M.5 single-seat flying-boat fighters. In 1921 a later member of this family of flying boats, an M.7, won the Schneider Trophy at 117.86 mph (189.67 km/h).

In the mid-1920s Mario Castoldi became Macchi's chief designer. His red-painted M.39 floatplane racer won the 1926 Schneider Trophy at a record speed of 246.49 mph (396.68 km/h), and the Macchi-Castoldi MC.72, powered by a double Fiat AS.6 24-cylinder engine able to sprint at 3,100 hp, set a world speed record of 440.69 mph (709.07 km/h) in 1934. Two years later Castoldi designed a landplane fighter, the MC.200 Saetta ("Thunderbolt"). Its small 840-hp Fiat engine resulted in mediocre performance, but it possessed excellent maneuverability; armament consisted of two .50 machine guns and racks for two bombs of up to 352 lb. (160 kg) each. Together with the Fiat C.R.42 biplane, the MC.200 formed the mainstay of Italian fighter strength at the outset of World War II, and about 1,000 were built. Use of the 1,075-hp Daimler-Benz–Alfa Romeo engine in the redesigned MC.202 Folgore ("Lightning") fighter resulted in one of the best aircraft to operate in the Mediterranean. The MC.202 first saw combat in 1942; the following year the MC.205 Veltro ("Greyhound") began active service. This was essentially an MC.202 equipped with a 1,465-hp Daimler-Benz–Fiat engine, which gave it a maximum speed of 391 mph (630 km/h). Allied pilots considered it the equal of the P-51D Mustang, but only 262 MC.205s were built.

After World War II Aeronautica Macchi's products included the MB.308 lightplane, the twin-engined B.320, and, most notably, the MB.326 jet trainer, first flown on December 10, 1957. The MB.326 seated pupil and instructor in tandem and was an attractive and successful machine. Powered by a Viper engine of 2,500 lb. (1,130 kg) thrust, it went into production in 1960, and has since been built in many versions in Italy and under license in Brazil, South Africa, and Australia. Several models are equipped for tactical support missions, and some single-seaters intended primarily for this role are powered by the Viper 632, rated at 4,000 lb. (1,830 kg) thrust. A later Aermacchi product is the MB.339 multirole trainer/attack aircraft, with a Viper 632 engine, a restressed airframe, and redesigned cockpits incorporating a higher rear seat to give the instructor or navigator an improved view.

Mach number

A measure of airspeed in relation to the speed of sound, named after the Austrian physicist Ernst Mach (1838–1916). The speed at which sound waves travel varies according to the conditions (principally temperature) of the air through which they pass. At sea level the speed of sound is approximately 760 mph (1,223 km/h), while at 37,000 ft. (11,277 m) it falls to about 660 mph (1,062 km/h).

The Mach number of 1 is assigned to an airspeed equal to the speed of sound at the altitude the aircraft is flying. An aircraft flying at Mach 0.5 is flying at half the speed of sound; at Mach 2.0 it is at twice the speed of sound, and so on.

Mannock, Edward "Mick"
(1887–1918)

Leading British fighter ace of World War I, credited with 73 victories. Beginning the war in the Medical Corps, but later commissioned in the Royal Engineers, he almost immediately applied for pilot training with the Royal Flying Corps. He went to France in March 1917 to join 40 Squadron, flying NIEUPORT scouts. His first victory was on May 7, 1917, the day his idol, Albert BALL, was killed. By January 1, 1918, he was credited with over 20 victories, and after two months in England on enforced leave, he joined 74 Squadron as a captain and flight

Macchi M.5

commander, flying SE5As. A ruthless, determined, and completely dedicated fighter pilot, his total of combat victims rose rapidly and he was given command of 85 Squadron.

On July 26, 1918, he took off at dawn with Lt. D. C. Inglis, and they destroyed a German aircraft over the front-line trenches. Mannock's SE5A was hit by ground fire and it quickly crashed, killing Mannock, whose body was never recovered. He was awarded a posthumous Victoria Cross.

Marauder

The first Martin B-26 Marauder medium bomber flew in 1940. This twin-engined, shoulder-wing monoplane was capable of carrying a 4,000-lb. (1,800-kg) bomb load. Powered by 2,000-hp Pratt & Whitney radial engines, maximum speed was in excess of 300 mph (480 km/h). Wingspan of late models was 71 ft. (21.64 m) and length 56 ft. 1 in. (17.10 m).

A series of crashes early in the Marauder's history led to official doubts about continuing production, but an investigation found that training techniques were at fault. Early B-26s acquitted themselves well in combat in the Pacific, though when first committed to fighting in Europe their reputation was further marred when all 10 aircraft on an early mission from England failed to return. A change from low- to medium-altitude operations totally altered the aircraft's fortunes, and it eventually produced the lowest loss per sortie record of all U.S. bomber types operating in Europe. A total of 5,157 B-26s of all types were built.

Mariner

Martin PBM Mariner twin-engined flying boats entered service with the U.S. Navy late in 1940. The Mariner was notable for its gull wings, designed to keep its propellers well clear of the water. The first model of the Mariner had floats that retracted into the wings, but all later aircraft used fixed floats. The major production model, the PBM-5, was powered by Pratt & Whitney R-2800 radial engines. Wingspan was 118 ft. (35.97 m) and length 77 ft. 2 in. (27.52 m). Patrol-bomber Mariners were defended by eight .50 machine guns, three pairs of which were in power-operated turrets; the armament was deleted from those aircraft built as transports. Altogether 1,289 Mariners of all types were produced during World War II, a further 36 amphibious versions were built in 1948–49 for rescue duties. Mariners served with Navy squadrons until after the end of the Korean War.

Mars

The largest military flying boat ever built in the United States, the original Martin XPB2M Mars was intended as the prototype of a very long-range patrol bomber for the U.S. Navy. First flown on July 3, 1942, the Mars had a wingspan of 200 ft. (60.96 m) and length of 117 ft. (35.66 m). Four Wright R-3350 radial engines gave a top speed of 121 mph (355 km/h). The Mars design was not adopted for its original intended role, but a limited production of cargo versions was ordered under the JRM designation. Five were completed in

Martin Mariner

1946, and four served with Squadron VR-2 on naval transport duties for some 10 years.

Marseille, Hans-Joachim (1919–1942)

German fighter ace of World War II, credited with 158 victories, Marseille was the highest-scoring Luftwaffe pilot in service elsewhere than on the eastern front. Opening his career along the coast of the English Channel, Marseille destroyed several RAF aircraft over the sea, but not until he was assigned to *Jagdgeschwader* 27 in North Africa in April 1941 did he begin to acquire the status of a celebrity. Rommel's winter offensive of 1941–42 gave him the opportunity to become a leading ace; by June 18, 1942, he had shot down more than 100 aircraft. In one day alone, September 1, 1942, he claimed 17 victories, of which at least a dozen were verified. He had his own personal BF 109, "Yellow 14," and became renowned for his marksmanship—on one occasion he used only ten 20-mm cannon shells and 180 rounds of machine-gun ammunition to destroy six aircraft. On September 30, 1942, his fighter caught fire in the air due to a mechanical fault. Marseille bailed out, but was thrown against the tailplane by the slipstream and killed.

Martin Marietta

Martin Marietta's aerospace activities are now restricted to tactical and strategic missiles, equipment for the space program, and a certain amount of subcontracting work for other manufacturers, but one of its original component companies, Glenn L. Martin, was long in the forefront of the American aviation industry.

Glenn L. Martin (1886–1955) had built one of the world's first large twin-engined bombers for the U.S. Army in 1913, and a year later completed an Army trainer and established a factory at Santa Ana, Cal. For the next three years Martin built military trainers, but also supplied large bombers used in the 1916 expedition against Pancho Villa. On American entry into World War I Martin joined with Wright interests in forming Wright-Martin Co., which produced trainers and a large number of Hispano-Suiza engines.

By 1918 Glenn L. Martin Co. was again an independent manufacturer, and at the request of the Army Air Service produced the first American-designed bomber, the MB-1, which was powered by two 400-hp Liberty engines. It made its first flight in August 1918, but because of the end of the war only 10 were built. Martin also constructed two MBT torpedo bombers for the Navy, and several transport versions for the Post Office, some of these being converted into T-1 Army transports. In 1920 delivery began of the MB-2, used by Billy Mitchell to demonstrate the airplane's ability to sink warships. Martin produced 20 MB-2s as NBS-1 (night bomber short-range), and other manufacturers including Curtiss also

Martin Mars

Martin 130 flying boat of Pan American Airways

built the aircraft. All were called "Martin Bombers," and the MB-2 served as the standard Army heavy bomber until 1928; the MT was a Navy torpedo version. A long series of important single-engined machines followed. These began with the T3M torpedo bomber designed by the Navy. Martin built 102 T4M-1 torpedo bombers, powered by the new Pratt & Whitney Hornet engine, and a small series of BM-1 dive-bombers. Soon afterward came the large PM-1 and PM-2 patrol flying boats, followed in 1932 by the Model 123 bomber, used by the Army Air Corps as the B-10 and developed into the B-12 and B-14 and export models 139 and 166. This was an extremely advanced aircraft, with maximum speeds of 207 to 268 mph (330–430 km/h), at a time when a typical single-seat fighter could only achieve 175 mph (281 km/h). Powered by two Cyclone or Hornet radial engines, the B-10 and its successors were all-metal cantilever monoplanes with stressed-skin construction, retractable landing gear, flaps, variable-pitch propellers, turrets, internal bomb bays with powered doors, and numerous other pioneering features.

In 1935 Martin built the Model 130 *China Clipper* flying boat for Pan Am. The *China Clipper* and its other sister ships could carry 40 passengers nonstop to Hawaii, and via Midway, Wake, and Guam, to Manila. Martin's factory at Baltimore went on to build the even bigger Model 156, followed eventually by the XPB2M-1 MARS. Another flying boat, produced from 1937 until 1949, was the Model 162 MARINER, which

served as patrol, antisubmarine, and transport aircraft during World War II. From the Mariner was developed the P-5 Marlin, of which 196 were manufactured between 1951 and 1960.

Martin's wartime bomber production began with the Model 167 twin-engined medium bomber, which went into service in France in October 1939, and served with the RAF as the Maryland. A more powerful development was the Baltimore, which also never flew for the United States, but was supplied to British, South African, and Italian forces. Martin's main wartime product was another twin-engined bomber, the MB-26 (Model 179), which first flew in November 1940. Martin chose a wing designed primarily for efficiency at cruising speeds, which meant that the B-26 landed fast, a feature that initially caused many crashes and earned the machine an undeservedly bad reputation; later, as the MARAUDER, it was to give outstanding wartime service.

After the war, Martin built the three-jet XB-51, six-jet XB-48, and Navy AM-1 Mauler, all attack bombers that were not produced in quantity. Only 19 piston/jet P4M Mercators, long-range Navy patrol bombers, were built. The Model 202 and 404 civil airliners—an unusual departure for Martin—enjoyed reasonable success, but license-production of the CANBERRA jet bomber as the B-57 was virtually the last aircraft built by Martin.

The company had entered the rocket-propulsion field soon after the war. In 1949 it launched the first Viking research rocket, having previously built Gorgon ramjet test vehicles and KDM-1 target

drones. Later came the B-61 Matador pilotless jet bomber, the TM-67B Mace of 1959, the Lacrosse (a tactical battlefield missile launched from a large truck), the Pershing (able to deliver a nuclear warhead over ranges up to 400 mi., 640 km), and the Titan (an Air Force intercontinental ballistic missile).

Maxim Gorki

The ANT-20, designed by A. N. TUPOLEV, was a giant Russian aircraft with a wingspan of 206 ft. 8 in. (63 m), constructed in 1934 for propaganda work. It was powered by 875 hp engines in the wings and two similar power units in a pod above the fuselage. The reputed cost of six million rubles was paid for by workers' donations. Inside the fuselage, which measured 107 ft. 11 in. (32.90 m), was a radio station, photographic darkroom, sleeping chambers, movie theater, and newspaper office with printing press. When flying at night, political slogans could be flashed on the underside of the wings, and at stopovers propaganda leaflets were distributed. There was accommodation for a crew of 23 and some 75 passengers.

On May 18, 1935, the *Maxim Gorki* was carrying 36 "shock workers" (some with their families) and was accompanied by a camera ship and a light aircraft to provide a size contrast. The small plane pulled a loop too close to the eight-engined giant and collided with one of the *Maxim Gorki's* wings. The massive aircraft disintegrated killing all on board.

McConnell, Joseph, Jr. (1922–1954)

Highest-scoring American fighter pilot of the Korean War, McConnell had first seen combat flying B-24 Liberators with the Eighth Air Force in the last years of World War II. He became part of the second F-86 SABRE wing to enter combat in the Korean War, and as a captain with the 16th Fighter Squadron had destroyed his fifth MiG-15 by February 16, 1953. On April 12 he bailed out from a badly damaged F-86F but was rescued from the Yellow Sea. On his 106th and last mission on May 18, McConnell destroyed three MiGs to achieve the record score of 16. After his return to the United States McConnell was killed while testing the F-86H on August 26, 1954.

McDonnell Douglas

McDonnell Douglas was formed in 1967 as the result of the acquisition of DOUGLAS by McDonnell Aircraft Corporation (MAC). McDonnell was created in 1939 and received a contract for the XP-67 bomber destroyer in September 1941, and later produced Fairchild AT-21 trainers under license. In January 1943 it was chosen to build the world's first carrier-based jet fighter, the XFD-1 (later FH-1) Phantom for the U.S. Navy. Flown for the first time in January 1945, the Phantom I was powered by two Westinghouse J30 engines of 1,600 lb. (725 kg) thrust. In January 1947 the F2H (later F-2) Banshee appeared, powered by larger J34 engines. Nearly 900 Banshees of four main types were delivered by 1956; many were long-range naval night fighters with Westinghouse or Hughes radar and four 20-mm cannon. In 1951 the F3H (later F-3) Demon made its maiden flight; 519 of these large and complex transonic naval fighters were eventually built.

After the XP-67, MAC next built for the Air Force the XF-88 Voodoo, an advanced long-range fighter (1948), following this with the larger F-101 Voodoo of 1954. With two Pratt & Whitney J57 engines of 14,800 lb. (6,700 kg) thrust, the Voodoo was the most powerful fighter of its day; all-weather interceptors, tactical fighter-bombers, and long-range reconnaissance versions were developed. Immediately after the war MAC also built the XHJD-1 Whirlaway helicopter, and the XF-85 Goblin, a very small fighter that was designed to be carried inside a B-36 bomber, from which it would fly off to engage enemy fighters. Still other machines that did not enter production included the Model 120

flying-crane helicopter, the XV-1 Army CONVERTIPLANE with rotor and propeller, the Little Henry ramjet helicopter, and the Model 119/220 business jet (the first executive jet ever certificated, but ahead of its day).

In 1958 MAC flew the first F4H (later styled F-4) PHANTOM II, which started to enter service in 1960 and became one of the most successful of all supersonic fighters. The F-15 EAGLE, an air-superiority fighter for the U.S. Air Force, began production in the early 1970s.

In the late 1950s McDonnell also entered the aerospace industry. It built the Mercury and Gemini capsules used in early American manned spaceflights. It was also a pioneer firm in the field of rocketry and guided missiles. It produced the Gargoyle glide bomb (1944), Katydid pulse-jet drone (1945), Kingfisher air-to-underwater missile (1949), Talos long-range ramjet-powered ship-to-air missile (1952), GAM-72 Quail bomber-launched electronic decoy (1958), ASSET test vehicle flying at Mach 18 (1963), Alpha Draco with a lifting body flying at Mach 5 in the atmosphere (1959), and M-47 Dragon infantry antitank missile (1968).

McDonnell acquired control of the ailing Douglas Aircraft to form McDonnell Douglas Corporation on April 28, 1967. McDonnell and Douglas retain their separate identities although products bear the corporate title. McDonnell Douglas produced the Model 188 (Breguet 941) STOL transport in the United States, and also designed the YC-15 military STOL transport. Major military projects include the AV-8B, an advanced development of the HARRIER with a super-critical wing giving twice the load range of the original British version, and the F-18, a multirole tactical fighter and attack aircraft for the U.S. Navy and Marines, which was designed by Northrop, the principal associate contractor.

McGuire, Thomas B., Jr.

Second-highest scoring American fighter ace of World War II, McGuire was sent to Australia in March 1943 and assigned first to the 49th Fighter Group and then, in July 1943, to the 475th Fighter Group. Flying P-38 LIGHTNINGS with the 431st Fighter Squadron and later with the Group Headquarters, he shot down a total of 38 Japanese aircraft in the Pacific. McGuire was awarded the Congressional Medal of Honor for gallantry displayed in fighter sweeps over the Philippines on December 25 and 26, 1944, when he shot down seven enemy aircraft. McGuire was killed when his aircraft stalled and crashed during a low-altitude dogfight on January 7, 1945.

Me 163. See KOMET

Me 262

The world's first operational jet-powered aircraft, the Messerschmitt 262 was designed as a fighter, but large numbers were converted on Hitler's own orders into fighter-bombers. The prototype made its maiden flight on April 4, 1941, powered by a nose-mounted Jumo piston engine; the initial jet-powered flight was made on July 18, 1942. Production Me 262A-1a aircraft had a wingspan of 40 ft. 11 in. (12.48 m) and length of 34 ft. 9 in. (10.60 m). They used a tricycle undercarriage and two Jumo 004B turbojets, and had been equipped with four 30-mm cannon in the nose, but conversion of all the initial Me 262s from fighters into Me 262-2a fighter-bombers delayed the aircraft's trial appearance in battle until late June 1944. Eventually the Luftwaffe received permission to use a small number of Me 262s as interceptors. In the last six months of the war they had a fair measure of success against

Messerschmitt Me 262A-1a

Messerschmitt Me 209 (109R) world speed record holder of 1939

Allied heavy bombers. As a fighter the Me 262 lacked acceleration and was unable to make a tight turn, but it possessed sufficient speed to break off combat at will. Allied fighters were avoided if possible.

Additional variants produced included a night fighter with radar (Me 262B-1a/U1) developed from the two-seat trainer (Me 262B-1a), and the Me 262 (Mark 114) bomber-destroyer with a 50-mm cannon. Fighter-bombers carried two 550-lb. (250-kg) bombs, and the Me 262A-1b was fitted with 24 rocket projectiles below the wings.

Messerschmitt

Willy Messerschmitt (1898–) was one of the leading German aircraft designers before and during World War II. While still at school Messerschmitt had constructed a glider, and by the early 1920s he and Friedrich Harth had built and flown several designs. In 1923 Messerschmitt established his own aircraft factory and produced the tiny M17 lightplane (generally equipped with a 32-hp Bristol Cherub engine), which Messerschmitt himself flew across the Alps to Rome, and the M18 airliner of 1925, of which 26 examples with various engines were sold. In 1927 Messerschmitt AG came to an agreement with BFW (Bayerische Flugzeugwerke): BFW would build only Messerschmitt designs, while Messerschmitt would give BFW an option on each aircraft. The united companies established themselves at Augsburg, with Messerschmitt as chief designer.

Despite Messerschmitt's ability to produce successful designs, BFW went bankrupt in 1931. Messerschmitt himself had managed to sell an M23 to the Nazi leader Rudolf Hess, and after the Nazis came to power in 1933 Messerschmitt was able to resume business. Despite the enmity of Air Minister Milch and of Ernst HEINKEL, whose aircraft Messerschmitt was told to build, the firm produced the Bf 108 Taifun cabin monoplane (1934), the outstanding BF 109 fighter (1935), which remained in production throughout World War II, and the Bf 110 ZERSTÖRER (1936), a twin-engined fighter escort and interceptor. In 1938 the company once more took the name of Messerschmitt AG. It was soon the largest German aircraft manufacturer, and its designs now bore the prefix "Me." One of the first aircraft to do so was the Me 209, a diminutive contender for the world speed record built around the DB 601 engine. As the "Me 109R," this machine set a world speed record of 469.22 mph (754.97 km/h) in the hands of Fritz Wendel in 1939. Subsequent attempts to turn it into a fighter, with wider-span wings and an enlarged tail fin, came to nothing, and the Me 209 II was an entirely different airplane embodying Bf 109G airframe components, a DB 603 engine, and a wide-track undercarriage. Proposals to fit the Me 209 II with a Jumo 213 engine and to develop a high-altitude interceptor version (the Me 209H) were canceled, and the DB-605 engined Me 309, with a tricycle undercarriage and pressurized cockpit, proved to have unduly heavy control forces. Only four prototypes of the Me 309, intended as a replacement for the Bf 109, were constructed.

Planned as a successor to the Bf 110 Zerstörer, the Me 210 of 1939 was a snub-nosed, twin-engined heavy fighter. Substitution of a single fin and rudder for its original twin tail assembly did little to alleviate the aircraft's dangerous handling characteristics. Me 210A machines had DB 601 engines, two machine guns and two cannon in the nose, and remote-controlled barbettes on either side of the fuselage each mounting a single rearward-firing machine gun. Unfavorable Luftwaffe reports on these first production aircraft led to the introduction of leading edge slots and a new rear fuselage, but the Me 210 was replaced in 1943 by the Me 410 Hornisse—a reworking of the earlier design which overcame the handling problems of the Me 210. The Hornisse was manufactured as a heavy fighter, photo-reconnaissance machine, antishipping aircraft, and high-speed bomber.

Other wartime Messerschmitt designs included the Me 163 KOMET, the only operational rocket-powered interceptor ever built; the twin-engined long-range Me 261 Adolfine, which was intended to carry the Olympic flame to Tokyo on a record-breaking flight (only three prototypes were constructed); the world's first operational jet aircraft, the ME 262; the four-engined Me 264, with which it was originally planned to bomb New York, but which was later intended to be a maritime reconnaissance aircraft (only two prototypes were completed); the Me 321 military transport glider and its six-engined development, the Me 323 Gigant; the abortive Me 328 "piloted missile" with twin pulse jets (vibration from the power units proved an insuperable problem); the Me/Bv 155 high-altitude fighter; and the experimental P.1101 SWING-WING design of 1945 upon which the Bell X-5 (see X-1) was based.

Willy Messerschmitt worked for a period after the war in Argentina before reestablishing himself in Germany in the mid-1950s. Messerschmitt is now part of Messerschmitt-Bölkow-Blohm.

Metal fatigue

Metal failure is caused by repeated stress. At low stresses a ductile metal such as steel or aluminum stretches, but when the stress is removed, the metal returns to its original state. Beyond a certain stress, however, the deformation is permanent, and eventually a point is reached when the metal will fracture. Mild steel, for example, will fail when a stress of some 56,000 lb./sq. in. (386,106 kilopascals) is applied to it.

But mild steel may fail under much lower stress, if that stress is applied first in one direction, then in another, repeatedly. This is fatigue failure, and it produces a brittle fracture.

Fatigue failure almost invariably begins at irregularities on the surface of the metal. These points act as stress raisers. Under constantly repeated stress the metal crystals start to break up, and submicroscopic cracks form and spread. They eventually link together and visible cracks appear, weakening the metal so much that it eventually fails. Fatigue strength decreases as the temperature increases. It is also adversely affected by corrosion and erosion, which provide potential sites for stress concentration.

Research done in the early 1950s demonstrated that typical metal parts can fail in fatigue at stresses roughly half as great as can safely be applied in the absence of repeated flexing. Previously it had merely been known that oscillating parts sometimes broke at what seemed "safe" stresses.

By 1930 fatigue was known and understood in the design of engines, and was countered by avoiding indentations, grooves, corners, scratches, angles or sudden changes in cross-section. Good parts had gentle curves, rounded inner corners, and a smooth polished surface.

Fatigue was largely ignored in the design of airframes, however, partly because an individual aircraft seldom flew more than a few hundred hours and partly because several authorities denied that aircraft were prone to fatigue at all. The subject was thrown into prominence in 1954 by investigations into the loss of two COMET jetliners over the Mediterranean. In the case of one it was proved that the whole fuselage had ripped open as the aircraft climbed to its cruising height. The repeated pressurization had caused small cracks to appear at the corners of cut-outs—windows and radio-compass aerials—which should have been more rounded instead of almost square. On each flight the fuselage had been once inflated and then deflated, and eventually one crack had suddenly turned into a catastrophic tear.

As soon as this was learned it became the practice to fatigue-test aircraft cabins by repeatedly pressurizing them thousands of times, while simultaneously flexing the wings and applying repeated shock loads from the landing gear. Today the most extreme care is taken to make aircraft either fatigue proof, often by specifying a "safe life" (perhaps 60,000 hours or 40,000 flights), or by making the structure "fail safe" so that, even if fatigue should cause any part to break, it will do so at a slow rate and the aircraft will still be able to land safely.

Meteor

The first RAF jet aircraft in squadron service, the twin-engined Gloster Meteor was the only operational Allied jet of World War II. The first Meteor (DG206) made its maiden flight on March 5, 1943, and in July of the next

year the first Meteor fighters entered service with the RAF. The 490-mph (788-km/h) Meteor F-3, with Derwent engines of 2,000 lb. (907 kg) thrust, served as the standard RAF fighter of the immediate postwar years. In 1949 the T7 trainer variant was introduced. From 1950 to 1955, the F8 fighter was the principal RAF day interceptor, equipping 19 regular squadrons and 10 reserve units. Powered by Derwents of 3,600 lb. (1,630 kg) thrust, the F8 had a wingspan of 37 ft. 2 in (11.33 m) and length of 47 ft. 7 in. (13.59 m). Photo-reconnaissance versions were also produced from 1950 onward; Armstrong-Whitworth Meteor night fighter variants began to reach the RAF in 1951.

On November 7, 1945, the Meteor F4 *Britannia* (EE454) flown by Group Captain H. J. Wilson established a world speed record of 606.38 mph (975.67 km/h), raised to 615.78 mph (990.79 km/h) by Group Captain E. M. Donaldson (EE594) on September 7, 1946.

Metropolitan

Among the most successful short medium-range airliners to appear in the years immediately following World War II were the twin-engined CONVAIR 240/340/440 series, often known as Metropolitans or Convairliners.

The series was begun with the 30-passenger Model 110, which first flew on July 8, 1946. Before the aircraft could enter production the design was revised in the light of prospective competition from the sleeker, higher-capacity Martin 202 and became the 40-passenger CV-240 of 1947. With two 2,400-hp Pratt & Whitney R-2800 radial engines giving a cruising speed

Gloster Meteor F8

Mi-12 "Homer"

of 270 mph (435 km/h) the Convair 240 joined American Airlines in January 1948 and had substantial subsequent success in the United States and overseas.

In 1951 came the CV-340, with a fuselage stretched to 79 ft. 2 in. (24.13 m) and seating for 44 passengers. Four years later the CV-440 Metropolitan made its first flight. This model could accommodate up to 52 passengers, and with 2,500-hp R-2800-CB17 engines the cruising speed was 289 mph (465 km/h), with a range of 1,270 mi. (2,045 km). Wingspan of the CV-440 was 105 ft. 4 in. (32.10 m) and length 81 ft. 6 in. (24.84 m). Metropolitans flew extensively with SAS, Sabena, Swissair, Braniff, Delta, and Eastern.

Many civilian Convairliners were later modified and equipped with turboprop engines (becoming 580s, 600s, and 640s) to serve into the 1970s with some smaller airlines. Military versions of the Convair 240/340/440 series included the C-131A air ambulance, T-29 navigational trainer, C-131B airborne electronic equipment test aircraft, and C-131D transport.

Mi-12

The world's largest helicopter, designed by the Russian M. L. MIL. When it first appeared at the 1971 Paris Air Show, Russia's mighty Mi-12 (codenamed "Homer" by Nato) dwarfed every exhibit except America's Lockheed C-5A Galaxy, parked alongside. The Western world had been aware of

its existence because from time to time the Russians had claimed world records for payload in the rotary-wing class. The most recent claim for the Mi-12 was its ability to lift a load of 88,636 lb. (40,200 kg) to a height of 7,398 ft. (2,255 m).

The Mi-12 was built to the specification of the Soviet Air Force, which wanted a helicopter to transfer loads from the cargo hold of the huge ANTONOV An-22 strategic freighter to the battlefield. Although the aircraft seen at Paris wore AEROFLOT markings, it may be considered as a military craft, since the airline and the air force serve each other.

Weighing some 231,000 lb. (105,000 kg) fully loaded, the Mi-12 is more than four times as heavy as America's biggest helicopters, the Boeing Vertol CH-47 CHINOOK and Sikorsky CH-53 (S-65), and about three times as big. With the recent cancellation of Vertol's half-completed XCH-62 HLH (Heavylift Helicopter), the Mi-12 is unlikely to be equaled for size and weight in the foreseeable future. Its twin intermeshing rotors, 220 ft. (67 m) in diameter, are powered by four 6,500-shp turbines and provide a maximum speed of 160 mph (257 km/h).

Despite its immense size, the Mi-12 is not based on any new technology but uses "building blocks" from other helicopters. In particular it employs the complete engine, mechanical transmission and rotors from a previous Mil helicopter, the Mi-6, itself twice the size of the largest U.S. machines.

Middle East Airlines

One of the most important international airlines of the Middle East, MEA was established in 1945 with three DH 89 biplanes, and soon received war-surplus DC-3s. Minority shareholdings were held successively by Pan American (1949–55) and BOAC (1955-61), and the airline was able to purchase a fleet of five Viscount turboprops. Pure jet services began in 1960 with Comet 4s leased from BOAC pending delivery of MEA's own Comet 4Cs in 1961. Merger with Air Liban in 1963 gave Air France a 30% interest in MEA and extended the route network. Comets were replaced by Boeing 720Bs on the West Africa and London services in 1966 and by a leased Boeing 707 in 1968. Lebanese International Airways was acquired in 1969.

Three Comets, a Boeing 707 and a leased Vickers VC-10 were destroyed in an Israeli attack on Beirut Airport in December 1966. These were replaced by Boeing 707s and Convair 990s, but in 1976 MEA suffered a more serious blow with the closing of Beirut Airport due to the Lebanese Civil War. Operations were transferred to London and then to Paris, but heavy financial losses were incurred. Before this disruption scheduled passenger services were operated throughout Europe, the Middle East, North and West Africa, and Asia. Route mileage totaled 38,756 mi. (62,358 km), and some 1,075,000 passengers were carried annually. The present fleet consists of 3 Boeing 747s, 3 Boeing 707s and 15 Boeing 720Bs.

Midway, Battle of

By the summer of 1942 the Japanese drive into the southern Pacific had been halted at the Battle of the CORAL SEA. It was evident, however, that a new Japanese thrust might be expected either against the Hawaiian Islands or in the Aleutians, and by June 2 the carriers *Enterprise*, *Yorktown*, and *Hornet* had rendezvoused northeast of Midway Island, some 1,250 mi. (2,000 km) west of Pearl Harbor, in anticipation of an enemy strike.

Soon after 9 a.m. on June 3, routine U.S. air patrols discovered part of a Japanese task force under Admiral Isoroku Yamamoto that comprised 350 vessels (including the aircraft carriers *Akagi*, *Kaga*, *Soryu*, and *Hiryu*) and an air amada of more than 1,000 combat planes. That same morning nine B-17 Flying Fortresses from Midway carried out a bombing attack that damaged a number of vessels (although not the carriers, which were still unlocated), and during the night a torpedo attack was made by Catalina PBY-5As.

At daybreak on June 4 American radar on Midway picked up an incoming raid comprising 36 Vals (dive-bombers), 36 Kates (torpedo-bombers), and 36 Zeros at a distance of 93 mi. (149 km) and an altitude of about 10,000 ft. (3,050 m). Twenty Brewster F2A-3 Buffalos and 7 Grumman F4F-4 Wildcats (both planes were inferior to the raiding craft, and the Buffalos especially so) were scrambled to intercept, but the formidable Zeros shot down at least 15 of them, Japanese losses apparently being only 2 Kates, 1 Val, and 2 Zeros (all probably to antiaircraft fire).

Retaliatory American strikes on the Japanese task force by B-26 Martin Marauder torpedo-bombers, TBF-1 Grumman Avengers, Douglas SBD-5 Dauntlesses, B-17s, and Vought SB2U-3 Vindicators achieved little and incurred heavy losses—only 1 of the 9 Avengers returned, and half of the 18 Dauntlesses were shot down.

Just after 9 a.m., torpedo-bombers from the American carriers made an attack on the four Japanese flat-tops. They were intercepted by defending Zeros and decimated: of the 41 aircraft sent in, only 6 returned.

With the Zeros fully engaged at low level, however, dive-bombers from the *Enterprise* were able to press home their attacks on the *Kaga* and *Soryu* unopposed until they began to pull out of their dives, and the *Akagi* was similarly pinpointed by planes from the *Yorktown*. Eighteen of the Dauntlesses were lost, but the three Japanese carriers had become burning hulks, two of them foundering within a few hours and the

MiG-3s of the Soviet Air Force

third finally dispatched by a torpedo from the U.S. submarine *Nautilus*.

The *Hiryu* escaped, and in the early afternoon sent off 18 Vals and 6 Zeros to attack the *Yorktown*. Defending Wildcats intercepted them 20 miles out and only 8 reached the U.S. carrier, which was hit by three bombs. Shortly afterward a second wave from the *Hiryu* comprising 10 Kates and 6 Zeros scored two torpedo hits on the *Yorktown*, which had to be abandoned 15 minutes later (she was taken in tow but torpedoed and sunk by a Japanese submarine on June 6).

Almost immediately after this Japanese success, however, 24 dive-bombers from the *Enterprise* attacked the *Hiryu* and reduced her to a smoldering wreck.

On June 5 and 6 any B-17s, Dauntlesses, and Vindicators that remained serviceable on Midway were sent out to attack the now-fleeing Japanese task force. They sank the heavy cruiser *Mikuma* and damaged several other capital ships.

At the end of the battle, the Americans had lost the *Yorktown* and about 150 aircraft, while the Japanese had sacrificed four carriers, two heavy cruisers, several destroyers, and some 275 aircraft. A major factor in the American victory was the availability of radar on U.S. ships; the Japanese were without this vital aid.

MiG

The great Soviet design bureau formed by the partnership of Artem Ivanovich Mikoyan and Mikhail Gurevich, whose aircraft are the backbone of the fighter squadrons of the USSR. Mikoyan was born in 1905 and joined POLIKARPOV's design team at the Central

Design Bureau to work on the I-153 in 1937.

Mikhail Iosifovich Gurevich graduated from the Kharkov Institute of Technology in 1923 and joined the flying boat design team at the Central Design Bureau in 1930. When he met Mikoyan, he was one of Polikarpov's Deputy Chief Designers.

In 1939 Mikoyan and Gurevich set up their own design bureau to develop the MiG-1 single-seat high-altitude interceptor which was completed at the beginning of 1940. Designed around the 1,350 hp AM-35 V-12 engine, the MiG-1 had a span of only 33 ft. 9½ in. (10.3 m) and length of 26 ft, (8.1 m). Maximum speed was 390 mph at 22,965 ft. (628 km/h at 7,000 m). The MiG-3 was similar but a number of minor improvements resulted in a very small increase in speed. The MiG-1 and MiG-3 were the only MiGs to be produced during the war, although the team was responsible for over a dozen prototypes during this period.

In 1946 the MiG team produced the MiG-9, one of Russia's first jet aircraft —the YAKOVLEV Yak-15 and MiG-9 made their first flights from the same airfield on the same day, April 24, 1946. The MiG-9 was a single-seat fighter, with an unswept laminar wing, powered by two RD-20 engines which was produced in small numbers; it had a maximum speed of 566 mph at 16,400 ft. (911 km/h at 5,000 m).

First observed at the 1948 Soviet Aviation Day display, the MIG-15 made its appearance over Korea in 1950. Since then it was produced in four countries and served with the air forces of at least eighteen. Powered by a Russian copy of the Rolls-Royce Nene, the first MiG-15s had a maximum speed of 647 mph (1,042 km/h) and were highly maneuverable.

The MiG-15 was followed by the MiG-17, a single-engined 700 mph (1,120 km/h) derivative of the MiG-15, and the MiG-19 850 mph (1,360 km/h) twin-engined development, the widely-used delta-winged MIG-21—which has a maximum speed of 1,320 mph at 41,000 ft. (2,125 km/h at 12,500 m)—the swing-wing MiG-23 (Foxbat), and the MIG-25 (Foxbat A) twin-engined reconnaissance fighter which established records of 1,808 mph (2,910 km/h) over a 621-mi. (1,000-km) closed-circuit course with a 4,409 lb. (2,000 kg) payload, and of 98,458 ft. (30,000 m) with the same payload.

MiG-15

The MiG-15 "Fagot" single-seat fighter marked the Soviet air force's real entry into the jet age, despite the earlier appearance of the German-researched MiG-9 and the Yak-15. It created a sensation when it appeared in Korea in 1950 and was widely used outside the USSR by at least 18 other air forces.

The MiG-15's first flight took place on December 30, 1947. The design owed much to Soviet and German research and the British Rolls-Royce Nene centrifugal jet engine, although the airframe was originally designed for a smaller diameter but less powerful German-inspired axial-flow engine. The MiG-15 was the first all-swept aircraft to be produced in the USSR. Powered by a 5,950 lb. (2,700 kg) thrust VK-1 Nene development, the MiG-15SD (or MiG-15*bis*) of 1949 had a span of 33 ft. 1 in. (10.08 m), a length of 36 ft. 4 in. (11.06 m), maximum takeoff weight of 12,756 lb. (5,786 kg) and a maximum speed of 668 mph at 39,370 ft. (1,076 km/h at 12,000 m). It was produced in standard and all-weather

versions, with bombs or rockets for ground attack and as a two-seater for training (MiG-15 UTI "Midget").

The MiG-15 was produced under license in Czechoslovakia and Poland. The MiG-17 was the definitive version, incorporating improvements derived from combat experience in Korea, where early MiG-15s failed to match the performance of American fighters.

MiG-21

The MiG-21 "Fishbed" single-seat fighter entered Soviet front line service in 1961 and has been exported to 22 countries. It first flew about 1956 and was the first Soviet production aircraft to have a delta wingform. It was also the first Soviet aircraft since World War II to be license-produced by a non-Communist country: in India, by the Hindustan Aeronautics Company at Nasik, near Bombay.

The MiG-21 has been produced in standard and all-weather versions, and has undergone only fairly conservative modification during its long life, apart from the experimental STOL which appeared in 1967 and the tailless Tu-144 analog. The early MiG-21F (1961) was powered by a 9,480–12,125 lb. (4,300–5,500 kg) thrust Tumanski R–11–300 engine, had a wingspan of 28 ft. $5\frac{1}{2}$ in. (7.15 m), a length of 51 ft. $8\frac{1}{4}$ in. (15.76 m), a takeoff weight of 15,543 lb. (7,050 kg) and a maximum speed of 1,352 mph at 36,090 ft. (2,175 km/h at 11,000 m). The MiG-21MF, with a 11,244–14,500 lb. (5,100–6,600 kg) thrust Tumanski R-13 engine, has a maximum takeoff weight of 21,605 lb. (9,800 kg).

MiG-25

The twin-engined MiG-25 appeared in public for the first time at the 1967 Soviet Aviation Day display. A single-seat all-weather fighter, the "Foxbat A"

is an interceptor version of the MiG-25 armed with four AA-6 "Acrid" missiles. It is believed to have been in service since 1972. A reconnaissance version of the MiG-25, Foxbat B, is known to have been over the Egyptian/Israeli border early in 1971. A two-seat trainer version has also been reported.

The MiG-25 is distinguished by its large rectangular ram air intakes and twin splayed vertical tail surfaces. Estimates of its maximum speed range between Mach 2.8 and 3.2. The span of the almost straight wings is about 45 ft. (13.7 m).

The current world records held by the MiG-25 are height—118,898 ft. (36,240 m; established in 1973); and speed over a 310 mi. (500 km) closed circuit course—1,852.5 mph (2,981.5 km/h; established in 1967).

Western experts were able to examine a MiG-25 ("Foxbat A") interceptor when one landed in Japan in September 1976.

Mil, Mikhail Leontevich (1909–1970)

Soviet helicopter designer, responsible for the giant craft produced in the postwar period. In 1931 Mil joined the rotary wing department at the Central Aero and Hydrodynamic Institute (TsAGI) where he worked on autogiros. In 1945 he became head of its Rotary Wing Scientific Research Laboratory, and in 1947 he established his own design bureau to develop the Mi-1.

The Mi-1 "Hare" became the first really successful Soviet helicopter and was widely produced in the Soviet Union and Poland. A small single-rotor helicopter with an anti-torque tail rotor (a configuration used for all subsequent Mil helicopters except the giant MI-12 "Homer"), it was followed by the Mi-4 "Hound," which bore a striking resemblance to the SIKORSKY S-55 (see HELICOPTER).

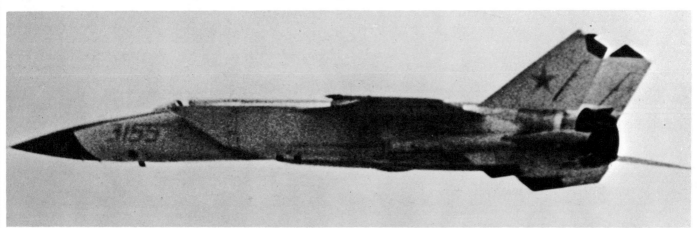

Mig-25

Mil Mi-10 "Harke"

The Mi-6 "Hook" made its first flight in 1957 and was then the largest helicopter in the world, establishing the pattern for the smaller eight-passenger Mi-2 "Hoplite" and 24-passenger Mi-8 "Hip." On all these helicopters the gas turbine engines were mounted above the cabin, which was left free at the center of gravity for payload. The Mi-2 and Mi-8, when they first appeared, made use of many Mi-1 and Mi-4 parts.

The Mi-10 "Harke" was a flying-crane version of the Mi-6. The huge Mi-12, first reported in 1966, is a four-engined twin-opposed rotor helicopter with a maximum takeoff weight of 231,485 lb. (105,000 kg) and maximum payload capacity of 88,636 lb. (40,200 kg). The Mi-24 "Hind" ATTACK HELICOPTER was first reported in 1973.

Military transport

The first airplanes used for the transport of troops and equipment were little more than adaptations of the civilian aircraft—or even the bombers—of their day. In the early 1920s the U.S. Army Air Service had already bought small numbers of Martin T-1 and Fokker T-2 aircraft, and one of their uses was in experiments with in-flight refueling. In the interwar years the RAF used bomber-transports for policing lawless areas in the Middle East, employing the VICKERS Vernon, Victoria, and Valentia—all biplanes powered by two Lion or Pegasus engines and able to carry up to 22 fully-equipped troops. They could use poor local airstrips, but had cruising speeds of under 90 mph (145 km/h).

These lumbering biplanes represented an old and clearly obsolete generation of transports when compared to the German JUNKERS Ju 52/3m of 1932, which remained a valuable transport throughout World War II. Although only able to carry 18 troops, the Ju 52/3m was sturdy and reliable and could fly from extremely short fields; it was particularly useful as a paratroop carrier. On the Allied side, the C-47, the military version of the Douglas DC-3, could carry up to 28 troops at speeds exceeding 180 mph (290 km/h); it was the workhorse of U.S. and British forces, and was even license-built in the Soviet Union. Over 10,000 military DC-3s went into U.S. service alone. For longer missions the Allies used the Douglas C-54 Skymaster, based on the design of the DC-4 airliner, the Curtiss C-46 COMMANDO, and variants of the Consolidated B-24 LIBERATOR such as the C-87, Liberator Express, and RY-2.

During 1940 a number of Germany's Blitzkrieg victories were gained by assault troops brought in by small gliders. Larger troop-carrying gliders soon figured importantly in the forces of several nations, but after the war interest rapidly waned. Two wartime military transports were, however, of considerable significance for the future. Both were German. The Me 323 Gigant was an enormous cargo carrier, converted from glider design and powered by six Gnôme-Rhône radial engines, with a laden weight of 99,000 lb. (45,000 kg; four times that of a C-47 or Ju 52). Huge double doors in the nose admitted large vehicles or an 88-mm gun, and its ten wheels and low-pressure tires allowed operation from soft airstrips. The Arado 232, though smaller, was in principle even more advanced. Its tail was carried on a slim boom, and at the rear of the freight hold were large doors that opened to the full section of the fuselage. The undercarriage was arranged to "kneel" the fuselage onto ten small wheels so that the floor was level and near the ground.

Beginning soon after the war Douglas built the large C-74 and double-deck C-124 Globemasters, while Boeing was

C-5A Galaxies (foreground) and C-141 StarLifters of the U.S. Air Force

C-130 Hercules

C-124 Globemaster

responsible for the C-97, used as a cargo aircraft and tanker. Marking a major step forward in the design of large transports was the Lockheed turboprop Hercules, which first flew in August 1954 and offered a high wing with integral tanks and notable efficiency, a large unobstructed fuselage pressurized throughout, full-section rear doors able to open for parachute drops, soft-field landing gear, and remarkable maneuverability. Lockheed also produced the four-turbofan C-141 StarLifter and the C-5A GALAXY, the largest aircraft now flying, while the ANTONOV An-12 and An-22 are the chief Soviet strategic transports. In recent years an increasing amount of military transport work, particularly in battlefield situations, has been done by large helicopters, such as the Soviet Mi-6 and U.S. Army Boeing-Vertol CH-47 CHINOOK.

Mirage IIIA, pre-production model of the famous Dassault fighter

Mirage

Series of highly successful French Mach-2 fighters which have been supplied to air forces all over the world and have proved both a principal export and a controversial political issue. Conceived as a lightweight delta-winged aircraft with a high rate of climb, the DASSAULT Mirage I first flew in June 1955; the Mirage III-C (the first of the series to enter service) made its maiden flight in October 1960. Models of this remarkably versatile fighter included the basic III-C Mach-2.2 interceptor (with twin 30-mm cannon), the ground-attack III-E, the photo-reconnaissance III-R, the Australian-built III-O, and the Swiss-produced III-S. All these were powered by the Atar-9C after-burning turbojet of 13,670 lb. (6,200 kg) thrust, giving a maximum speed of 1,460 mph (2,350 km/h). There was also a trainer variant, the III-B, with an afterburning Atar-9B of 13,225 lb. (6,000 kg) thrust. Length of the Mirage III was 49 ft. 3½ in. (15.03 m), and wingspan 27 ft. (8.22 m). It served with air forces that included those of France, South Africa, Pakistan, and Israel.

The Mirage IV twin-engined two-seat supersonic strategic bomber (1959) was a scaled-up Mirage III, designed to carry a nuclear weapon as part of France's *Force de Frappe*. The Mirage 5 (1967) was a simplified ground-attack III-E export version with an Atar-9C.

An orthodox nondelta configuration was adopted for the Mirage F-1 (1966), which was intended to replace the Mirage III. Powered by a single Atar-9K-50 engine of 15,873 lb. (7,200 kg) thrust, it entered service with the French air force in 1973 and was also built in South Africa. The multimission F1E (with an M53 engine of 18,646 lb.,

8,500 kg, thrust) first flew in 1974 and can carry over three tons of bombs in addition to its interceptor armament of twin cannon and air-to-air missiles. It is 49 ft. 2½ in. (15.00 m) in length and has a wingspan of 27 ft. 6¾ in. (8.40 m), and top speed of 1,450 mph (2,335 km/h).

The prototype of the variable-geometry Mirage G first flew in November 1967, and remained until early 1971 the only swing-wing fighter outside the United States and the Soviet Union. It has a length of 55 ft. 1 in. (16.79 m), a wingspan of 42 ft. 8 in. (13.00 m) with wings forward (24 ft. 3 in., 7.39 m, with the wings swept), and a top speed of 1,650 mph (2,655 km/h).

Missiles

Rocket (or occasionally jet) propelled weapons that fly through the air and are guided to their target. Ballistic missiles are normally launched on their tails and have a trajectory governed by aerodynamic fins or gimballed rocket nozzles during the initial powered part of their flight. After burnout or motor shut-down such missiles continue in free flight on a ballistic trajectory. Non-ballistic guided missiles are guided to, or home onto, their targets using a wide variety of systems, ranging from trailing wires to onboard radar or television equipment which locks onto the target.

The earliest missiles were used by the Germans in World War II. The main operational types were the so-called V-1 and V-2, but other Luftwaffe or army missiles included the Henschel Hs 293 radio-controlled air-to-surface missile, the Ruhrstahl/Kramer Fritz X (FX 100) radio-controlled glide bomb (which sank the battleship *Roma* and damaged her sister ship *Italia*), and

some 15 or so types of air-to-surface or surface-to-air missiles that did not reach the operational stage.

After 1945 missile research became a top priority in all major nations. The Soviet Union put most effort into large ballistic rockets, and has now deployed at least 19 types of ICBM (intercontinental ballistic missile) with nuclear or thermonuclear warheads, all recent types being either protected in deep silos in the ground or else moved on large transporters which never stay long in the same place. Of the nine types currently operational, SS.18 is the most formidable: with a length of 121 ft. (36.8 m) and diameter of 11 ft. (3.35 m), it can throw a payload of about 16,000 lb. (7,257 kg) (either a 50 megaton warhead, or eight independently targeted warheads) over a range of 10,500 mi. (16,900 km). In addition, the Soviet Strategic Rocket Troops have 990 SS.11 missiles (only slightly smaller) and 228 of the very large SS.9 type, as well as many thousands of smaller ballistic missiles.

In the United States the first ICBM was Atlas. It was developed by CONVAIR in 1956–60, and was unique in having an airframe made like a balloon from thin stainless steel, kept inflated under pressure and filled with liquid oxygen and kerosene. These 6,525-mi. (10,500-km) missiles were withdrawn in the mid-1960s, as were the 1,700-mi. (2,735-km) Thor and Jupiter, but 54 Titan II missiles still exist, together with 1,000 of the small silo-based Minuteman missiles. The only other long-range missiles are the SSBS of the French army, of which two squadrons are based in silos in southwest France, and others being deployed in China.

A further important series of strategic

missiles are used aboard submarines. The American Polaris is still used by the Royal Navy (four submarines, each with 16 missiles) but has been superseded in the U.S. Navy by Poseidon, which has much greater warhead power and enhanced precision over greater ranges.

The even bigger Trident missile is under development, and France is developing a series of improved versions of MSBS to arm six big missile submarines. The Soviet Union has about 750 modern submarine-launched missiles—both ballistic types and 264 N-3 winged cruise-type missiles. The N-8 submarine ballistic missile can fly missions of 4,900 mi. (7,885 km), and large numbers will be deployed aboard the enlarged Super Delta type of submarines.

There are over 120 kinds of tactical or long-range battlefield missile, ranging from the large U.S. Army Pershing II and the Soviet SS-12 ballistic missiles down to antitank missiles, which are guided by the operator through trailing wires, or by laser or radar homing, radar beam-riding, infrared homing, optical line-of-sight command, etc. Marine missiles usually finish their flight by skimming just above the tops of the waves at about the speed of sound and have some form of guidance to home onto their target. Modern ship-destroying missiles can be carried by small patrol boats and fired from below the horizon.

Almost all fighter aircraft carry air-to-air missiles, ranging from close-in dogfight missiles (Sidewinder, Magic, and the Soviet AA-8) through medium-range models (Sparrow, Skyflash, Anab, and R.530), up to the big Phoenix and AA-6 "Acrid," which are 20 ft. (6 m) or more in length and can kill aircraft at ranges up to 130 mi. (209 km). Similarly, most combat aircraft can carry missiles for use against surface targets: approximately 100 have been developed, using many kinds of guidance, including Martel (TV or radar-homing), Maverick (TV), Rb 08 (TV), Condor (electro-optical), Harpoon (sequence of guidances ending with radar lock-on), Shrike (radar-homing), Kormoran (radar-homing), and several large Soviet missiles.

A SRAM (Short-Range Attack Missile) has a range of 40 to 100 mi. (64 to 160 km), depending on attack procedure, and can deliver a thermonuclear warhead, yet it is so small that it can be fired from a large drum (like bullets from a revolver). Future cruise-type winged missiles will include the AGM-86, which fits the SRAM launcher but can fly 1,370 mi. (2,205 km), and the BGM-109 Tomahawk, which is only fractionally larger and can deliver a 1,000-lb. (453-kg) thermonuclear warhead over 2,240 mi. (3,605 km) at Mach 0.7.

The most numerous category of all is the surface-to-air missile (SAM) group, of which 12 designs were being developed in Germany by 1945, and some 200 models have been studied or deployed since. The first to enter extensive use was the U.S. Army Nike-Ajax of 1955, followed by the larger Nike-Hercules and Nike-Zeus, which in turn led to the Safeguard ABM (antiballistic missile) system. HAWK (Homing All the Way Killer) was built in considerable numbers as the first tactical army SAM, and the Soviet Union has ten major SAM types, of which most are built into mobile vehicles with good cross-country performance. Typical of modern systems are the Franco-German Roland, fired from tank chassis and with various kinds of guidance, and the British Rapier, which can be used in a simple daylight system towed by a Land-Rover or a complex all-weather instant-readiness system in a tracked vehicle.

Mitchell

The twin-engined North American B-25 Mitchell, named in honor of General "Billy" Mitchell, was one of the best medium bombers of the war. It was a parallel project with the Martin B-26 MARAUDER and proved to be a highly versatile and dependable aircraft used by a number of nations. Powered by Wright R-2600 radial engines, late versions of the Mitchell could reach speeds of 272 mph (438 km/h) and carry 4,000 lb. (1,814 kg) of bombs. Wingspan was 67 ft. 7 in. (20.60 m) and length 52 ft. 11 in. (16.73 m). Some versions were produced with both bombardier and gun noses, the latter usually with eight .50 machine guns. The B-25G and B-25H carried a 75-mm cannon, the largest caliber gun used operationally by American warplanes for ground-attack purposes. The Mitchell was the first U.S. bomber lend-leased to the Soviet Union; other aircraft went to Britain, Brazil, and China. A total of 9,815 were produced.

Mitchell, Reginald Joseph (1895–1937)

British aircraft designer who designed the SPITFIRE. Mitchell joined SUPERMARINE in 1916; by 1919 he was the firm's chief designer, and in 1920 its chief engineer. During the 1920s he designed a series of civil and military flying boats, including a number of British entrants for the Schneider Trophy races which eventually led to the S4, S5, and S6 series of racing floatplanes. On the basis of his experience with the highly successful S6B, he persuaded Supermarine to let him design as a private venture a fast monoplane fighter for the RAF. The result was the Spitfire, which first flew on March 5, 1936. Only a year later, on June 11, 1937, Mitchell died of cancer, but he left his country a superb fighter that became a legend in aviation

B-25J Mitchell

and played a major part in Britain's survival during World War II.

Mitchell, William (1879–1936)

American protagonist of an independent air force and pioneer military air strategist. Born in Nice, France, the son of a U.S. senator, William E. Mitchell enlisted in the army as a private at the turn of the century.

He served in Cuba, the Philippines, and Alaska, and as a General Staff Officer studied strategy; he was asked to prepare a report on Army aviation in 1915, and although as a colonel of 36 he was too old and senior to qualify for pilot training, he learned to fly at his own expense. In 1918 he was in France, in charge of the U.S. Army Aviation Service forces under General Pershing, where his courage, dash, and impatience with administrative formality endeared him to all save his seniors. After the Armistice, Mitchell was appointed, as a general, second-in-command of the Air Service and almost singlehanded sustained the young service against political attack and financial denial. In 1921 he demonstrated the effectiveness of aircraft against large warships by sinking the former German battleship *Ostfriesland* with bombs dropped by Martin MB-2s and in 1923 gave a further demonstration by sinking two old U.S. battleships (*Virginia*, *New Jersey*). Despite Mitchell's popularity with press and public the Navy secured the backing of Congress, and gradually turned Mitchell into an impatient and angry man whose outspoken comments finally earned him a courtmartial which suspended him for five years. He resigned, to lecture, write, and campaign for air power, finally dying of a heart attack in 1936.

In 1946 he was posthumously awarded the Congressional Medal of Honor.

Mitsubishi

Mitsubishi Jukogyo K.K. (Mitsubishi Heavy Industries) entered the field of aircraft production in the years immediately following World War I. Mitsubishi had sent Dr. Kumezo Ito to Europe to study aerial warfare in 1918. Two years later an aviation subsidiary was created, moving in 1921 to Oye-cho near Nagoya, where Mitsubishi's principal aircraft factory still operates. The manufacture of foreign designs under license began at once; among the overseas companies involved were Blackburn, Curtiss, and Hanriot. By the early 1930s Mitsubishi had gained sufficient design strength to be independent of external advice or licensing. More than 2,000 aircraft were delivered to the Army and Navy, but the Western world's first sight of a Mitsubishi product only occurred when the mailplane *Divine Wind* arrived in London in April 1937 after a $94\frac{1}{2}$-hour flight from Tokyo. From this single-engined machine, with glasshouse canopy and spatted wheels, stemmed the Ki 5 reconnaissance aircraft and the Ki 30 bomber, used widely in operations in China. The standard Army bomber of World War I, the Ki 21, nicknamed SALLY by the Allies, was produced up until 1944; Mitsubishi made 1,713 and NAKAJIMA 351. Another twin-engined aircraft was the Ki 46 ("Dinah"), a high-speed long-range reconnaissance machine, capable of 375 mph (604 km/h). Considered one of the finest Japanese aircraft of the war, many Ki 46s also saw service as night fighters and trainers. From the Ki 30 was developed the single-engined Ki 51 ("Sonia") attack/reconnaissance aircraft of which almost 2,400 were produced between 1939 and 1945. Mitsubishi also built 507 Ki 57/MC.20 transports (which were similar to the DC-2), 698 Ki 67 ("Peggy") high-speed heavy bombers (combat ready only in the summer of 1944), and 22 Ki 109 fighters with a single 75-mm gun. Mitsubishi also produced a wide range of naval aircraft. These included the K3M high-wing crew trainer, of which 624 were delivered between 1930 and 1941, and the F1M biplane seaplane, capable of dive-bombing. Between 1936 and 1942 the Navy's main fighter was the highly maneuverable A5M ("Claude") series of low-wing open-cockpit machines.

Mitsubishi was also responsible for the G3M ("Nell"), the Navy's main heavy bomber from 1935 to 1943, and its successor, the G4M BETTY. Probably best known of all Mitsubishi's wartime products was the ZERO fighter, built in far greater numbers than any other Japanese warplane; over 10,400 between 1939 and 1945. Other Mitsubishi wartime products included the J2M RAIDEN Interceptor, the A7M Reppu (intended as a replacement for the Zero but not developed in time to see action), a total of 515 dual trainer versions of the Zero, and (in 1945) seven J8M Shusui ("Swinging Sword") rocket-powered interceptors based on the Me 163 KOMET.

Allowed to reenter the aviation field in 1952, Mitsubishi began by overhauling American fighters, and then went on to build 300 F-86F Sabres under license, followed by 230 F-104 Starfighters and 120 F-4EJ Phantoms. Other license-production has included the Sikorsky S-55, S-58, and S-61 helicopters. Mitsubishi's first postwar design of its own, the twin-turboprop high-wing Mu-2, a seven-seat multirole utility aircraft, first flew in 1963; nearly 500 were sold to military and civil customers. Eight years later the XT-2 supersonic jet, powered by two Adour afterburning turbofans, made its maiden flight, and Mitsubishi has since produced two developments in quantity for the Japan Air Self-Defense Force, the T-2 dual trainer and FS-T2-KAI single-seat close-support fighter, both of which carry a 20-mm Vulcan multibarrel cannon and can be armed with a wide range of bombs, rockets, and missiles.

Monocoque

Literally meaning single-shelled, a monocoque structure is exemplified by a lobster claw or a tin can. Most early aircraft were made by fixing many parts together and gaining strength from a network of bracing wires throughout

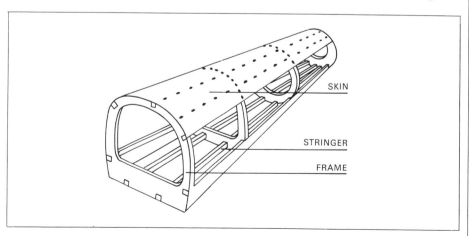

A monocoque fuselage structure consists of a stressed skin stiffened by stringers that run lengthwise through the fuselage and are spot-welded or riveted to the skin. This strong shell is supported at intervals by transverse frames or rings

the interior of the fuselage and surrounding the wings and tail. By 1909 several engineers, especially Ruchonnet and Louis Bechereau in Switzerland, had tried to approach the superior monocoque form, in which there is nothing but the outer skin. The technique developed swiftly from 1910 to 1912, and was first seen applied in the DEPERDUSSIN racers. After 1920 the development of metal stressed-skin construction extended the technique to the wings and tail, though some interior spars and ribs were needed. Even in modern thin-winged supersonic aircraft a high proportion of the stresses are carried in the strong surrounding skin, which is usually still fastened to a skeleton underneath.

Monoplane

An airplane with only a single pair of wings. Almost all modern aircraft are of this configuration; the wing loading is higher than that of a biplane but drag is substantially reduced and performance consequently much superior.

In 1912–13 there was a series of accidents to monoplanes in England and Europe caused by structural failures. As a result the British government banned their use by the Royal Flying Corps. Although it was subsequently determined that there was nothing inherently dangerous about monoplanes, the prejudice against them persisted for some years. The RAF's first monoplane fighter, the HURRICANE, did not enter service until 1937, and classic biplanes such as the FAIREY Swordfish achieved notable successes in World War II. However by 1930 the great majority of new aircraft, both civil and military, were monoplanes.

There are three types of monoplane. Most large and medium sized commercial transports are "low winged" (the wing is mounted on the bottom of the fuselage). Most warplanes, and many light single-engined airplanes are "high winged" (with the wing on top of the fuselage). The third type, the mid-wing monoplane, is less often used.

Montgolfier Brothers

French inventors of the hot-air balloon, and builders of the first man-carrying balloon. Joseph Michel (1740–1810) and Jacques Étienne (1745–99) were mill owners who conducted scientific experiments in their spare time. In 1782 they began experimenting with paper bags filled with hot air, which they sent aloft. On June 5, 1783, they gave a public demonstration of a

The Montgolfière, the first man-carrying balloon

hot-air balloon, 30 ft. (9.15 m) in diameter. Made of linen lined with paper, and inflated with hot air from a furnace, it reached a height of 6,000 ft. (1,830 m). On September 19, 1783, at Versailles, the Montgolfiers demonstrated a larger balloon which carried a sheep, a duck, and a rooster in a basket suspended underneath, the first living creatures to make a balloon ascent. The flight lasted about eight minutes. While Joseph worked (inconclusively) on the problems of control and propulsion, Étienne built a much larger balloon which carried its own fire and could therefore stay in the air for longer periods. It lifted the first

human to leave the Earth (the physicist de Rozier) on October 15, 1783, in a tethered flight, and achieved the first free flight on November 21 (see BALLOONS). Étienne never made an ascent, and Joseph made only one (which nearly ended in disaster), but the brothers were widely acclaimed and honored for their work, which paved the way for the development of hydrogen balloons.

Morane-Saulnier

The French aircraft manufacturer Morane-Saulnier was renowned, from its foundation in 1911, as a producer of high-performance monoplanes. One of the very first Moranes won the Paris–Madrid race of 1911 in the hands of Jules Védrines in 14 hours 6 minutes, and Gustav Hamel chose a similar high-wing Morane for racing and aerobatic displays, in which he daringly kept cropping the wings for increased speed. In 1913, Roland GARROS, Morane's demonstration pilot, made the first crossing of the Mediterranean, 453 mi. (770 km)—between France and Tunisia, taking 7 hours 53 minutes in a 60-hp Gnôme Morane monoplane.

At the start of World War I large numbers of Moranes entered service, some being biplanes (Type G), although most were mid-wing, high-wing or parasol monoplanes. Garros himself flew a parasol Type L with bullet deflectors on the propellers allowing a machine gun to be fitted for the first time between the spinning blades, and quickly ran up several victories before being forced to land behind the German lines (see INTERRUPTER GEAR). Another Type L was used by R. A. J. Warneford of the British Royal Naval Air Service when he managed to drop six 20-lb (9-kg) bombs on the Zeppelin LZ-37 near Ghent, the first Zeppelin ever to be brought down by enemy action (June 1915). At the outbreak of World War II the MS. 406 was the only modern fighter serving in large numbers in the French air force. A low-wing monoplane made of a metal/wood sandwich, the 406 was underpowered with an 860-hp Hispano-Suiza engine, and poorly armed with only a single 20-mm cannon and two machine guns, but no fewer than 1,081 were delivered by the French collapse in June 1940. A few MS. 4060s saw action in Indochina, and several hundred others were ordered for export, but the only deliveries were to Switzerland (where the type initiated the Swiss-produced D-3800 series produced throughout World War II), Turkey, and Finland. After 1945 only three important Morane production models appeared: the MS. 733 trainer/attack aircraft, the Paris MS. 760 executive jet, and the Rallye lightplane (which survived Morane's bankruptcy in 1963 and is still produced by a subsidiary of AÉROSPATIALE).

Mosquito

One of the most versatile warplanes used by the RAF in World War II, the de Havilland DH 98 Mosquito was an immensely fast and versatile aircraft, initially a light bomber, built of wood. The prototype flew for the first time on November 25, 1940, and 105 Squadron received the first Mosquitoes in May

De Havilland Mosquito

1942. A large number of variants were soon produced, which included night-fighters (with four 20-mm cannon and four machine guns), photo-reconnaissance aircraft, fighter-bombers, ship-destroyers (the Mark XVIII, with a 57-mm gun), pathfinders and escort fighters. Capable of carrying a 4,000-lb. (1,800-kg) bomb, and powered by two Merlin engines giving a top speed (in late models) of 408 mph (656 km/h), Mosquito bombers ranged across Europe virtually unchallenged. Wing-span of most variants was 54 ft. 2 in. (16.51 m) and length 40 ft. 6 in. (12.43 m). Total production was 7,781 machines.

Moth

A very popular sport biplane that helped to popularize private flying in Britain and gave DE HAVILLAND its initial success as a company. The original DH 60 Moth of 1925 was powered by a specially designed 60-hp four-cylinder Cirrus engine giving a maximum speed of 91 mph (146 km/h);

later Moths were equipped with 75-hp Genet, 85-hp Cirrus II, 90-hp Cirrus III, or 105-hp Hermes I engines and had speeds around the 100-mph (160 km/h) mark. The later DH 60G Gipsy Moth (1928) employed a Gipsy engine of 100 or 120 hp and was used for many outstanding flights, including those of Amy JOHNSON and Francis Chichester; total British production was 595 aircraft.

Still later Moths included the DH 60GIII Moth Major of 1931 (Gipsy III or Gipsy Major engine; 112 mph, 180 km/h), the DH 60M of 1928 with a metal fuselage and its military trainer DH 60T derivative, the four-seat DH 75 Hawk Moth of 1928 (240-hp Lynx), and the DH 80A Puss Moth high-wing monoplane of 1930 (128 mph, 206 km/h, from a 130-hp Gipsy Major).

The original Tiger Moth was the DH 71 racing monoplane (1927), of which only two were built (one took the lightplane speed record at 186 mph, 299 km/h), but the DH 82A Tiger Moth two-seat biplane (1931), with a

DH 60 Moth

Gipsy Major engine, became the RAF's standard World War II primary trainer. Production eventually amounted to some 7,290 aircraft, including license-production abroad. The DH 83 Fox Moth of 1932, with a four-seat cabin, used as many Tiger Moth components as possible.

To replace the Puss Moth, the three-seat DH 85 Leopard Moth monoplane began production in 1933 and won that year's King's Cup air race; the DH 87 Hornet Moth (1934) was intended as a replacement for the Gipsy Moth. The DH 94 Moth Minor was a two-seat low-wing monoplane flown in 1937, planned as a successor to the original DH 60 Moth. British production was terminated by the outbreak of World War II, although manufacture was continued briefly in Australia.

Mozhaiski, Alexander Fedorovich (1825–1890)

Designer of what Soviet historians claim to have been the first airplane to fly, Mozhaiski built a monoplane with a wingspan of 74 ft. 10 in. (22.8 m) and chord of 46 ft. 7 in. (14.20 m). It was powered by two British-built light-weight steam engines. Mozhaiski's monoplane was completed about 1882, and is believed to have left the ground for the first and only time in the fall of 1884 when it was launched after running down a ramp. It flew some 80 ft. (25 m) before heeling over and crashing.

Although Mozhaiski had carried out numerous experiments into drag and angles of attack, his aircraft can hardly be said to have achieved sustained powered flight—the propulsive force was that of the down-ramp run. Furthermore, the Frenchman Félix du Temple (1823–90) had already carried out similar experiments with a steam-powered airplane running down a ramp in 1874.

Mustang

One of the outstanding fighters of World War II, the NORTH AMERICAN P-51 Mustang owed its origin to the presence in America during April 1940 of a British Air Purchasing Commission. The president of North American Aviation offered to produce a new fighter for the RAF powered by the Allison V-1710-39 engine and agreed to a time limit of 120 days for completion of the prototype.

The result was the NA-73X, which made its first flight on October 26, 1940. The British government ordered 320 aircraft, the first of which arrived in England in November 1941. The new fighter was heavily armed (four .50 and four .30 machine guns) and possessed good low-level performance; its laminar-flow wing was a significantly advanced feature. Unfortunately, the Allison engine developed insufficient power at height for the Mustang I to find any application as a fighter, and the RAF fitted cameras in the aircraft and used them for military liaison work.

The RAF's Mustang IA, armed with four 20-mm cannon, received the American designation P-51; the A-36 was a dive-bomber variant used in Italy by American units during 1943. The last Allison-powered Mustang was the P-51A (RAF Mustang II), which began reaching service units in March 1943.

Four Mustang Is were experimentally fitted with Merlin engines by Rolls-Royce in England during the fall of 1942, and this supercharged power unit radically improved the aircraft's performance at altitude. A year later the first Merlin-engined Mustangs were delivered to U.S. combat units; the P-51B was built at Inglewood and the identical P-51C at Dallas. With a maximum speed of 440 mph (708 km/h) at 25,000 ft. (7,620 m), and a range (with drop tanks) that enabled it to escort British-based U.S. bombers the full distance to Berlin and back, the Mustang revolutionized the air war over Germany.

Final version of the Mustang, the P-51H

The P-51D, with a teardrop canopy, was issued to squadrons in both Europe and the Pacific from mid-1944 onward. Nearly 8,000 P-51Ds were built, and by the end of the war all but one of the U.S. 8th Air Force's fighter groups had been equipped with Mustangs. The P-51D had a wingspan of 37 ft. (11.28 m) and length of 32 ft. 3 in. (9.83 m). A final version, the P-51H, was capable of 487 mph (780 km/h), but saw service only against Japan and, later, in Korea.

The Mustang served with U.S. units into the 1950s, supplemented by the P-82 long-range fighter (which consisted of two P-51 fuselages joined together). Many of the world's smaller air forces still retained P-51s in the 1970s, and in addition to the 14,000 built during the war by North American, further Mustangs have subsequently been assembled by Cavalier Aircraft Corporation, some as two-seaters (for civil use or—with dual controls—for training duties).

Super-Mystère with its full complement of weapons

Mystère

The first French Mach-1 jet fighter was the DASSAULT Mystère. Dassault's first jet fighter had been the Nene-engined Ouragan (1949), which entered service early in the 1950s. By using a slimmer wing section and a greater degree of sweep, Dassault produced the three Mystère I prototypes (1951). In 1952 a Tay-powered Mystère II became the first French aircraft to break the sound barrier, and Atar-engined versions of this machine were soon equipping French service units. The Mystère II was 38 ft. 6 in. (11.73 m) in length, with a wingspan of 38 ft. 1 in. (11.60 m). Its successor, the Mystère IV (1952), was flown by the French, Israeli, and Indian air forces.

The Mystère series culminated in the Super-Mystère of 1955, powered by an afterburning Atar 101G engine of 9,920 lb. (4,500 kg) thrust. It was the first French service aircraft capable of Mach 1 in level flight and served with both the French and the Israelis until the early 1970s. Maximum speed was 743 mph (1,195 km/h) above 36,000 ft. (11,000 m), and its service ceiling was 55,775 ft. (17,000 m). It was 46 ft. $1\frac{1}{4}$ in. (14.05 m) long, with a wingspan of 34 ft. $5\frac{3}{4}$ in. (10.51 m).

N

Nakajima

During World War II Nakajima Hikoki KK (Nakajima Aircraft Co.), established in 1917, built more aircraft than any other Japanese manufacturer. In the early 1930s it had produced the first important all-Japanese fighter, the Army Type 91, a parasol monoplane with a 500-hp Nakajima Kotobuki engine (a license-built Bristol Jupiter). The highly maneuverable Ki.27 low-wing fighter which followed had a 650-hp Nakajima Ha.la engine, and could reach 286 mph (460 km/h). During the so-called Nomonhan Incident Ki.27s claimed nearly all the 1,252 reported victories over Soviet aircraft against the loss of only 100 of their own number. The Ki.43 HAYABUSA, manufactured throughout World War II, succeeded the Ki.27 as the standard army fighter, with nearly 6,000 built. The Ki.44 Shoki or "Demon" was a tricky, small-winged fighter armed with various combinations of machine guns and 20-mm or 30-mm cannon that enjoyed a production run of 1,225, while the versatile Ki.84 HAYATE was one of the best Japanese fighters of the war. Other Nakajima army aircraft included the Ki.49 Donryu or "Storm Dragon" (code-named "Helen" by the allies). The Ki.49 was a heavy bomber and electronic-warfare aircraft that saw action between 1942 and 1944. Nakajima also produced the Ki.115 Tsurugi or "Sabre" KAMIKAZE plane.

Nakajima's aircraft for the Japanese navy began with license-production of the GLOSTER Gambet. After this came the A2N and A4N biplanes, followed by the B5N ("Kate") three-seat low-wing monoplane. The B5N was Japan's standard carrier-based torpedo and bombing aircraft between 1936 and 1942; deliveries totaled 1,149. Although outwardly conventional, it was in fact very advanced for its time, with stressed-skin structure, and hydraulically folding wings. The "Kate" made its mark at Pearl Harbor and went on to sink the *Lexington*, *Yorktown*, and *Hornet*.

Nakajima was also responsible for the E8N reconnaissance floatplane (755 were built between 1934 and 1940), and for the A6M2-N seaplane version of Mitsubishi's ZERO fighter (327 built between 1941 and 1943). Another major Nakajima product was the J1N Gekko ("Moonlight") of 1941, a twin-engined machine of high performance and exceptional range. Designed as a fighter, it began life as a shore-based reconnaissance aircraft, but later served as an escort or night fighter, and in one form (J1N1-C-KAI) was the first night fighter with oblique guns, two cannon firing up and two firing down. The successor to the B5N "Kate" was the B6N Tenzan ("Jill"), a low-wing machine with a 1,800-hp radial engine, a long glasshouse cockpit canopy, and a top speed of nearly 300 mph (480 km/h). The B6N inflicted heavy losses on Allied ships in conventional and Kamikaze attacks around Okinawa in 1945. A somewhat similar but even faster machine was the C6N Saiun or "Painted Cloud," which could reach 379 mph (610 km/h) with a 2,000-hp Homare engine. Code-named "Myrt," the C6N was designed as a carrier-based reconnaissance aircraft, but was later pressed into use as a night fighter with two oblique cannon.

Nakajima also produced a number of technically advanced aircraft that never saw service. Among them was the G5N Shinzan ("Mountain Recess") bomber, evolved from the prototype Douglas DC-4 (sold to Japan in 1939). It had a wingspan of 138 ft. (42 m) and a crew of ten. The later G8N Renzan ("Mountain Range") heavy bomber was capable of 368 mph (592 km/h) on four 2,000-hp Homare engines, but only four were completed by VJ-day. Japan's only jet aircraft was also a Nakajima product. This was the Kikka ("Orange Blossom"), which resembled a lower-powered ME 262, and had Ne-20 axial engines of 1,047 lb. (475 kg) thrust. Two were built and one flown by the end of the war.

National Airlines

National Airlines began operation on October 15, 1935, when it opened a St. Petersburg–Daytona Beach service with three small Ryan monoplanes. Lockheed Model 10 Electras were acquired and service extended to Miami in 1937. National was the only major U.S. airline never to operate the twin-engined Douglas DC-2 and DC-3. A major event in the airline's history was the award of a Jacksonville–New York route in February 1964, thus breaking Eastern Air Lines' long-held monopoly on New York–Florida services. A southern transcontinental route was awarded to National in March 1961 and in 1970 the airline began scheduled flights between Miami and London.

The present route system consists of a network along the eastern seaboard linking Florida with Washington, Philadelphia, New York, and Boston; and a Gulf coast route to New Orleans and Houston with westward extension to Las Vegas, Los Angeles, San Diego and San Francisco.

The present fleet consists of 38 Boeing 727s, 15 DC-10s, and 2 Boeing 747s. Some 4,300,000 passengers are carried annually.

Navigation

The science of finding the way. All early aerial navigation was by direct "pilotage," the aviator simply map-reading his way with reference to features on the ground. The only navigational instrument was a magnetic COMPASS, standard on most aircraft from 1910. By 1914 pilots had begun to complain that their compasses were erratic in clouds, and eventually the cause was traced to a combination of factors, one of which was that the aircraft was changing direction un-

known to the pilot. Another was an inherent compass error due to motion of the aircraft. In 1917 S. Keith-Lucas in England solved this problem with a spherical bowl compass which contained fluid to damp out the spin of the magnet/needle system. This allowed aviators to use their compass when on northerly courses (where previous readings had been shown to be particularly confusing), even in cloud.

Still, the only way pilots could navigate was to work out basic "triangle of velocities" calculations from knowledge of airspeed and compass course and a guess at the wind velocity. In a few cases they carried a sextant and could navigate at night by means of stars, as in a ship at sea. When ALCOCK AND BROWN flew the North Atlantic in 1919 they had no means of navigating except by setting an approximate compass course; they expected to make landfall somewhere between Scotland and the Bay of Biscay.

While the new compass improved pilot navigation, the airborne D/F (direction finding) loop laid the foundation for radio "navaids" (navigational aids). A British invention, devised by Dr. James Robinson, this at first comprised a large coil of wire taped to the upper and lower wings and interplane struts. The pilot had to turn his aircraft until the received signal in this coil was at minimum in order to find the bearing of the ground transmitter (whose signal identified its location). This allowed a pilot, in theory, to fix his position by obtaining two such bearings in quick succession. By 1932 the D/F loop had been reduced to a coil only 12 in. (30.48 cm) in diameter, capable of being rotated accurately without having to turn the aircraft, and by 1938 it had shrunk further, with increase in radio frequency, so that it could be placed in a small streamlined fairing above the fuselage. Continued development led to today's sophisticated radio compass, which gives instant bearing on any of a large number of ground stations and needs only a small aerial that does not project outside the aircraft. An incidental effect of radio direction finding was that interference from engine magnetoes led to the gradual perfection of "screened" ignition systems, which suppressed radio interference.

Another development of the mid-1920s was the ground radio beacon. In its first form this sent out a radio beam which rotated in azimuth once a minute, with a characteristic signal toward due north. Thus an airborne navigator could listen in his headphones and with a stopwatch time the interval

between the null point (absence of signal) and the north signal and determine the bearing from his aircraft of the ground station. In the United States the Radio Range had come into use. The range station consisted of two transmitters pumping out A–A–A... in morse code, and two others transmitting N–N–N... The A ($\cdot$—) and N (—$\cdot$) signals merged to form a clear steady note along exact pathways through the sky. On either side of this invisible pathway the pilot or navigator heard an A sequence or N sequence. In 1930 there were 50 of these LF (low-frequency) range stations in the United States, increasing to 250 by 1940. However, Radio Range was less useful than the ground radio beacon since it could be used only along narrow tracks in the sky.

1941 a radically new navaid was developed for the RAF under the name of Gee. It consists of a Master ground transmitter station and two Slave stations. The Master continuously emits a series of coded pulses, controlled by an electronic clock of high accuracy; one Slave repeats part of this signal, and the other repeats the remainder, filling the ether with successions of circular broadcast signals, each like the ripples from a point on a pond. The points where the "ripples" intersect lie on curves called hyperbolas, and so Gee was the first of the "hyperbolic" navaids. The navigator in the distant aircraft watches a cathode-ray tube and, by noting time-differences between the Master and two Slave signals at his location, can see exactly where he is by referring to a Gee chart (covered with a hyperbolic lattice which remains fixed in space). Gee enabled British bombers to navigate accurately over Germany, and in modified forms is still used.

LORAN (long-range navigation) is an American hyperbolic navaid using LF, and thus able to cover vast areas at some expense in accuracy. The DECCA NAVIGATOR is a quite different British hyperbolic navaid using not pulses but CW (continuous-wave) radio signals. At first confined to Europe, and needing complex maps and dial instruments, the Decca system later evolved into automated navaids able to pinpoint aircraft position anywhere, even down to ground level where other radio aids cannot operate.

In 1958 an international conference recommended the universal employment of a radio range development called VOR-DME (very-high-frequency omni-range/distance-measuring equipment). The VOR station is rather like a radio lighthouse which, figuratively, flashes each time its rotating beam sweeps past due north. The VOR cockpit instrument indicates aircraft bearing from the selected VOR station (or, by dialing a number from 000 to 359, from any chosen airway heading) while the separate DME receiver indicates miles-to-go to the next "waypoint" or destination. There are now many hundreds of VOR/DME stations, but they can only be used by aircraft flying directly from one to the next. Recently RNav (area navigation) equipment has been introduced which can process VOR data and display it to the pilot in such a way that he can fly direct to his destination, without the inconvenience of having to "dog-leg" between VOR beacons.

Apart from navaids relying on devices outside the airplane, there is also doppler, which measures the exact phase-shift (or change in frequency) of four radio beams sent out to four different places on the Earth's surface, enabling a calculation to be made of the ground-speed and drift due to wind, and INS (INERTIAL NAVIGATIONAL SYSTEM).

Nieuport

Among the earliest aircraft produced by the French manufacturer Nieuport were the 28-hp Type IIN (which set a world speed record of 74.42 mph, 119.74 km/h, in 1911) and the two-seat Type IVG (used by the Italian army). Nieuport monoplanes again broke the world speed record (82.73 mph, 113.11 km/h) later in 1911 and took the altitude record (20,079 ft., 6,120 m) in 1913.

One of Nieuport's military two-seaters had been used by the armed services of France, Italy, and Russia immediately before World War I, and under the direction of Gustave Delage, designer and successor to the Nieuport Brothers, the racing seaplane intended for the Gordon Bennett Trophy contest was completed as the Nieuport 10 SESQUIPLANE. It had a machine gun armament and served with the French, British, Italians, Belgians, and Russians in the first years of the war. The smaller, lighter 80-hp Nieuport 11 "Bébé" (1915) replaced it as a scout, and was followed in production by the Nieuport 12 (110-hp Clerget engine), which was a two-seat fighter-reconnaissance plane and trainer, and the Nieuport 14 and 15, both single-engined bombers. The Nieuport 16 was a development of the 11 with a 110-hp engine which proved distinctly nose-heavy, and it was superseded by the Nieuport 17, with the same engine but more wing area to improve handling.

Although widely used by the French in 1916, the Nieuport 17 was outclassed on the western front by the following summer, and it was replaced by the Nieuport 23, with improved synchronization for its Vickers gun and a 120-hp engine. This, too, had become obsolescent by 1918, while the Nieuport 17*bis* with a 130-hp Clerget engine and fully faired fuselage sides saw limited service in 1917 but proved too heavy.

Neither the Nieuport 24 (with twin machine guns, a rounded fuselage section, and a fixed tail fin) nor the similar 27 (with revised armament) saw much operational service, and the

Nieuport 28 (with the unreliable 160-hp nine-cylinder Gnôme Monosoupape engine) was initially used principally by American forces during 1918. Due to defective glue, the fabric tended to pull away from the wings of Nieuport 28s during dives or violent maneuvers, and by the time the problem was cured the U.S. squadrons had changed to Spads.

After the war Nieuport absorbed the Astra airship firm and adopted the title Nieuport-Delage. Its products included the Type 29 fighter (300-hp Hispano-Suiza engine) and the Nieuport 62 sesquiplane interceptor, together with the Loire-Nieuport LN401-411 dive-bombers, used by the French navy early in World War II. The company was nationalized in 1936 to become part of Société Nationale de Constructions Aéronautiques de l'Ouest.

Night Fighter

Fighters capable of locating and shooting down hostile aircraft at night were only slowly developed during World War I. Although the earliest bomber sorties against England were in daylight, the more vulnerable Zeppelins came by night, when they could not be intercepted. Attacks by German bombers, over both Britain and France, were eventually mounted entirely by night, except for tactical sorties over the western front. By 1917 a semblance of a defense system had been set up over Britain; the night fighters were principally BE2C and 2E aircraft (their limited maneuverability was of little consequence at night), Bristol F.2Bs,

Nieuport Scout

Camels, and DH 4s. The first night victory was actually gained by a French Morane, which dropped small bombs on a Zeppelin over Belgium. In France, the RAF's 151 Squadron gained at least 21 confirmed victories using Sopwith Camels, in which the fixed Vickers guns directly ahead of the cockpit were usually replaced by sloping Lewis guns that did not blind the pilot when fired.

The World War I night fighter pilots relied on their eyes. The first night fighters specially equipped to "see" in darkness did not appear until July 1939, when the first primitive fighter radar, called AI III (AI = airborne interception), was cleared for use in the RAF's Bristol Blenheim IF light-bomber conversion. In 1940 the better AI IV was fitted to the greatly superior Beaufighter, and this gradually mastered the German night bombers during the BLITZ. The technique was for each Beaufighter to be directed (vectored) by a ground controller to a position behind an enemy aircraft and at about the same height. The observer in the fighter would then watch two cathode-ray tubes, which indicated whether the enemy was to the left or right, above or below, and thus guided the pilot until the "hostile" could be sighted visually and shot down in the conventional manner. Abortive experiments of 1940–42 included the Turbinlite, in which a night fighter attempted to illuminate a hostile bomber so that Hurricane day fighters could shoot it down, and the LAM (long aerial mine), in which the night fighter trailed an explosive charge on a long cable.

By 1943 radar of extremely short wave length had become available. This gave much better definition and could be used at low altitude. The aerial system was mounted on a single parabolic reflector that searched the sky ahead and later could be automatically locked onto a target. The fairing over this aerial gave night fighters a distinctive "thimble" nose, although today's fighter radomes can be made pointed in shape and capable of withstanding flight through rain at supersonic speed.

One of the most successful World War II night fighters was the MOSQUITO, which had both AI IV and the new centrimetric radar (British AI VIII and American AI X, called SCR-720 in U.S. service and fitted to the Northrop P-61 Black Widow). Luftwaffe night fighters included the Bf 110G ZERSTÖRER, Dornier 217J and N, and Ju 88C and G, as well as limited numbers of HENSCHEL 219s and ME 262Bs. Many German night fighters were equipped with extremely effective Schräge Musik

(slanting, or jazz, music) armament, which incorporated heavy cannon mounted in the top of the fuselage to fire obliquely upward. Their AI radars, with clumsy external aerial arrays, were backed up by passive sensors that emitted no signals and thus did not betray the fighter's presence. These sensors included Naxos, which enabled the crew to home on the RAF bombers' navigation radars, and Flensburg, which homed on the "Monica" radars specially added to RAF bombers to protect them against the same enemy night fighters.

Today fighter aircraft are almost universally capable of operating at night or in bad weather and all are radar-equipped. In 1948–50 the U.S. Air Force ordered from Hughes Aircraft the first of a more advanced kind of fighter guidance system in which radar was linked with an autopilot so that the interceptor could be flown without the pilot's aid to destroy a hostile aircraft. This was coupled with new armament systems involving first FFAR (folding-fin aircraft rockets) and later air-to-air missiles (initially the Falcon family and the medium-range Sparrow).

Noise suppression

Aircraft noise is a major problem, since most large civil airports are located near densely populated areas. Minimum Noise Routing (MNR) can be used to keep aircraft away from major urban concentrations, providing the procedure does not impose unacceptable economic penalties (fuel, time) on the operators. Silencers for jet engines are available, but fitting them to the older types of airliners involves withdrawing the aircraft from service, installation costs, and a weight penalty that reduces performance. Much of the objectionable noise created by turbofan-powered aircraft is caused by the fan itself, and the single-stage fans without inlet guides introduced in 1968 are less noisy than earlier two-stage systems.

ICAO recommendations for aircraft noise restriction do not apply to older airliners. For modern subsonic airliners, however, they propose a maximum noise level of 108 EPNdB (Effective Perceived Noise in decibels, a measure which takes into account the high-frequency emissions of jet engines to which the human ear is particularly sensitive).

For takeoff, readings are measured 21,000 ft. (6,500 m) from the start of the takeoff run and also at a point 2,100 ft. (650 m) to one side of the runway; during landing approaches the

reading is taken 390 ft. (120 m) below the 3° descent path.

American and British airport noise regulations are based on these recommended values. To reduce noise, flight profiles may involve using less than maximum power, adopting a steep takeoff gradient so that a high takeoff thrust may be employed, or climbing out at a low airspeed. Movements of jet aircraft at night are restricted at many airports and prohibited at some. Servicing areas should be soundproofed, with special mufflers for engine run-ups.

A supersonic airliner presents even more serious problems, since at 60,000 ft. (18,300 m) it leaves a "boom path" 30 mi. (48 km) wide, and the Concorde presents an example of how the problem of aircraft noise may actually become a factor of significance in relations between nations.

North American Aviation

One of America's principal aircraft manufacturers, whose name is associated with some of the best World War II and Korean War warplanes and with advanced designs in the postwar period.

Established in 1928 as a holding company, North American Aviation was reorganized in 1934. A new Manufacturing division included the Fokker Aircraft Corporation, in which General Motors had a 40 percent holding.

The first NAA product was the General Aviation GA-15, which went into production as the North American 0-47 observation aircraft for the U.S. Army. It was an all-metal stressed-skin machine of advanced design. The NA-16 of 1935 was a low-wing trainer with tandem open cockpits; it led to the TEXAN series, including the NA-50 fighter, which was intended for delivery to Siam but reached the U.S. Army in small numbers as the P-64 trainer.

During World War II North American produced the F-51 MUSTANG fighter and the B-25 MITCHELL bomber. In 1947 it flew America's first jet bomber, the B-45 Tornado. Powered by four J47 engines of 5,800 lb. (2,627 kg) thrust, the Tornado served mainly in a reconnaissance role, under the designation RB-45C. The first NAA jet fighter, the FJ-1 Fury, which was built for the U.S. Navy, first flew in November 1946, and 100 were delivered. In the course of its design German swept-wing data became available, and the Air Force version, the XP-86, was quickly redesigned to become the F-86 SABRE, the hero of the Korean War.

The Navion lightplane was a commercial disaster, but continued in production with other manufacturers and was eventually built in large numbers. The AJ-2 was the chief production version of the U.S. Navy Savage, planned as a carrier-based attack bomber with the exceptional speed of 471 mph (756 km/h) using two R-2800 Double Wasp piston engines together with a J33 turbojet in the tail. It was eventually employed mainly as a tanker and reconnaissance machine. The T-28 Trojan/Fennec trainers were powered by either 800-hp Wright R-1300 engines or 1,425-hp R-1820 Cyclones, and many are still used as utility and light attack aircraft, rebuilt machines having 2,450-hp turboprops and heavy weapon loads. The world's first supersonic fighter was the NAA swept-wing F-100 Super SABRE, while the extremely complex A5 Vigilante, first flown 1958, led to the Mach-2 RA-5C carrier-based strategic reconnaissance machine, featuring extensive boundary-layer blowing and powered by two J79 engines.

Only two XB-70 Valkyries built by North American actually took to the air, but these six-engined Mach-3 bomber prototypes provided a valuable means of testing the newest technology. The X-15 hypersonic research aircraft produced by the company for what became known as NASA, with Air Force and Navy funding, was constructed of a special high-nickel alloy and attained Mach 6.72. The T-2 Buckeye trainer began with a single J34 turbojet and remained in production in the mid-1970s with two J60s or J85s. In 1967 North American merged with Rockwell Standard Corporation, becoming part of ROCKWELL INTERNATIONAL in 1973.

Northrop

John Knudsen Northrop, who had been associated with Lockheed Aircraft in its earliest days, and played a major role in the design of the record-breaking VEGA, and the later Altair, Sirius, and ORION, became vice-president and chief engineer of Northrop Aircraft in 1929. There he continued to build advanced, stressed-skin monoplanes, beginning with the Alpha single-engined low-wing transport, followed by the Beta, Gamma, and Delta. While the Gamma and Delta were developed and placed in commercial service, Northrop produced a military successor which went into large-scale production for the U.S. Army in 1934 as the A-17 attack bomber. The later A-17A had a more powerful (825-hp) Twin Wasp Junior engine and retractable landing gear. The RAF and the South African Air Force used the aircraft under the name of Nomad; other export orders included 150 of the original fixed-undercarriage model for China.

In 1939 John Northrop resigned when Douglas purchased the Northrop Corporation, reestablishing independent operations as Northrop Aircraft, at Hawthorne, Cal. The first new product was the N-3PB patrol seaplane, which was derived from the A-17A, but none reached the purchaser, Norway, before the German invasion, and the aircraft instead equipped a Norwegian RAF squadron based in Iceland. Northrop's biggest task in World War II was subcontract manufacture, but it also designed and produced the chief U.S. night fighter, the large and extremely complex P-61 Black Widow. The P-61 was powered by two 2,000-hp Double Wasp engines and equipped with SCR-720 radar and an armament of four 20-mm cannon and (in some models) a remote-controlled dorsal turret carrying four .50 machine guns.

In 1940 Northrop flew the first of a series of tailless, flying-wing aircraft, the N-1M. The "all-wing" formula was extended to the XP-56 pusher fighter, capable of over 400 mph (640 km/h), the twin-jet XP-79B fighter, and a design for a gigantic long-range bomber, the XB-35, for which the N-9M aircraft were twin-engined one-third size flying scale models. The B-35 was intended as a rival to the CONVAIR B-36 as prospective equipment for the future Strategic Air Command, and the first prototype flew on June 25, 1946. It was a true flying wing, without fuselage or tail. It had a span of 172 ft. (52.42 m) and was propelled by four 3,000-hp Wasp Major engines driving the trailing edge driving pusher propellers. The Air Force purchased 14 YB-35s, and in 1947 flew a jet-propelled version, the YB-49, powered by eight J35 axial turbojets. Although a six-jet reconnaissance version was also produced, the entire flying-wing project failed to secure a production contract.

Northrop did, however, build a strategic missile with no horizontal tail, the Air Force's SM-62 Snark, as well as the F-89 Scorpion night fighter, of which ten versions were produced. A small number of C-125 three-engined STOL transports were manufactured, but the company's future was assured by the development of the T-38 Talon supersonic trainer, first flown on April 10, 1959: 1,187 Talons had been built by 1972. The T-38 design had originally been intended to be developed as a small fighter, and on July 30, 1959, Northrop flew at its own expense a prototype of this interceptor called the N-156F FREEDOM FIGHTER. Eventually this was adopted by the U.S. Air Force as the F-5 and supplied to many foreign nations.

Northrop's next-generation fighter, the YF-17, led to the Navy F-18 multi-role fighter, in which Northrop became the principal associate to prime contractor McDonnell Douglas. Northrop is itself continuing with land-based developments of the YF-17/18, and also has other divisions diversified in advanced-technology fields, one of them the Ventura Division which produces RPVs (Remotely Piloted Vehicles) and targets and their engines; the Aircraft Division also makes almost the entire fuselage of the Boeing 747 and parts of other aircraft.

Nowotny, Walter (1920–1944)

Austrian-born Luftwaffe ace of World War II, Nowotny was the fifth-highest scoring pilot of the war, credited with 258 confirmed victories (all but 3 on the eastern front). Nowotny was assigned to *Jagdgeschwader* 54 in February 1941; he was to assume command of the unit in October 1942. On August 4 of that year he claimed 7 Russian aircraft in one day, and in the month of June 1943 scored a total of 41 victories (10 on June 24 alone). In February 1944 Nowotny left the eastern front to take command of the special unit, Kommando Nowotny, testing the ME 262 jet fighter in Luftwaffe service. After achieving 3 more victories, he was killed when his Me 262 crashed at Achmer airfield on November 8, 1944, after engaging an American bomber raid (the cause of the crash is unknown).

Nungesser, Charles Eugene Jules Marie (1892–1927)

Credited with 45 victories, Nungesser was the third-ranking French fighter ace of World War I. Born in Paris, Nungesser went to South America at the age of 17, became a racing driver, and later learned to fly. He returned to France in 1914, joined the army, and obtained a transfer to the Flying Service. In 1915 he was a reconnaissance pilot, but received an assignment to a fighter squadron in November 1915. Most of his victories were achieved in NIEUPORTS, and although wounded on innumerable occasions he continued to fly until the end of the war, eventually having to be carried to his aircraft.

After the war Nungesser ran a flying school in France and later barnstormed in America. He disappeared over the Atlantic on May 8, 1927, in an attempt to fly from France to the United States.

Northrop YB-35

Northrop P-61 Black Widow

O

Ohain, Hans-Joachim Pabst von
(1911–   )

A pioneer German turbojet designer, von Ohain was responsible for the engine used in the world's first jet-powered aircraft, the He 178. While still a student of applied physics and aerodynamics at Göttingen University in the 1930s, von Ohain obtained patents on a centrifugal turbojet design. He joined the aircraft manufacturer Heinkel and in 1937—the same year that Britain's Frank WHITTLE tested his pioneer turbojet—von Ohain demonstrated his first jet engine, which developed 550 lb. (250 kg) thrust.

A power unit for installation in a Heinkel airframe was tested in 1938 but proved unsatisfactory: in order to hold down the diameter of the engine, the compressor had been made too small and the combustion chambers were cramped. A redesigned version, the He S-3b, with 1,000 lb. (454 kg) thrust, was successfully used to power the Heinkel He 178 in the world's first jet-propelled flight (August 27, 1939). Von Ohain went to the United States after World War II, where he was employed on government research projects. (See also JET ENGINE.)

Olympic Airways

Olympic Airways was founded in April 1957 by Aristotle Onassis, as the successor to the former Greek National airline TAE, formed in 1951. Olympic was given a 50-year concession from the Greek government, but following heavy financial losses, Onassis withdrew in December 1974 and the airline was re-organized and became government-owned. Olympic Airways has extensive networks of domestic and European passenger and cargo services. It also operates to New York, Montreal, Chicago, Nairobi, Johannesburg, Singapore, Melbourne, and Sydney. In addition it serves Cyprus, Tel Aviv, Cairo, Benghazi, Kuwait, Dhahran, and Dubai.

Olympic has used an unusually wide range of aircraft including the Soviet Yak-40, and is the only European airline to have operated Japanese aircraft (YS-11s) on scheduled services. The main fleet consists of 2 Boeing 747s, 8 Boeing 707s, 6 Boeing 727s, 7 Boeing 720Bs, and 7 YS-11s. Some 2,800,000 passengers are carried annually.

Orion

Thirty-five examples of the Lockheed Orion single-engined six-passenger airliner were built in the early 1930s, together with five additional aircraft produced by conversions. The Orion had a wingspan of 43 ft. 10 in. (13.05 m) and length of 27 ft. 6 in. (8.38 m); they were the first Lockheed aircraft to feature a fully retractable undercarriage. Engines fitted included the 450-hp Wasp (Models 9 and 9E), 550-hp Wasp (9D, equipped with wing flaps), 575-hp Cyclone (9B), 645-hp Cyclone (9F), and 650-hp Cyclone (9F-1). Maximum speed was 224 mph (360 km/h). Flown by a number of U.S. airlines (notably American Airways, Varney Air Service, Northwest, Trans-Continental and Western, and Pan American), the Orion was also used by three Mexican operators and by Swissair, which had two machines on its Zurich–Munich–Vienna route.

A later Lockheed product, the P-3 reconnaissance aircraft, also bore the name Orion. It was developed from the Lockheed ELECTRA airliner, and deliveries to the U.S. Navy began in 1962.

Ornithopter

Building a machine that would soar like a bird by flapping its wings is one of the oldest proposals for a way that man could fly. Such a machine is known as an ornithopter. Although experiments continue to be made, human muscles alone are too weak to flap wings large enough to support a man's weight, and mechanical power—everything from clockwork to steam engines has been tried—has so far not been successfully applied to this problem. The most promising route to man-powered flight seems to lie through the PEDAL PLANE.

P

P-1/6. See HAWK

P-26. See PEASHOOTER

P-36. See HAWK

P-38. See LIGHTNING (LOCKHEED)

P-39. See AIRACOBRA

P-40. See WARHAWK

P-47. See THUNDERBOLT

P-51. See MUSTANG

P-59. See AIRACOMET

PBM. See MARINER

PBY. See CATALINA

Pan American

The oldest international airline in the United States, formed in 1927 by Juan C. Trippe as Pan American Airways. The tiny company secured a post office contract to fly mail between Key West and Havana in 1927, which was originally carried in a borrowed Fairchild FC-2 floatplane flown by Cy Caldwell. Fokker F.VII-3m trimotor aircraft were used to initiate services throughout the Caribbean, followed by F.Xs (1929). As Pan Am routes extended down to South America, a Sikorsky S-36 flying boat was used for route surveying, followed by S-38s.

Certain South American services were run by Pan-American Grace Airways (PANAGRA), in which Pan American had a 50 percent interest along with W. R. Grace, a major South American shipping company, until 1966, when it was sold to Braniff. The NYRBA (New York, Rio, and Buenos Aires Line) was taken over by Pan Am in 1930, which thus acquired Commodore flying boats.

The S-40 *American Clipper* joined Pan Am in 1931 and became the first of the line's flying clippers. The four-engined S-42 flying boat (1934) served in the Caribbean and pioneered routes for Pan Am across both the Pacific (1935–37) and the Atlantic (1937–39), but in January 1938 the company's chief pilot, Edwin C. Musick, was lost with all his crew in the *Samoan Clipper*. The first Martin 130 flying boat (*China Clipper*) joined Pan Am in 1935, and a route to the Orient was opened in 1936 (Martins from San Francisco to Manila, an S-42 shuttling to Hong Kong).

In 1937 DC-3s replaced S-42s on the east coast of South America; the Boeing 314 was to prove the last clipper flying boat. By 1945 faster, more economical Constellations and DC-4s were available, with Stratocruisers in 1949 on the "President" service across the North Atlantic. American Overseas Airways was absorbed by Pan Am in 1950, and the DC-6B appeared on the North Atlantic run in 1952 ("Rainbow" service), followed by the DC-7B (1955) and DC-7C (1956—when Pan Am became the first airline to carry 200,000 passengers in one year on this route).

Boeing 707 jets entered service in 1958, the 747 Jumbo Jet (which Pan Am was the first to use) arriving in 1970. In 1976 a New York–Tokyo service over the North Pole was opened with Boeing 747SP aircraft.

By the mid-1970s the jet clipper fleet consisted of 130 aircraft, and some 9,600,000 passengers were carried annually.

Parachute

A parachute consists of a canopy to which rigging lines and a harness are attached. A number of panels (gores) are sewn together to form the canopy; the traditional silk is now largely replaced by synthetic fabrics.

Parachutes may be used for bailing out in emergencies; for paratroop operations (large canopies with a slow rate of descent); for supply dropping; for spacecraft recovery; for braking high-speed aircraft after touchdown; and for sport, in which case the canopy has one or more openings to make the parachute "steerable" during descent.

Flat parachutes open quickly and have a central vent with the rigging lines passing completely over the canopy, which will lie flat when spread out on the ground. Shaped parachutes (sometimes almost hemispherical) cannot be laid out flat.

The deployment of parachutes may be carried out by an auxiliary parachute, by a pilot parachute and sock, or by a ripcord (operated manually, by an automatic timing device, or by a static line attached to the aircraft). EJECTION SEAT parachutes are opened by drogue and designed to withstand high-speed ejection.

The first successful descent by parachute was made in 1783 by a Frenchman, Louis-Sébastien Lenormand, from a tower. The first descent from an airplane was made by U.S. Army Captain Albert Berry in 1912.

Pattle, Marmaduke Thomas St. John (1914–1941)

The top-scoring RAF fighter ace of World War II, Pattle was officially credited with 41 victories. Born in South Africa, he went to England in 1936 and joined the RAF. He was sent to the Middle East in 1938 and his squadron moved to Greece in November 1940. Flying a Gloster Gladiator II biplane he ran up his tally of enemy aircraft to at least 24 by the end of the year. Changing to a Hawker Hurricane he increased his score to more than 40 successes before being shot down by a Bf 110 Zerstörer in a dogfight over Eleusis Bay on April 20, 1941. Due to lack of documentation during the British retreat in Greece at the time of Pattle's death some of his victories were probably never recorded.

Paulhan, Louis (1883–1963)

Paulhan was one of the best known pioneer aviators, and one who gained enormous public enthusiasm for the fledgling flying machine. Gaining aviator's certificate No. 10 in 1909, he took part in the Reims meeting that summer and in several other early shows, but achieved prominence in April 1910, when he won the prize of £10,000 offered by the London *Daily Mail* newspaper for the first flight between London and Manchester completed within 24 hours. The distance was 198.8 mi. (317 km). Paulhan took half the allotted time, including a night stop. In the years before World War I he made many other notable flights, including one in which he established a height record of 4,495 ft. (1,370 m) over Los Angeles. During 1916–18 he flew French fighting scouts, adding further to his decorations, which included that of a Commander of the Legion of Honor.

Pearl Harbor

The principal U.S. naval base in the Pacific and scene of the opening Japanese strike that brought America into World War II. In 1941, Japan's plan to extend its empire included a surprise attack on the Hawaiian Islands base of Pearl Harbor, designed to destroy the U.S. Pacific Fleet. Six Japanese aircraft carriers, supported by a large force of warships, sailed to within 200 mi. (320 km) of Oahu and at 6 a.m. on December 7, 1941, launched the first of two air task forces totaling 351 aircraft. These consisted of approximately equal numbers of Aichi D3A "Val" dive-bombers, Nakajima B5N "Kates" (both bomb- and torpedo-armed), and ZERO fighters for escort and strafing. The attackers found 96 vessels at Pearl Harbor, including eight battleships and nine cruisers.

Achieving complete surprise, the Japanese met little effective resistance and in under two hours had sunk three battleships and two other warships. Strafing and bombing attacks destroyed 167 U.S. aircraft on the ground, while another 10 were shot down. Japanese losses were 29 aircraft, mostly as a result of antiaircraft fire. American casualties amounted to 2,322 military personnel and 68 civilians killed, and 1,178 persons wounded. With the exception of the battleship *Arizona*, which exploded, all ships sunk at Pearl Harbor were later salvaged and many returned to active duty. The attack was successful in restricting U.S. Navy action during the Japanese invasions of the Philippines and Dutch East Indies, but no American aircraft carriers were in Pearl Harbor on December 7.

Peashooter

In the early 1930s biplanes were still the standard fighters of the world's major air forces. In the United States, however, the Boeing Model 248 of 1932 marked the beginning of a new generation of monoplane interceptors. Designated the P-26, the "Peashooter," as it was soon nicknamed, was the first all-metal U.S. fighter. It had a nine-cylinder Pratt & Whitney radial engine giving a top speed of 234 mph (377 km/h) and was armed with two machine guns. The undercarriage was still not retractable and the cockpit was unenclosed. Wingspan was 28 ft. (8.53 m) and length 23 ft. 7 in. (7.19 m).

Production Peashooters began to reach Army Air Corps units early in 1934 and saw service in the continental United States and in Hawaii and the

Boeing P-26A Peashooter

The Jupiter, *a British-built pedal plane. During actual flight the pilot is enclosed by a canopy*

Panama Canal Zone. Between 1938 and 1940 it was replaced by the P-35 and P-36, but a dozen aircraft were flown against the Japanese by the Philippine air force in December 1941, and the export (Model 281) Peashooters delivered to the Chinese in 1934–36 also saw action. The Guatemalan air force used seven Peashooters as fighters until 1946 and as trainers into the 1950s.

Pedal plane

A pedal plane, rather than an ORNITHOPTER with its birdlike flapping wings, appears to be the most practical means of achieving sustained and successful man-powered flight. The source of power is a pedal-driven propeller. A man pedaling can generate about 2 hp for a brief period (sufficient for take-off), and can then maintain about $\frac{1}{2}$ hp over a longer period. In order to generate enough lift, the wingspans of pedal planes have been up to 82 ft. (25 m), with a 15:1 aspect (length to width) ratio, resulting in considerable weight and drag.

Lightweight wood has most frequently been chosen for the wing structure, generally with a light fabric covering supported by foam plastic or impregnated paper honeycomb. To keep propeller turbulence away from the wings a pusher layout is common; the propeller itself requires a diameter of 6 to 9 ft. (1.8–2.7 m) for maximum efficiency. A twin-boom layout, a tailless configuration, or a canard may be employed, or the propeller may be located above or below a normal fuselage. Chain or belt drive is the usual choice for power transmission. For takeoff it is necessary to drive the undercarriage along the ground as well as supply power to the propeller. The torque transmitted to the propeller is inevitably erratic, due to the limitations of muscle power. There is some evidence that the use of two men would be mechanically advantageous, despite the additional weight.

Success in the field of man-powered flight has been limited, although a large number of pedal planes have been constructed. Two British designs with wingspans of about 82 ft. (25 m) were flown in the 1960s, and one was airborne for over half a mile. The gross weights of these two aircraft (including pilot) were 250 lb. (113.5 kg) and 268 lb. (121.6 kg).

On June 29, 1972, RAF Flight Lieutenant John Potter covered 4,065 ft. (1,239 m) in the *Jupiter*. A further development of this design is the Prestwick Dragonfly. In 1977 a new record for man-powered flight was claimed by Takashi Kato of Nihon University, Japan. His *Stork B*, with a wingspan of 68 ft. 9 in. (20.95 m) and unladen weight of 79 lb. (35.8 kg) covered a distance of 6,870 ft. (2,094 m).

A large cash prize still awaits the first pedal plane to achieve a controlled figure-of-eight maneuver without losing height.

Percival

Formed in 1932 by Capt. E. W. Percival, Percival Aircraft made a name for itself as one of Britain's leading manufacturers of light aircraft and military trainers. Its first design was the three-seat Gull, a low-wing cabin monoplane flown for the first time in 1931. The Gull Major employed a four-cylinder Gipsy Major 130-hp engine; the Gull Six of 1932 employed a six-cylinder 200-hp Gipsy power unit which enabled it to achieve record-breaking long-distance flights in the hands of Jean BATTEN and Amy JOHNSON. The Mew Gull (1934) also had a 200-hp Gipsy Six and was capable of 250 mph (400 km/h), winning the King's Cup air race in 1937, 1938, and 1955 and breaking the England–Cape Town record in 1939 (Alec Henshaw, 3 days 6 hours 58 minutes).

The Vega Gull four-seat cabin monoplane (1935) could achieve 170 mph (275 km/h) on a 200-hp Gipsy Six, and won the King's Cup in 1936; a military version (the Proctor) served as a liaison aircraft during World War II, and the twin-engined Q.6 of 1937 also saw RAF use.

In 1946 a three-seat primary trainer (the P.40 Prentice) was produced, followed by the P.48 Merganser and its development, the P.50 Prince, a high-wing eight-passenger feederliner with two 520-hp engines that found employment in charter, airfreight, and survey work. Naval (Sea Prince) and RAF (Pembroke) versions were also developed. The P.56 Provost two-seat basic trainer of 1950, with a 550-hp Leonides engine, was succeeded by the P.84 Jet Provost, powered by a Rolls-

Royce Viper engine, which made its first flight in 1954.

Percival was acquired by the Hunting Group in 1954 and became known as Hunting Percival. In 1957 it became Hunting Aircraft, and was later one of the companies that amalgamated to form the BRITISH AIRCRAFT CORPORATION.

Petlyakov, Vladimir Mikhailovich (1891–1944)

Designer of the Pe-2, one of the most versatile aircraft of World War II, the Soviet designer V. M. Petlyakov became a member of the aircraft design section at the Central Aero and Hydrodynamics Institute (TsAGI) in 1920. He played a leading role in the wing design of TUPOLEV's TB-1 and thereafter was in charge of wing design at TsAGI. Petlyakov was responsible for the initial design studies on the ANT-20 MAXIM GORKI when Tupolev was abroad. He was subsequently extensively involved in plans for the ANT-42/Pe-8 four-engined strategic heavy bomber, and later took charge of its development. In 1938, however, Petlyakov was one of the many aircraft designers and scientists arrested during Stalin's purges, and the Pe-2, for which he is best known, was designed when he was in prison.

Originally known as Project 100, the Pe-2 was a fast twin-engined monoplane which was originally conceived as a high-altitude long-range fighter. During construction of the prototype it was decided that the aircraft should become a high-altitude bomber, and in the course of the war it even saw service as a dive-bomber and ground-attack machine. The Pe-2 was powered by two 1,000-hp M-105R liquid-cooled engines. Wingspan was 56 ft. 2½ in. (17.13 m) and length 41 ft. 6 in. (12.66 m). The maximum speed was 336 mph (540 km/h) at 16,400 ft. (5,000 m) and up to 2,640 lb. (1,200 kg) of bombs could be carried. A total of 11,427 Pe-2s were built.

Phantom

One of the most versatile and rugged of all jet fighters, the Phantom was probably the most successful warplane of the 1960s—4,000 were completed by 1970. Originally designed for the U.S. Navy, the two-seat McDonnell Douglas F-4 Phantom II general-purpose fighter first flew on May 27, 1958. It proved so successful that it was adopted for use by the U.S. Air Force and by the Marine Corps, and it was employed by numerous foreign air forces—including those of Britain, West Germany, Iran, Israel, Japan, South Korea, Spain, Greece, and Turkey.

With twin J79 turbojets giving a maximum speed of 1,584 mph (2,550 km/h) and combat ceiling of 71,000 ft. (21,640 m), the Phantom can carry a warload greater than that of the World War II B-29 bomber. It had attachment points for up to 16,000 lb. (7,250 kg) of bombs, drop tanks or air-to-surface missiles. Its main armament initially consisted of air-to-air missiles alone (F-4A, B, C, and D), but a multibarrel 20-mm cannon was introduced on the F-4E.

Developments of the F-4B, which was 58 ft. 3 in. (17.76 m) long with a wingspan of 38 ft. 5 in. (11.71 m), were the RF-4B (Marine Corps) and RF-4C (Air Force)—both equipped for photo-reconnaissance duties—the F-4G and J (U.S. Navy), the Spey-engined F-4K and F-4M (British Royal Navy and RAF respectively), and the F-4N (U.S. Navy). The RF-4E is used by Germany, and the F-4EJ by Japan.

An earlier aircraft to bear the name Phantom was the McDonnell FH-1, the U.S. Navy's first operational carrier-borne jet fighter.

Piaggio

The major Italian industrial firm of Piaggio entered the field of aviation during World War I. Most of its early products were foreign designs built under license, notably the Dornier WAL flying boat and the Bristol and Gnôme-Rhône radial engines. It also made a series of SAVOIA-MARCHETTI S.55 twin-hull flying boats. For the 1929 Schneider Trophy race it produced, however, a technically outstanding seaplane of its own design, the P.7. The aircraft had no ordinary floats and rested in the water on its sealed fuselage. To take-off, the pilot accelerated using a conventional marine screw propeller. A small arrangement of hydrofoils then gradually lifted the fuselage out of the water, and the engine's power was transferred to the aircraft propeller. Potentially the fastest of all competitors, the P.7 failed, however, to take part in the race.

A stream of small single-seaters, cabin machines, flying boats, and bombers designed by Piaggio over the succeeding years all failed to see major production. Among the bombers, notable types included the P.32, a very advanced twin-engined monoplane with

Petlyakov Pe-8 heavy bomber of World War II

U.S. Navy F-4B Phantom

Piaggio P.136-L of the Italian air force

patented double flaps; the twin-finned 350-mph (560-km/h) P.123, with three 1,500-hp Piaggio P.XIIRC35 radial engines and an extremely slender fuselage; and the four-engined P.50 with a very large single fin.

A critical shortage of engines stood in the way of production of the P.50, but since the Italian air force had no strategic bomber it was decided in 1938 to rework the P.50-II into the P.108, which became Italy's only four-engined heavy bomber of World War II. Similar to the B-17 FLYING FORTRESS in size, the P.108 first flew in 1939, and the 108B bomber saw action in the Mediterranean and Russia in 1942–43; a total of 163 were built. Along with the 15 108C airliners, most of the surviving bombers were taken over by the Luftwaffe as transports in 1943.

In 1948 the P.136 amphibian, with two pusher 215-hp engines, began a long and successful series of aircraft that included the P.166 high-wing landplane, powered by two 340-hp Lycoming pusher engines. More than 100 have been built for training, ambulance missions, liaison and antisubmarine work. In 1951 the Italian air force chose the single-engined P.148 as its primary trainer. From it was developed the more powerful P.149, of which 72 were built by Piaggio and 190 by Focke-Wulf in Germany. Another recent Piaggio product is the PD-808 executive jet and navigational trainer, designed by Douglas but developed and built by Piaggio, which made its first flight in 1964.

Pilcher, Percy Sinclair (1866–1899)

Scottish aviation pioneer who, but for his death in a flying accident, might have achieved sustained powered flight some four years before the Wright Brothers.

Pilcher's interest in heavier-than-air flying was inspired by Otto LILIENTHAL. In 1895, when he was a lecturer at the University of Glasgow, Pilcher built his first Lilienthal-type glider, the *Bat*, and flew it on Scottish slopes with increasing success. He twice visited Lilienthal in Germany, but recognized that sustained flight could be achieved only with an engine, and regarded gliding as merely an essential intermediate stage during which he would learn how to design a controllable aircraft and pilot it correctly. In 1896 he built a much better glider, the *Hawk*, which was superior in many respects to any constructed by Lilienthal. A hang-type monoplane, it was structurally sound and had landing wheels. In 1897 he made glides with the *Hawk* up to a record-breaking distance of 750 ft. (228 m). By this time he had begun the design of a powered aircraft and a lightweight engine. The result would have been more advanced than the tail-first Wright FLYER biplane of 1903, and in many ways resembled the Blériot monoplanes of nearly a decade later. Tragically, Pilcher was killed in September 1899, while flying the *Hawk* under tow. His powered machine remained unfinished, and since he was working in isolation he made little contribution to the progress of aviation.

Pioneer flights

By the end of World War I aircraft had become reliable machines, capable of meeting exacting performance specifications (for the first days of flying see EARLY AVIATION). The challenges of time and distance, however, had still to be vanquished. The pioneer flights of the 1920s and 1930s overcame one by one the obstacles of ocean, desert, jungle, and polar ice.

The first challenge that attracted aviators in 1919 was the North Atlantic. The German airship *L 72* had been intended to bomb New York during World War I (as the French *Dixmude* she disappeared over the Mediterranean in 1923 with a loss of 53 lives), but no flying machine had actually spanned the 2,000 mi. (3,200 km) of stormswept Atlantic, despite the London *Daily Mail*'s offer of a $50,000 prize for the first flight, in either direction, between the Old World and the New in less than 72 hours. Given the prevailing westerly winds, a west-to-east crossing was the obvious choice, and during 1919 Newfoundland found itself colonized by an itinerant flying population all seeking to be "first across."

Four Curtiss NC flying boats had been developed during World War I with the purpose of flying to Europe to join the Allied air forces. They were now prepared for a peacetime flight across the Atlantic, though they did not register as contestants for the *Daily Mail*'s prize. NC-1 (Lt. Cdr. P.N.L. Bellinger), NC-3 (Cdr. J. H. Towers), and NC-4 (Lt. Cdr. A. C. Read) left Trepassey Bay, Newfoundland, at 6 p.m. on May 16 for the Azores, guided by U.S. destroyers located at 50-mi. (80-km) intervals across the Atlantic with instructions to fire star shells every five minutes after dark. The dawn brought rain, mist, and heavy cloud, but the NC-4 made it to Horta, on Fayal Island in the Azores, flying on to Lisbon on May 27 to complete the first-ever Atlantic crossing by air.

The NC-1 came down 100 mi. (160 km) short of the Azores, and her crew was rescued by a Greek freighter; the NC-3 had to taxi the last 200 mi. (320 km) to the Azores after landing on the sea for a radio position check.

In Newfoundland, the landplane aspirants, taking a more northerly route, had less success. On hearing of NC-4's arrival in the Azores, H. G. Hawker and K. Mackenzie Grieve took off in the Sopwith *Atlantic*, but a clogged cooling filter boiled the engine; the plane was forced down in the ocean and the flyers were fortunate to be picked up by a tramp steamer. F. F. Raynham and W. F. Morgan's Martinsyde crashed on takeoff, but on June 14 ALCOCK AND BROWN made a successful attempt in their VIMY, landing in Ireland 16½ hours later.

In July the British dirigible *R-34* flew the first round-trip across the North

U.S. Navy Curtiss NC-4 flying boat, first aircraft to cross the Atlantic

Atlantic, Scotland–New York–England, and at the end of 1919 a Vimy made the first England–Australia flight, piloted by Ross SMITH and his brother (November 12–December 10; Hounslow–Darwin). The following year another Vimy attempted the first flight from London to Cape Town (Lt. Col. Pierre van Ryneveld/Sqn. Ldr. Quintin Brand), but crashed at Wadi Halfa; a replacement Vimy was wrecked at Bulawayo, and the journey was completed in a DH 9.

A prize of $25,000 was now offered by Manhattan hotel proprietor Raymond Orteig for a nonstop flight between New York and Paris. French fighter ace René FONCK crashed his Sikorsky S-35 taking-off from Roosevelt Field in 1926, killing two crewmen. N. Davies and S. H. Wooster lost their lives when the Keystone Pathfinder failed to get airborne at Langley Field, Va., on April 26, 1927, and NUNGESSER'S Levasseur *L'Oiseau Blanc* vanished in the Atlantic after leaving Le Bourget on May 8, 1927. LINDBERGH finally succeeded in taking the Orteig Prize on May 20–21 of that year, beating C. D. Chamberlin and C. Levine in the Bellanca *Columbia* (New York–Eisleben, Germany; June 4–5) and R. E. BYRD, B. Balchen, B. Acosta, and G. O. Noville in the Fokker trimotor *America* (New York–Ver-sur-Mer, Normandy; June 29–July 1).

The Douglas WORLD CRUISERS completed the first round-the-world flight in 1924, and the same year Lt. Georges Pelletier d'Oisy flew from Paris to Tokyo, although his original aircraft was destroyed in a crash at Shanghai and he went on to Japan in a borrowed plane.

An Italian aircraft piloted by Lt.

Arturo Ferrarin had previously flown from Rome to Tokyo in 1920 (the only survivor of 11 that had started out), and in 1925 Cdr. Francesco de Pinedo of the Italian air force made another Rome–Tokyo flight, via Australia and the Philippines, returning direct to his starting point.

T. Vanderhoop flew a Fokker from Amsterdam to Batavia, in the Dutch East Indies, in 1924, but a crossing of the South Atlantic by two Portuguese naval officers (Arturo de Sacadura Cabral and Gago Coutinho) required three planes and two months to get them to Pernambuco in 1922, two machines being lost at intermediate stops. Ramón Franco of Spain flew from Palos to Brazil in 1926 in the Dornier WAL *Plus Ultra*, and in 1927 the twin-hulled Savoia-Marchetti flying boat *Santa Maria* (de Pinedo again) flew from Dakar to Pernambuco, going on to tour South America, the West Indies, and the United States, where it was destroyed by fire at New Orleans when fuel spilled during an engine run-up. (De Pinedo was killed in 1933 taking-off from New York on a Baghdad flight.)

Saint-Roman and Mounèyres vanished in their Farman GOLIATH seaplane in 1927 on a Senegal–Brazil attempt, but a wing and landing gear from a Goliath found in the Amazon estuary may indicate that it was a success that cost them their lives. Capt. Dieudonné Costes and Lt. Joseph Le Brix made the first nonstop South Atlantic crossing on October 14, 1927, flying from St. Louis, Senegal, to Porto Natal, Brazil, in a single-engined Breguet landplane. They flew on to the United States, shipped the plane to Tokyo, and returned home via Hanoi, Calcutta, and

Karachi to Paris. A nonstop Rome–Porto Natal flight was made by Ferrarin and Carlo Del Prete in a Savoia-Marchetti (1928), and Bert Hinkler made the first eastbound crossing of the South Atlantic in November 1931—solo in a PUSS MOTH.

The east-west route over the North Atlantic was pioneered by the single-engined Junkers monoplane *Bremen*, flown by Baron von Huenefeld, Maj. James Fitzmaurice, and Hermann Koehl (Baldonnel–Greenly Island, Labrador; April 1928); the first Paris–New York nonstop flight was achieved by Dieudonné Costes and Maurice Bellonte in September 1930 flying the scarlet-painted single-engined Breguet *Question Time*.

Many flyers vanished over the Atlantic. H. C. MacDonald disappeared in a Gipsy Moth in 1928 (a rusty can washed up in Ireland a year later contained a note "Going down in mid-Atlantic, engine trouble," with a signature that looked like "McDonnel"); F. F. Minchin, L. Hamilton, and the 62-year-old Princess Anne Lowenstein-Wertheim vanished on September 1, 1927; and Oskar Omdahl and Mrs. Frances Grayson were lost in December 1927 (a bottle washed up at Salem, Mass., in 1929 contained a message "We are freezing. Gas leaked and we are drifting off Grand Banks. Grayson.").

The Pacific challenge was overcome by KINGSFORD SMITH in May-June 1928, flying the Fokker F.VII/3m *Southern Cross* from Oakland to Honolulu, then taking off from Kauai for Suva in Fiji and reaching Brisbane and Sydney. Kingsford Smith subsequently took the *Southern Cross* around the world on a two-year tour, and in 1934 flew the Altair *Lady Southern Cross* from Brisbane to Oakland via Suva and Honolulu.

Wiley POST and Harold Gatty set a new round-the-world time in 1931, and while trying unsuccessfully to beat this record Clyde Pangborn and Hugh Herndon made the first nonstop trans-pacific flight (Tokyo–Wenatchee, Washington; October 1931) in the Bellanca *Miss Veedol*.

Arctic exploration got under way in 1925, when Roald Amundsen and Lincoln Ellsworth mounted an expedition with two Dornier Wals, both of which came down some 136 mi. (219 km) from the North Pole. One plane was eventually able to get off (with all six explorers aboard) and reached Spitsbergen. The following year BYRD and Floyd BENNETT reached the Pole in a ski-equipped Fokker F.VII/3m *Josephine Ford*, and in 1928 George Wilkins and Ben Eielson made the first heavier-than-air crossing of the Arctic Ocean, flying in a Lockheed VEGA.

The South Pole was reached by air for the first time on November 29, 1929, when Byrd, Balchen, Harold June, and Capt. Ashley McKinley reached 90°S. in the Ford trimotor *Floyd Bennett*. A transantarctic flight was achieved in 1935, when Ellsworth and Herbert Hollick-Kenyon flew the Northrop Gamma *Polar Star* from Dundee Island to Little America.

Piper

Piper Aircraft owes its origin to the partnership of W. T. Piper and C. G. Taylor, who together established Taylor Aircraft Company in 1931. Their first product was the CUB, and when Piper bought Taylor out in 1936 (see TAYLORCRAFT) this same basic design continued to be produced, along with the 75-hp Cruiser and 100-hp Super Cruiser. In 1937 the firm was reorganized as Piper Aircraft Corporation, and during World War II built the L-4 Grasshopper observation plane (a military development of the Cub with a 65-hp engine), the AE-1 Navy ambulance (a modified Super Cruiser), and the PT-1 low-wing primary trainer (1943).

After the war the Cub series continued, together with the four-seat Clipper (1949) and its development the Pacer (1950), with an engine of 115, 125, or 135 hp. Piper moved into the multi-engined market with the four-seat Apache of 1953, while another advance was represented by the tricycle undercarriage of the high-wing Tri-Pacer. Agricultural aircraft (PA-25 Pawnee), were successfully introduced into the Piper range which was progressively extended with the single-engined PA-24 Comanche, the PA-23 Aztec twin, the two-seat PA-28 Cherokee (1961), the PA-30 Twin Comanche (1962), the six-seat PA-32 Cherokee-Six (1963), the PA-31 Navajo twin (1964), the six/seven-seat PA-34 Seneca, with two 215-hp engines (1971), the 17-passenger PA-35 Pocono, with two turbosupercharged 500 hp Lycomings (1968), and the twin-turboprop Cheyenne (1969).

The 1976 single-engined Cherokees had engines from 150 hp (Cruiser) to 235 hp (Pathfinder) in power, and the Cherokee Arrow II featured a retractable undercarriage. Navajo models included the stretched 10-passenger Chieftain, with turbosupercharged 350-hp engines.

Piston engine

Reciprocating engines, developed from automobile engines in the early days of powered flight, are today used almost exclusively in LIGHT AIRCRAFT. There are two basic types: RADIAL ENGINES, which are air-cooled and have the cylinders arranged radially around a central crankcase; and in-line engines, with the cylinders often in V, H, or broad-arrow formation. Larger in-line power units are normally liquid-cooled, using ethylene glycol. Lubrication is generally on the dry-sump principle in order to ensure an adequate oil supply in any flight attitude. Sleeve-valve engines are smooth in operation and retain positive timing at high revolutions. In very powerful engines the heat to which exhaust valves are subjected is a limiting factor, but sleeve valves can be used in units developing over 3,000 hp.

SUPERCHARGERS are fitted to boost the power output for takeoff and at high altitude. They are usually either of the centrifugal type (one- or two-stage, one- or two-speed), or are exhaust-driven (turbosuperchargers).

Aircraft engine carburetors may be required to handle 200 gallons (757 liters) of fuel an hour. Fuel injection into the supercharger is frequently used to ensure engine operation during negative-G maneuvers.

Piper Cherokee

Cutaway view of the Rolls-Royce V-12 Merlin, a famous supercharged piston engine of World War II turning out nearly 2,000 hp. It powered Britain's Spitfire, Mosquito, and Lancaster, and (under license-production by Packard) transformed the American P-51 Mustang into one of the outstanding fighters of the war

Pitot tube

A device for measuring the flow velocity of air or water, the Pitot tube is named after its originator, Henri Pitot (1695–1771). An aircraft Pitot tube (or pressure head) projects forward into the airflow. As the aircraft moves through the atmosphere, air is driven into the Pitot tube and passes into a capsule within the airspeed indicator (see INSTRUMENTS). Expansion of this capsule, modified by the actual external atmospheric pressure as measured by the STATIC TUBE, moves the needles on the dial of the airspeed indicator.

Po-2

Designed by N. N. POLIKARPOV in 1926, the U-2 biplane was redesignated the Po-2 as a memorial to Polikarpov when he died in 1944. The U-2 was a two-seat single-engined aircraft originally intended as a trainer, but over a lifespan of some 40 years it had a remarkably varied career. Altogether some 33,000 were built; they were used as air ambulances, glider tugs, and agricultural aircraft; they dropped parachutists, carried out forestry and fishery patrols, and did aerial survey work, and

during World War II they saw valiant service as night bombers.

Originally an angular-looking aircraft with interchangeable wings and tail surfaces, the U-2 was redesigned in 1927 because of poor performance. Some aircraft were produced with float undercarriages and a number featured enclosed cabins, but the majority had two open cockpits (about 800 were built with three cockpits in tandem). The U-2 was powered by a 100-hp M-11 air-cooled radial engine. Wingspan was 37 ft. 5 in. (11.40 m) and length 26 ft. 8 in. (8.13 m). Loaded weight was 1,962 lb. (890 kg) and maximum speed 93 mph (150 km/h).

Pokryshkin, Alexander Ivanovich (1913–   )

Second highest-scoring Soviet fighter pilot of World War II, Pokryshkin was credited with 59 victories achieved during 137 engagements in 550 sorties. With I. N. KOZHEDUB and Marshal G. K. Zhukov, Pokryshkin shared the honor of being the only "three-times Hero of the Soviet Union."

Pokryshkin began his military service in 1934, and after training at an air

force technical school, qualified as a mechanic. He learned to fly at a local flying club, and in September 1938 went to the Kasha air force flying school near Sevastopol. In 1941 Pokryshkin was a flight commander in the 55th Fighter Aviation Regiment equipped with I-16 RATA and MiG-3 fighters on the Romanian frontier. He scored his first victory on June 22, 1941, while flying an MiG-3, but on July 20 was himself shot down by ground fire: he crash-landed his MiG and, after destroying it, walked back to his unit through the enemy lines. Pokryshkin went on to fly most of the major Soviet fighter types, but he was probably best known as Russia's leading AIRACOBRA pilot. He later headed a squadron of the 16th Guards Fighter Aviation Regiment, and at the end of the war he commanded the 9th Guards Fighter Aviation Division.

Polikarpov, Nikolai Nikolaevich (1892–1944)

N. N. Polikarpov was a Soviet aircraft designer responsible for more than 40 basic designs, among them the I-16 RATA monoplane fighter of the

Spanish Civil War, and the U-2 (PO-2) biplane trainer, used with great effect during World War II as a night bomber.

Polikarpov began his professional career at the Russo-Baltic Company's aircraft division, where he worked on the Ilya Mourometz, the world's first four-engined bomber. After the Russian Revolution he went to the Duks factory (later State Aircraft Factory No. 1) in Moscow, and in 1922–23 designed the IL-400 monoplane fighter. Appointed director of the department of landplane construction in 1926, he was responsible for the development of nine fighter, trainer, reconnaissance, and bomber aircraft up to 1929, but was then imprisoned for allegedly holding up the development of new types of aircraft and thereby sabotaging the country's security.

Polikarpov worked for nearly a year under detention and earned his release by producing the successful I-5 biplane fighter. In 1932 he was made responsible for fighter design at the Central Design Bureau (TsKB), and produced the I-15 and I-153 biplanes and the I-16 Rata monoplane.

From the I-16 a less successful series, the I-180/185 was developed. A number of fatal crashes occurred during tests, largely as the result of engine trouble, for which members of Polikarpov's team (though not himself) were held responsible. Other aircraft designs produced during the early part of World War II included the TIS heavy escort fighter, ITP high altitude interceptor, NB(T) night bomber, and BDP glider, but none of these reached production status. The Malyutka rocket fighter had not progressed beyond the project stage before Polikarpov's death.

Post, Wiley (1900–1935)

Born in Texas, Wiley Post was working as an oilfield roughneck in 1919 when he was so impressed by a barnstorming flying circus that he began to follow the stunt flyers from airport to airport, picking up odd jobs. He returned to oil drilling in Oklahoma but was blinded in one eye by a splinter of metal and received $1,800 compensation, which he used to buy a secondhand Canuck in 1924.

Six years later Post won the Chicago–Los Angeles Air Derby, and in July 1931 he completed a 15,500-mi. (25,000-km) flight around the Northern Hemisphere (New York–Sealand–Berlin–Irkutsk–Alaska–New York) at the controls of the Lockheed VEGA *Winnie Mae* in 8 days 15 hours 51 minutes, accompanied by Harold Gatty. In 1933 Post flew the same trip solo, using

a gyro control, and cut his time to 4 days 19 hours 36½ minutes.

He then made plans to undertake a series of high-altitude flights, but in August 1935 set out to fly from New York to Moscow in an extensively modified Lockheed ORION with humorist Will Rogers as a passenger. The first stage of the journey was to Alaska, with floats being fitted en route at Seattle. On August 15 they left Fairbanks but had to force-land on a small lagoon near Barrow, where Post made some adjustments and asked for directions.

Shortly after takeoff the engine failed and the Orion crashed from an altitude of only 50 ft. (15 m) killing both occupants. It was later determined that the aircraft was nose-heavy as a result of modifications effected before the flight and the use of floats, and this would have made control almost impossible without engine power.

Winnie Mae is now housed in the aviation collection of the National Air and Space Museum in Washington.

Pratt & Whitney

U.S. aircraft engine manufacturer which is one of the world's three largest suppliers of power plants for both civil and military airplanes.

Originally established in 1925, and today a part of United Technologies Corporation, Pratt & Whitney specialized in air-cooled radial engines, its earliest designs being the 9-cylinder Wasp of 1926 (600 hp; BOEING 247), the 9-cylinder Hornet of 1927 (750 hp; SIKORSKY S-42), and the R-985 Wasp Junior (450 hp; Lockheed Model 12, ELECTRA). World War II engines included the 14-cylinder R-1830 Twin Wasp (1,200–1,450 hp; CATALINA, LIBERATOR, WILDCAT), and the 18-cylinder R-2800 Double Wasp (2,000 hp; MARAUDER, HELLCAT, THUNDERBOLT, CORSAIR).

Reciprocating-engine development culminated in the 28-cylinder 3,500-hp Wasp Major (CONVAIR B-36, STRATOCRUISER); early turbojet work included license-production of the Rolls-Royce Nene and Tay. The JT3C two-spool turbojet (18,000 lb., 8,154 kg, thrust with afterburner) is installed in the B-52 STRATOFORTRESS, Super SABRE, early BOEING 707s and DC-8s, DELTA DAGGER, and Voodoo, and the JT3D turbofan (18,000 lb., 8,154 kg, thrust) powers the Lockheed StarLifter and later Boeing 707s and DC-8s. The JT9D is used in the BOEING 747 and DC-10, while the JTF 10A (TF 30) powers the F-111 (with afterburner) and CORSAIR II. The JTF 22 is employed in the EAGLE, TOMCAT and F-16.

Pressurization

At high altitudes, aircraft cabins must be pressurized to protect crew and passengers from the dangerous effects of low pressure and the oxygen deficiency that accompanies it. Pressurization is also important during descent to prevent pressure inequalities causing discomfort or damage in ears, sinuses, and intestines. Without pressurization, oxygen starvation causes loss of efficiency at altitudes above 10,000 ft. (3,048 m), while at 20,000 ft. (6,096 m) unconsciousness may occur within a few minutes (see AVIATION MEDICINE).

Airliner cabins are usually pressurized to the equivalent of an altitude of about 5,000 ft. (1,524 m), air being continuously pumped in and leaked out to maintain its freshness. The pressurized air may be obtained either from a mechanically-driven compressor (the normal system in piston-engined planes), or from the compressor system of one or more of the jet engines. Since engines may only be idling under certain conditions (during descent, for example), the air supply from the compressor may be inadequate. The engine from which the cabin air is drawn may therefore have to be speeded up, or facilities may be provided to draw air from three or four engines. Before being fed to the cabin the air must be cooled by expanding it through a turbine and passing it into a heat exchanger.

Propeller

The airplane propeller, or airscrew, obtains thrust from moving a large mass of air at low velocity; a JET ENGINE's thrust is obtained from a small mass of gas at high velocity. At speeds up to about 500 mph (800 km/h), propellers provide more economical propulsion than jets, and many smaller transports have TURBOPROP engines, while LIGHT AIRCRAFT almost universally employ propellers driven by PISTON ENGINES. A typical propeller has two or more blades (often three or four) radiating from a central hub and forming part of a spiral surface, like segments of a screw thread. They are twisted along their length and strike the air at a specific angle of attack. As they revolve their airfoil cross-sections they develop thrust in much the same way as an airplane WING develops LIFT, creating reduced pressure in front and increased pressure behind. The angle at which the blades are set on the hub (the pitch) determines how far the propeller screws itself through the air on each revolution. A low or fine-pitch setting makes best use of

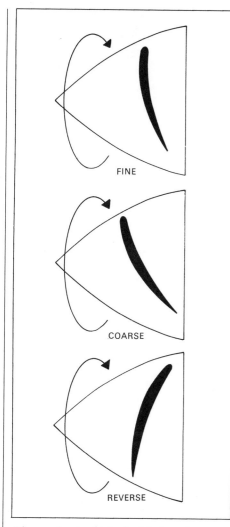

The angle at which the blades of a variable-pitch propeller are attached to the hub can be altered in flight. A fine-pitch setting is most useful at takeoff and landing; a coarse pitch is employed at cruising speed. The pitch can also be reversed to provide braking on landing

engine power at low speeds (during takeoff, for example), while a high or coarse-pitch setting is most economical at high speeds. A fixed-pitch propeller can be at its most efficient only at a single airspeed, and variable-pitch propellers were therefore developed, with either manual or automatic adjustment. The pitch can also be reversed to provide braking on landing.

To counteract the torque of a propeller, multiengined aircraft normally have "opposite-handed" propellers, or else counter-rotating propellers on a single hub (also fitted to powerful single-engined airplanes).

Pulse jet

A pulse jet is a simple JET ENGINE in which there are no rotating parts. Air is sucked into the combustion chamber through a number of valves, which are spring-loaded in the open position. Fuel injected into the combustion chamber is ignited, and the resulting expansion of gases forces the air-intake valves shut, allowing the exhaust only one means of exit (backward down the jet pipe), thus creating forward thrust. When the combustion chamber is emptied, the spring-loaded intake valves reopen and the cycle begins again.

A pulse jet was used to propel the German v-1 "flying bomb" of World War II, and has since found limited application driving the rotors of helicopters (an engine is attached to the tip of each blade). By careful design of the air intake duct it is possible to control the changing pressures of the resonating cycle without using valves, but pulse jets have not been widely used because of their high fuel consumption, vibration, and inferior performance when compared with a conventional jet engine.

Pup

Described by its British pilots as "the perfect flying machine," the Sopwith Pup (an unofficial nickname) was a tiny World War I biplane fighter, powered by an 80-hp Le Rhône or Clerget rotary engine, and armed with one synchronized fixed machine gun. Wingspan was 26 ft. 6 in. (8.07 m) and length 19 ft. 3 in. (5.86 m). Delightful in control response, the Pup was highly maneuverable, and when it first appeared it could fight its opponents on almost equal terms. The first Royal Flying Corps unit to receive Pups was 54 Squadron, which arrived at the front on December 24, 1916. By late 1917, however, the 112-mph (194-km/h) Pup was outdated for combat duties, and was replaced largely by Sopwith CAMEL. Most Pups were then relegated to training duties in Britain, where they also figured in experimental landings on aircraft carriers. In all, 1,770 Pups were manufactured, of which over 800 were still on RAF strength in November 1918.

Pusher

Pushers are aircraft whose engines are arranged to drive a propeller (or propellers) located at the back of the fuselage, the trailing edge of the wing, or the rear of an outrigged nacelle. From this position behind the engine the propeller pushes the aircraft through the air. The more conventional front-mounted propeller is termed a "tractor:" it pulls the plane forward. In the aircraft with a single engine, a pusher propeller gives the pilot an unimpeded forward view from his cockpit, and this was consequently a popular arrangement in the early phases of World War I, because it gave the air gunner a good field of fire.

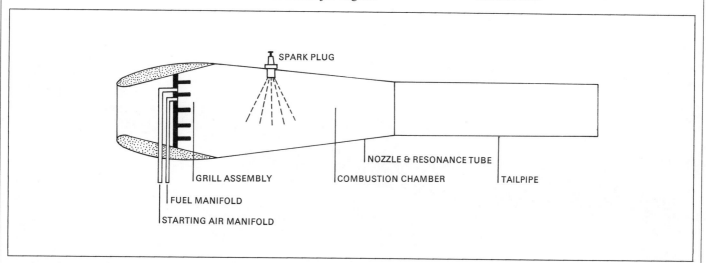

Details of a pulse jet

Classic pusher biplane of World War I: a Voisin Scout

In the multiengined aircraft pushers located at the back of the wing tend to incur a lower drag penalty than forward-facing propellers.

Q

QANTAS

The Australian national airline was formed in 1920 at Longreach, Australia, as Queensland and Northern Territory Aerial Services, using a war-surplus Avro 504K and a BE2E. The airline opened its first scheduled service, Charleville–Cloncurry, on November 2, 1922, flying three wartime FK-8s, and it acquired four-passenger DH 50s in 1924. Two flying schools were opened with DH MOTHS in 1926, and the airline helped to establish the FLYING DOCTOR SERVICE in 1928. Reconstituted as Qantas Empire Airways Ltd. in 1934 to operate the Singapore–Darwin–Brisbane sec-

tion of Imperial Airways England–Australia route, the airline introduced Short C-class flying boats in 1937 and ran a nonstop Australia–Ceylon courier service with CATALINAS between 1942 and 1945.

The Australian government became the airline's sole owner in 1947. Sandringham flying boats were introduced on a service to Fiji in 1950, and the Avro Lancastrians flown on the London route since 1945 were replaced in 1954 by Super Constellations, which also operated a transpacific service. Boeing 707 jetliners were introduced on the Sydney–San Francisco route in 1959, and the company's title was changed to Qantas Airways Ltd. in 1967. Scheduled passenger and cargo services now cover New Zealand, the Pacific islands, and San Francisco, the Far East, India, and European capitals.

Some 1,400,000 passengers are carried annually, and the fleet consists of 11 Boeing 747s, 11 Boeing 707s, 2 DC-4s, and 1 HS-125.

R

R-34

The largest airship of its day, the British *R-34* was the first dirigible to cross the Atlantic. The *R-34* was 643 ft. (196 m) long, with four 1,250-hp engines that gave a top speed of 50 mph (80 km/h). Its design was based on that of a German Zeppelin (the L33) which came down in England during World War I.

In July 1919 (less than a month after ALCOCK AND BROWN's flight) the *R-34* became the first aircraft of any type to make a round-trip crossing of the Atlantic, from Scotland to Mineola, Long Island, and back to England. Strong winds over Newfoundland on the outward journey from East Fortune, Scotland, made it difficult for the ship to reach Roosevelt Field. Time for the westward journey was 108 hours 12 minutes; the return trip to Pulham in Norfolk took 75 hours 3 minutes. The total distance covered was 6,330 mi. (10,187 km).

The *R-34* was wrecked in a gale while moored at Howden in the winter of 1921.

Racing

From almost the earliest days of aviation prizes and competitions stimulated the production of faster aircraft and the achievement of longer flights. Particularly in the years between the two world wars a series of international races had an enormous effect on the progress of aviation.

At the Reims aviation meet of 1909, the first great international event in the history of aviation, the first race for the James Gordon Bennett Aviation

The R-34, *first airship to cross the Atlantic*

Cup, was staged, with Glenn CURTISS winning at 47.65 mph (76.68 km/h). Subsequent Gordon Bennett races were won by Grahame-White in a Blériot (1910 in New York) and Charles Weymann in a Nieuport monoplane (1911 in England), and DEPERDUSSIN monoplane racers carried off the last two events that preceded World War I. A revival was staged as a time-race in France in 1920, but only two competitors completed the course, Sadi Lecointe's Nieuport averaging 168.50 mph (271.16 km/h) to win the trophy outright with a third consecutive French victory.

The Schneider Cup for seaplane racers, presented by industrialist Jacques Schneider, enjoyed a longer history, the first event being held at Monaco in 1913, where a Deperdussin emerged victorious. A British Sopwith won in 1914, and after the war Italian successes in 1920 and 1921 were followed by a British victory in 1922 and two consecutive American wins (1923 with Navy Curtiss CR-3 first and second, 1925 James DOOLITTLE in a Curtiss R3C). The Italians took the 1926 contest with the Macchi M.39 monoplane, but in 1927 and 1929 British Supermarine floatplanes entered by the RAF were successful and the 1931 contest was an easy victory for the S6b at 340.08 mph (547.23 km/h), making Britain the outright winners of the Trophy with three consecutive victories.

The Pulitzer Trophy was originally conceived in 1919 by the newspaper-owning Pulitzer Brothers as a prize for a long-distance race, but it became a closed-circuit event at the U.S. National Air Races in 1920, when it was won by an Army Verville VCP-R. The next year's victory went to a factory-entered Curtiss CR-1. In 1922, 1923, 1924, and 1925 military pilots were successful, but the American armed services thereafter lost interest in racing as a means of developing pursuit planes and the contest lapsed.

The Curtiss Marine Flying Trophy was contested 10 times between 1915 and Glen Curtiss' death in 1930, five victories being achieved by Curtiss seaplanes.

In England an Aerial Derby around London took place in 1912 and was repeated in 1913 and 1914. Revived in 1919 as a race over two circuits of the city, the event continued until 1923, but had to be abandoned in 1924 because there were no British racing planes capable of competing. Instead, King George V presented the King's Cup for a handicap contest, and this became the principal British air race.

In 1930 the Thompson Trophy closed-circuit unlimited-class race was intro-

Supermarine S6b racer

duced at the U.S. National Air Races, and the following year the Vincent Bendix Trophy 2,000-mi. (3,200-km) cross-country race was added to the NAR program. Gee-Bees won the Thompson in 1931 and 1932, but Wedell-Williams racers took first prize in 1933 and 1934, as well as in the 1932, 1933, and 1934 Bendix events. The last two Thompsons before the war went to Roscoe Turner in the Laird-Turner L-RT, while Seversky SEV-S2 aircraft took three Bendix races in a row (1937, 1938, 1939).

After World War II, modified ex-combat planes dominated American racing, Goodyear-built Corsairs proving quicker than Mustangs in the Thompson, although P-51s monopolized the Bendix. Jet divisions of both races were initiated, but these all-military affairs ended when the Air Force and Navy were barred from competing against each other, and the Air Force pulled out of low-level races after 1949.

By the 1970s racing had ceased to be of any technical interest to manufacturers, and modified World War II fighters remained the principal unlimited-class participants at events held at Reno, when American air racing was revived in 1964. Racing for smaller types of aircraft still produces keen competition but has limited spectator appeal, although it served to maintain American interest in air racing through the 1950s.

International long-distance races were frequently promoted in the 1920s and 1930s. They included the infamous 1927 Dole Derby from Oakland to Honolulu in which 10 lives were lost when three planes came down in the Pacific, and

the 1934 McRobertson England–Australia race (won by a de Havilland COMET), but the 1953 England–Australia contest in which a CANBERRA finished first was the last major event of this type.

Radar

Although the term radar (RAdio Direction And Range) was coined in America, the technique was being developed in Britain and Germany as well. Conventional radio direction finding used before radar required the operation of at least two intersecting beams if range as well as direction was to be obtained. With radar, the direction and range not only of physical features on land but also of aircraft can be detected by a single transmitter/receiver. Furthermore, radar operates without the knowledge or cooperation of the target, whereas ordinary radio direction finding requires the target to transmit for a fix.

Radar employs very short-wave radio transmissions, which are reflected back from the target, the time taken for this double journey being registered on receiving equipment that incorporates a cathode ray tube. The receiver is calibrated to indicate how far away the target is, depending on how long the radar echo takes to return to its source.

In navigational radar, the Plan Position Indicator (PPI) display is used. This comprises a scanner (aerial) electrically coupled to coils around the neck of a cathode ray tube. The center of the screen represents the position of the aircraft, and as the scanner rotates so the beam oscillates backward and forward to the margin of the screen in

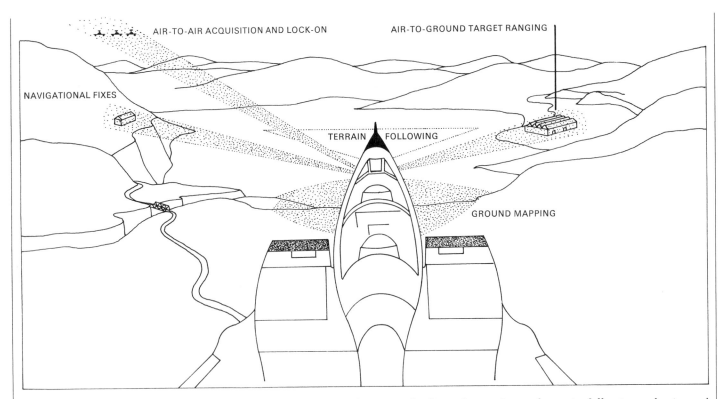

AIR-TO-AIR ACQUISITION AND LOCK-ON   AIR-TO-GROUND TARGET RANGING

NAVIGATIONAL FIXES

TERRAIN   FOLLOWING

GROUND MAPPING

Radar fulfils a variety of functions in a modern multirole combat aircraft. Ground-mapping and terrain-following radar is used for navigational fixes in blind conditions as well as for low-level attack; radar provides air-to-air search, acquisition, and lock-on, while air-to-air and air-to-ground ranging for missiles, bombs or cannon is also radar-operated

concert with the direction in which the scanner is pointing. The image of the echo appears on the screen, which because of its fluorescence retains a fading afterglow until the next sweep of the scanner reactivates the reflection.

Cumulo-nimbus and cumulus clouds contain heavy concentrations of rain drops, hail, and snowflakes, which will return an echo. Aircraft radar is used to detect these bad-weather clouds so that the pilot can avoid them.

Radar is an essential feature of air traffic control systems, with processed digital displays now eliminating the rotating beam and providing a constantly illuminated image incorporating superimposed digital information.

As the density of air traffic increased, secondary surveillance radar (SSR) was developed for airport controllers. This system is a communication channel which receives coded emissions from a transponder in an aircraft. The identity of the airplane can thus be established as well as its range, azimuth, and height. Additional information from the aircraft can also be transmitted and thrown onto the radar screen, special positions on the display being available for encoded notification of in-flight emergencies (including hijacking) or communications failure.

In addition to its basic function as a navigational aid, radar is used by com-

bat planes both to locate targets and to provide warning of an impending attack by hostile interceptors or missiles. Forward-Looking Radar (FLR) provides data on the slant range to a target for processing by an aircraft's computer, sends pitch command signals to an attitude direction indicator, and presents information either as a ground map or a PPI display on the radarscope. Additional functions include terrain-following when flying at a pre-set altitude, terrain avoidance (by showing terrain at or above the height of the aircraft's track within a five- or ten-mile range), ground mapping (including a high-resolution capability at low altitude), and a closed-circuit television display for weapon aiming. Sideways-Looking Radar (SLR) can be used to produce a photographic filmed radar map of the terrain on either side of an aircraft. Airborne attack radar is integrated with other avionics as part of the equipment of strike aircraft and of such new supersonic strategic bombers as the B-1 (see BOMBING).

Helicopters use tactical radar for search and rescue operations, and in 1976 "rodar" was being investigated— a system that uses the rotor arms as an aerial and provides a high definition short-range picture that would enable survivors in the sea to be discerned.

Radial engine

Aircraft PISTON ENGINES in which the cylinders are arranged in circular fashion around the crank-case are known as radial engines. Up to nine cylinders can be incorporated in a single circular row; one of the connecting rods is a strongly constructed master rod that receives the thrust delivered from the other connecting rods. A two-throw crank can enable two rows of cylinders to be employed.

Radial engines are air-cooled. They first came into prominence in the 1930s, when their high power outputs became appreciated. Power can be increased even further if sleeve valves are employed, since these have a higher maximum operating temperature.

The principal disadvantage of radial engines is their large frontal area, which imposes much greater drag than a comparable liquid-cooled in-line power unit.

Radio

Prior to the outbreak of World War I aircraft had carried radio receivers and transmitters on an experimental basis, mainly with a view to their eventual use as aids to army or navy reconnaissance.

All the early airborne communications were by W/T (wireless telegraphy) in which Morse signals were superimposed on carrier frequencies in the MF (medium-frequency) band around 500–1,500 kHz. The equipment was, by modern standards, extremely bulky, and even a simple airborne T/R (transmitter/receiver) often weighed 250 lb. (113 kg) and needed a large current supplied from a heavy lead-acid battery charged by a small windmill-driven generator. A further problem was that with such modest frequencies an aerial some 300 ft (91 m) long had to be extended with a lead weight on the end, and reeled in before landing (a hazard in formation flight).

The obvious need was for shorter wavelengths, and as a result of major developments by radio manufacturers, coupled with the introduction of "screened" ignition systems which did not emit interfering waves on the same frequencies, HF (high-frequency) R/T (radio telephone) communications in the 6,000 kHz band came into use for civil and military aviation. It was now possible to use ordinary voice speech and to fit the aerials into or around the aircraft. Continued fundamental development had led by 1938 to a completely new system of VHF (very high frequency) radio communications in the 100–150 MHz band. This again reduced the size and weight of airborne equipment, made it possible to use aerials forming part of the aircraft surface or projecting as a small blade (with no external wires at all), and dramatically improved the clarity of speech. The first VHF radios were in RAF service before World War II, and a visiting U.S. mission in 1939 copied the equipment for American use pending development of U.S. counterparts.

During 1944–57 further basic research produced still shorter wavelengths, resulting in the introduction in the late 1950s of UHF (ultra-high frequency) radio in the 225–470 MHz band. Today VHF and UHF are used by military aircraft, lightplanes, and helicopters, and also for local voice contact in transport aircraft. At the other end of the spectrum there has been development toward much longer wavelengths in the VLF (very low frequency) region where "survivable communications" is an area of intense interest in the 14–60 kHz. band.

A typical advanced aircraft flying a demanding combat mission will now have 16 CNI (communications/navigation/identification) subsystems covering frequencies from 10 to 4,300 MHz. The CNI avionics (aviation electronics) will in turn be part of the total on-board ICNI (integrated CNI) installation, which will be totally linked with from 4 to 18 other subsystems for reconnais-sance sensing, weapon control, weapon aiming, flight control, information management, information presentation, data storage, data processing, and many other functions. The installation must be of a digital nature and compatible within itself, with ground stations, with weapons, satellites, various kinds of ground and airborne recorders, airfield aids (e.g. for blind landing), and, in the case of combat aircraft, with other friendly forces while preserving security from the enemy.

Radio communications have also been used for the remote piloting of RPVs (remotely piloted vehicles) and radio-command drones, targets, and missiles. Other radio uses include NAVIGATION, and indication of aircraft height above terrain, the radio altimeter (arguably, the radar altimeter) using extremely high frequency in various ground-reflection modes.

As radio frequencies increase, they become less susceptible to interference (e.g. from sunspots), but at the same time they become more directional. Thus VHF cannot, unlike HF, be "bent" to communicate with vehicles beyond the horizon. A satellite communication system, known as Aerosat, is being designed to work at L-band frequencies (1,500 MHz), and will provide a quality of voice communication superior even to UHF, but with a range of many thousands of miles.

Radio direction finder

A device used to determine the bearing of a radio signal. It consists of a conventional radio receiver with a directional antenna, usually in the form of a loop which can be rotated about its vertical axis. By turning the antenna to find the points of strongest and weakest reception, the direction of the transmitter can be found.

The radio direction finder is used in NAVIGATION, either as a compass, using just one bearing, or as a means of fixing position, in which case bearings are taken on two transmitters. The system has the advantage of simplicity, since no special equipment is needed, but its range is limited. The radio direction finder can also be used from a shore station to take bearings on aircraft—thus the importance of maintaining "radio silence" in military operations (see RADIO).

Raiden

Developed for the Japanese Navy as a pure interceptor with a premium on speed and climb, the MITSUBISHI J2M Raiden or "Thunderbolt" (code-named "Jack" by the Allies) first flew as the J2M1 on March 20, 1942. Underpowered with its original 1,460-hp Kasei 13 (Mars) radial engine, J2M2 initial production aircraft had instead a methanol/water-injection Kasei 23a of 1,820 hp, which gave a top speed of 367 mph (590 km/h) at 20,000 ft. (6,100 m).

The Japanese Navy began to operate the four-cannon J2M3 Raiden at the beginning of 1944, but a number of aircraft were lost when they disintegrated in midair, probably as a result of the engine vibration that had plagued the design since its inception. Wingspan of the J2M3 was 35 ft. 5 in. (10.80 m) and length 32 ft. 8 in. (9.94 m).

The last Raiden to attain production status was the J2M5, with a supercharged Kasei 26a engine and armament reduced to two 20-mm cannon. Only 30 to 40 of these reached operational units, but with a maximum speed of 380 mph (610 km/h) and excellent maneuverability, this final variant of the Raiden proved a formidable B-29 SUPERFORTRESS-interceptor, capable of climbing to 26,250 ft. (8,000 m) in less than 10 minutes. Total production of the J2M series was only about 500 machines.

Ramjet

The ramjet is the simplest form of JET ENGINE. It possesses no moving parts, and is dependent on the forward motion of the aircraft itself to "ram" air into the engine. The air is forced through an intake incorporating single- or double-cone center bodies designed to decelerate and compress the flow into a combustion chamber, where fuel is injected to provide an inflammable mixture. Initial ignition is generally achieved by means of a pyrotechnic flare, and is then stabilized by a baffled section at the rear of the combustion chamber. Forward thrust is generated as exhaust gases pass backward down the nozzle.

A ramjet cannot operate from rest; forward motion must first be imparted to it so that air enters the intake. The AFTERBURNER present in some high-performance turbojet engines operates on the ramjet principle, but use of a ramjet as a main power plant has been limited largely to missiles, despite experiments by René Leduc. Ramjet engines are thus only designed for short periods of operation, during which they run at full power.

Rata

The Russian POLIKARPOV I-16 of 1933

was the first monoplane fighter with a retractable undercarriage and enclosed cockpit to enter service anywhere in the world. The I-16 first saw combat in the SPANISH CIVIL WAR; the initial shipment of 105 aircraft together with pilots and personnel arrived by sea in September 1936, and an eventual total of 475 I-16s were supplied. The aircraft's nickname of Rata ("Rat") was given by the opposing Nationalists; the I-16 went on to fight in the Russo-Finnish War, but by the time of the German invasion of Russia in 1941 even the considerable improvements that had been made to the original design were inadequate to enable the aircraft to stand up to the Luftwaffe with any great success. The I-16 was a parallel development with the I-15, I-15*bis* and I-153 biplane interceptors, which also served in Spain and were to prove even more outdated against the Luftwaffe in 1941.

Late versions of the I-16 had a 1,000-hp nine-cylinder air-cooled engine giving a top speed of 326 mph (525 km/h) and a ceiling of 29,530 ft. (9,000 m). Wingspan was 29 ft. 6 in. (9.00 m), and length 20 ft. 1 in. (6.13 m). Although highly maneuverable, the Rata lacked longitudinal stability and had very heavy elevator controls. Armament consisted of two 20-mm cannon in the wings and two .30 machine guns in the top of the engine cowling.

Reconnaissance aircraft

The first military reconnaissance aircraft were balloons, the earliest of all being that of Captain J. M. J. Coutelle at the Battle of Fleurus on June 26, 1794, where his constant signals played a significant part in the French victory. Reconnaissance balloons were widely used in the American Civil War, but the modern reconnaissance mission can be said to date from February 24, 1912, when Captain Piazza of the Italian army air detachment photographed Turkish positions in Tripolitania from a Blériot. A little later, in April 1912, an Italian airship took reconnaissance motion pictures.

Over the western front in 1914 aerial reconnaissance was the primary mission of most early airplanes, though they seldom carried cameras. By May 1915 the first aerial camera, the Thornton-Pickard Type A, designed by a Royal Flying Corps team in France, had begun to show the kind of fine detail that could be achieved with suitable equipment. By 1918 hundreds of reconnaissance machines were taking up to 25,000 photographs per week, the chief aircraft types being the British RE 8, French Breguet 14A.2 and Salmson 2,

I-16 Rata of the Soviet Air Force

Italian Pomilio, and various German and Austro-Hungarian "C-types." Maritime reconnaissance for the Allies was carried out by SHORT seaplanes from a variety of tenders.

By 1939 specialized reconnaissance and observation aircraft were taken for granted. Among them were the WESTLAND Lysander, HENSCHEL 126, Douglas 0-46, and Hanriot NC.530. Such aircraft in fact played little part in World War II. Reconnaissance was accomplished partly by versions of high-performance medium bombers, such as the B-24 LIBERATOR and B-25 MITCHELL, and partly by camera-equipped versions of high-speed fighters, such as the SPITFIRE, LIGHTNING, A-61 Black Widow, and MUSTANG. The MOSQUITO was in fact originally ordered for long-range high-altitude reconnaissance, but Spitfires often flew at tree-top height and secured clear pictures from point-blank range by using oblique cameras with the film moving at a speed linked to the aircraft's ground speed to avoid blurring. Such a sortie secured the first close-up pictures of German ground-based detection radar. Another technique was the mapping mosaic, in which vertical or near-vertical cameras of long focal length took a carefully sequenced series of overlapping prints which when fitted together formed a continuous map.

The German's favorite World War II reconnaissance plane was the Focke-Wulf Fw 189 and Fw 200 KONDOR, while the Japanese used the excellent Mitsubishi Ki.46 "Dinah" and the Russians the versatile Pe-2 light bomber.

In the period following the war radar and electronic warfare resulted in a much more complicated reconnaissance aircraft. The main equipment of the U.S. Air Force in the 1950s consisted of the four-jet RB-45, the RB-47 STRATO-JET and the Convair RB-36, all able to

fly thousands of miles (the RB-47 was specially fitted for in-flight refueling). For more limited, tactical use, the RF-84F and RF-101C showed how special reconnaissance aircraft carrying multiple cameras could be developed from jet fighters.

The Russians, during this period, used a variety of converted aircraft, but developed flying boats, such as the BERIEV Be-6, Be-10, and Be-12, for maritime reconnaissance. Nato maritime reconnaissance used aircraft such as the Lockheed P-2H Neptune, the Avro Shackleton (later replaced by the Nimrod), the Franco-German Breguet Br.1150 Atlantique, the Lockheed P-3C Orion, and the S-3A carrier-borne Viking. Notable Japanese reconnaissance craft were the Kawasaki P-2J and the SHIN MEIWA PS-1.

A totally different kind of camera platform was the Lockheed U-2, built to cruise at a height in excess of 70,000 ft. (21,000 m), at which it was thought—erroneously—that the aircraft could not be intercepted. The U-2, which made numerous flights over the Soviet Union and China between 1957 and 1963, was one of the new breed of reconnaissance aircraft whose function is to probe the enemy's electronic environment as well as survey his land surface. The U-2 has been replaced by the spectacular Mach 3+ SR-71A, which operates at 100,000 ft. (30,500 m).

Elint (electronic intelligence) has increased greatly in importance; by 1960 it had become even more vital than ordinary land/sea photography. It involves sensing, measuring, classifying, and recording every electromagnetic emission over hostile territory, to yield information on the enemy's communications, missile guidance, navaids, and other electronic equipment. It demanded bigger and heavier reconnaissance loads, and was soon backed up by

Lockheed U-2 high-altitude strategic reconnaissance aircraft

infrared (IR) Linescan and side-looking airborne radar (SLAR). IR Linescan presents a picture of the heat over the enemy's territory, and shows, for example, which engine of a line of trucks was running, or where a vehicle or aircraft had lately been parked. SLAR gives a continuous strip picture (sensed by radar instead of visible light) of the ground on each side of the aircraft's track.

Today most tactical missions are flown by a standard supersonic fighter or attack aircraft equipped with a multi-sensor pod containing cameras, radars, SLAR, Linescan, and other devices to bring back the most complete picture of all that is going on in and above the hostile country. No matter how sophisticated reconnaissance airplanes become, the largest share of reconnaissance work in the foreseeable future is likely to be done by artificial satellites rather than manned aircraft.

Regia Aeronautica

The Italian air force, known as the Regia Aeronautica between the 1920s and the end of World War II, had its origins in the Corpo Aeronautica Militare (CAM), established in 1912. A year before, a small army air detachment, equipped with an airship and airplanes of various types had seen service in Tripolitania against the Turks. There was no air opposition, but during this brief campaign Italian army officers flew what are now recognized as the world's first bombing and reconnaissance missions.

At first, most of the CAM's equipment consisted of foreign aircraft, or foreign designs built under license in Italy. A notable early exception was the series of CAPRONI heavy bombers. These gave Italy a force of true strategic bombers almost as early as Czarist

Royal Air Force Nimrod on Nato maritime reconnaissance duties, seen here investigating a Russian destroyer

Lockheed P-3C Orion patrol aircraft of the U.S. Navy, derived from the Electra turboprop airliner. The "Sting" at the tail encloses Magnetic Anomaly Detection equipment for locating submarines

Russia's SIKORSKY-designed "Squadron of Flying Ships." Moreover, unlike most other air services of World War I, the CAM undertook a prolonged and effective series of strategic missions unrelated to the immediate needs of the land war, but aimed against Austro-Hungarian cities, industries, and ports. The Ca 4-series triplanes, with bomb loads of up to 3,197 lb. (1,450 kg), together with the Ca 5-series biplanes, undertook repeated missions of up to seven hours' duration across the Alps into Austria or against naval bases along the Adriatic. The last of these raids was mounted against Pola on October 22, 1918, by 56 Capronis and 142 flying boats.

The most numerous of the Italian-designed machines that served in World War I were the Pomilio P-type biplanes, which equipped 30 squadrons of the CAM. These powerful armed two-seaters could reach speeds of over 120 mph (194 km/h) with the 300-hp Fiat A-12*bis* engine, and carried up to four Vickers, Revelli, or Lewis machine guns. SIA, a subsidiary of FIAT, built the SIA 7B and 9B, predecessors of the Fiat R.2 reconnaissance bomber, while the SAML S.1 and S.2 served widely as artillery spotters, reconnaissance machines, and light bombers. The Ansaldo SVA (Savoia/Verduzio/Ansaldo) series of fighter, reconnaissance, and bomber biplanes, with Warren (diagonal) braced wings, were among the fastest warplanes of World War I, reaching 143 mph (230 km/h). The slightly slower (137 mph, 220 km/h) Ansaldo A.1 Balilla ("Hunter") was the last CAM fighter of the war; 108 were built by the time of the Armistice.

The strength of the Italian air force was built up again after Mussolini came to power in 1923. Its most important early aircraft was the Ansaldo A.300 biplane, a robust single-engined machine, usually powered by a 300-hp Fiat A-12*bis*, which gave good service in many roles throughout Italy's colonial empire. Caproni produced an extensive series of large biplane bombers and colonial bomber-transports able to fly in both combat and utility roles in remote areas. SAVOIA-MARCHETTI's twin-hull S.55 flying boat, first flown in 1925, was later used by General Italo BALBO on the world's first long-distance formation flights, while a special Regia Aeronautica high-speed unit twice won the Schneider Trophy with its seaplane racers, the MACCHI M.7 in 1921, and the M.39 in 1926. In 1934 Francesco Agello set a world speed record of 440.69 mph (709.209 km/h) in a Macchi MC.72 floatplane, while later in the 1930s Italian officers set a number of world

altitude records, culminating in Mario Pezzi's 56,046 ft. (17,083 m) in 1938. (Still a record for a piston-engined aircraft.) Through the 1930s the Regia Aeronautica had seen more active combat than any other European air force, fighting in Albania, Ethiopia, Libya, and in the Spanish Civil War, in which it tried out many of its chief combat types, including the Fiat C.R.32 biplane fighter and Savoia-Marchetti SM.79 SPARVIERO bomber.

When Italy entered World War II in June 1940, the Regia Aeronautica had a high reputation. Indifferent equipment (it possessed no strategic bombers, aircraft carriers, or long-range escorts) and poor morale soon took their toll, however, and despite much individual courage, the air force had accomplished relatively little when Marshal Badoglio capitulated in September 1943. Most of the Regia Aeronautica was quickly taken over by the Luftwaffe, but a number of aircraft escaped to join the growing Co-Belligerent Air Force, which fought alongside the Allies. After the war the CBAF was constituted as the Aeronautica Militare Italiano (AMI), and since Italy's admission to Nato in April 1949 this has grown into a substantial force. Its chief modern aircraft include Aeritalia (license-built Lockheed STARFIGHTER) F-104S interceptors and F-104G strike/reconnaissance aircraft; Aeritalia (formerly Fiat) G.91 strike/reconnaissance machines; C-130, C-119, and G.222 transports; and a wide variety of helicopters and smaller aircraft.

Research centers

The world's oldest aircraft research center was established at Farnborough, in England, where the original Royal Engineers Balloon Factory became the Royal Aircraft Factory in 1912 and the Royal Aircraft Establishment in 1918. It continues to this day as the major British aerospace research body. The RAF operates several centers for research into such specific fields as guided weapons, propulsion, and armaments. In France major research centers are the Centre d'Essais en Vol (test-flight center) at Brétigny (south of Paris) and Istres (near the Mediterranean); Centre d'Essais Aéronautique de Toulouse, which is responsible for aerodynamic, structural, systems, and high-voltage testing; Centre d'Essais des Propulseurs, Saclay (engine testing); Centre d'Essais des Landes (rockets and weapons); Centre d'Essais de la Méditerranée (missiles); CAEPE (rocketry, spaceflight, and propulsion); Centre d'Expériences Aériennes Militaires,

Mont-de-Marsan (military aerospace); and LRBA/SEP/G2P/SNPE, numerous centers concerned with rocketry and spaceflight.

In the United States the National Aeronautics and Space Administration (NASA) ranks as the largest non-military research organization, with field stations concerned with manned flight in aircraft as well as rocketry and spaceflight. U.S. Air Force facilities include the Atlantic Missile Range (with headquarters at Patrick AFB near NASA's Kennedy Space Center) and the Pacific Missile Range (headquarters at Vandenberg AFB, Cal., with an anti-ballistic missile test site more than 4,000 mi., 6,400 km, downrange at Kwajalein Atoll). The biggest and longest-established U.S. aerospace laboratory is at Wright-Patterson AFB, Dayton, Ohio, which as Wright Field carried out all Army Air Corps research in the years before World War II. Eglin AFB, Fla., is a center for weapons trials and operational research, and at Edwards AFB, in the Mojave Desert of California, is the Flight Test Center (a vast facility also occupied partly by NASA). Arnold Engineering Development Center, in Tennessee, is the Western world's largest engine and propulsion test establishment. Among U.S. Navy test facilities are the Navy Pacific Missile Range, with headquarters at Point Mugu, Cal., and the Naval Weapons Center at China Lake, Cal., which designed and tested the Side-winder air-to-air missile.

The Russian equivalent of NASA is Central Aero- and Hydrodynamic Institute, established in 1918 in Moscow at the initiative of N. Y. ZHUKHOVSKI. Today it exists at three locations: the original headquarters, new institute premises at Zhukovskaya outside Moscow, and the Institute of Hydrodynamics at Novosibersk.

Richthofen, Manfred von (1892–1918)

Germany's legendary "Ace of aces" in World War I, known as the "Red Baron" for his bright-red Fokker Dr.I triplane. Manfred von Richthofen began his military life at the age of 11 at the Wahlstatt Military School and in 1911 joined a cavalry regiment. He transferred to the air service in May 1915, initially as an observer, but won his pilot's badge on December 25, 1915. In August 1916, while serving on the Russian front, he was selected by BOELCKE to join the newly-forming *Jagdstaffel* 2 in France. The first of his eventual 80 confirmed victories was achieved on September 17, 1916; his 11th on November 23 was no less a pilot than the Royal Flying Corps' contemporary leading ace, Major L. G. Hawker. After his 16th victory, von Richthofen was appointed commander of *Jagdstaffel* 11.

On June 26, 1917, he was given command of *Jagdgeschwader* 1 (four *staffeln*)—later to be named "Richthofen's Circus" by the Allies—but three weeks later was shot down and seriously wounded. Returning to operations, von Richthofen brought down his 79th and 80th victims on April 20, 1918, but next day was himself forced down during combat with Sopwith CAMELS of 209 Squadron. He was found dead of bullet wounds, but a question still remains as to whether he was hit in the air or by ground fire.

Rickenbacker, Edward Vernon
(1890–1973)

American World War I fighter ace and winner of the Medal of Honor. A successful racing driver in the United States until he traveled to France in 1917 as General Pershing's chauffeur, he transferred to the Aviation Section in August of that year, initially as an engineering officer. After learning to fly he joined the 94th Aero Squadron in March 1918.

Flying first NIEUPORT 28s and later SPADS he had scored 26 victories by the Armistice, despite spending July and August in a hospital for a mastoid operation. In September he became the squadron commander.

After the war he was active in the automobile and aviation industries in the United States, and during World War II undertook various special overseas missions (on one occasion he and his crew survived 21 days on a life raft after his plane ditched in the Pacific). In 1953 he became chairman of Eastern Air Lines, which he had helped to create.

Rocket

Both the jet engine and the rocket generate forward thrust as a reaction to a jet of exhaust gases streaming backward out of a nozzle, but the rocket obtains the oxygen necessary for maintaining combustion from its own fuel, whereas the jet engine relies on atmospheric oxygen. A rocket can operate therefore in a vacuum, and becomes progressively more efficient the thinner the atmosphere.

The rocket is believed to be an invention of the Chinese, and the first report of its use, in warfare, dates from 1232. Over the centuries it was used from time

Manfred von Richthofen

Captain Eddie Rickenbacker and his Spad 13

to time in European warfare, but by the 1800s rocket missiles had been developed, notably in Britain by William Congreve and his son, into quite an effective weapon.

In 1926 the first successful liquid-propellant rocket was fired by Robert Hutchings Goddard in the United States, using gasoline and liquid oxygen as propellants. In the 1930s German scientists were engaged in rocket development and in 1942 successfully fired the V-2 ballistic rocket.

Rocket motors have either solid or liquid propellants. Solid-propellant motors are the simplest, consisting of a casing filled with propellant, narrowing to a throat, and then expanding into a nozzle at the rear. The casing itself forms the combustion chamber. The liquid-propellant rocket motor is considerably more complex and consists of storage tanks for the propellants, thrust or combustion chamber, igniter, turbopumps, gas generator, and associated valves and control devices.

In most rockets bi-propellants are used, fuel and oxidizer being separate; liquid hydrogen or kerosene as fuel and liquid oxygen as oxidizer is a typical combination. Some rockets, however, employ mono-propellants, such as high-strength hydrogen peroxide or hydrazine, which are decomposed into a propelling gas by reaction with a catalyst. To get the propellants into the combustion chamber of small rockets, a pressurized inert gas such as helium or nitrogen is used. Large rockets have a turbo-pump with a turbine spun by a

gas generator. The propellants enter the combustion chamber through injectors and mix in a fine spray.

Liquid-propellant rockets are more powerful than solid-propellant types and can be shut off and re-started. Solid-propellant rockets lack this capability, but can be stored for long periods (a great advantage for military missiles), whereas most liquid-propellant rockets require filling with propellants just before takeoff.

Although the rocket has been used primarily for space travel and for missiles, there have been a number of rocket-powered aircraft. The Me 163 KOMET was an early example, and there were a number of experimental American rocket planes (X-1, X-15, etc.). However, the amount of fuel a rocket requires necessitates excessively bulky tanks for any useful flight duration. Jet engines now develop sufficient thrust to satisfy all requirements for flight within the atmosphere, and even the use of auxiliary rockets to provide a briefly enhanced performance in combat planes or to assist heavily-loaded takeoffs is seldom practiced.

Rockwell International

U.S. aircraft builder established in 1973 after the merger of Rockwell and NORTH AMERICAN AVIATION in 1967 to form North American Rockwell. Rockwell International's staple products are both designs taken over from NAA, the T-2 Buckeye carrier-borne jet trainer and the Sabreliner executive jet aircraft.

The Sabreliner flew for the first time in September 1958 and has appeared in several versions, including military T-39 military trainers and civilian transports with various fan engines. The North American-designed OV-10 Bronco, based on experience in Vietnam, was planned as a tactical multirole light-attack, reconnaissance, and casualty-evacuation machine, with good STOL performance on two 715-hp T76 turboprops. Versions include counterinsurgency, multiple-sensor and jet-boosted target tug models.

Another Rockwell International product, on quite a different scale, is the B-1 long-range supersonic bomber, the most expensive combat aircraft ever built (in terms of unit price). Powered by four F101 turbofans of 30,000 lb. (13,590 kg) thrust, the swing-wing B-1 can carry some 50,000 lb. (22,000 kg) of weapons including up to 32 Short Range Attack Missiles or Air Launched Cruise Missiles.

Rockwell general aviation products include various models of Commander high-wing twins, some with cabin pressurization and turboprop engines, and the Thrush Commander agricultural aircraft, produced in large numbers.

Roe, Alliott Verdon (1877–1958)

A pioneer of British aviation, Roe built and flew a 24-hp CANARD biplane in 1908, the first powered aircraft to fly in Britain, although only in short hops and not officially credited. A year later he built a machine with triplane wings and tail surfaces that enabled him to begin a series of successful flights, although its 9-hp engine remains one of the smallest motors with which controlled flight has ever been achieved. By 1910 this aircraft had been fitted with a 35-hp Green engine and was carrying passengers; it is generally recognized as the world's first fully successful triplane. In 1911 the first aircraft bearing the name AVRO (A. V. Roe, Ltd.) made its appearance. Roe sold his interest in AVRO in 1928 and bought a share in S. E. Saunders Ltd., forming SAUNDERS-ROE.

Rolls-Royce

Today the largest aircraft engine manufacturer in Western Europe, Rolls-Royce's activities in the field of aviation date back to 1915, when a liquid-cooled 60° V-12 engine was first bench-tested. Later known as the Eagle, and developed to give 360 hp, this unit equipped the FE2D, the HANDLEY PAGE 0/400, and the Vickers VIMY, as well as several flying boats and airships. Over half the

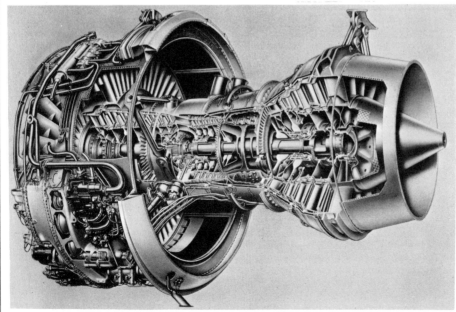

Cutaway view of the Rolls-Royce RB.211 engine, a huge but remarkably quiet high bypass ratio turbofan for the Lockheed TriStar. Unforeseen problems with the development of such a radically new engine, although finally overcome, were a major factor in the 1971 bankruptcy of Rolls-Royce

engines used in British aircraft during World War I were in fact manufactured by Rolls-Royce; they included 4,000 Eagles.

The V-12 layout was to be used on a long succession of Rolls-Royce engines. The FX engine of 1926, which later became known as the Kestrel, had aluminum alloy cylinder blocks and equipped the famous FURY and other Hawker biplanes of the 1930s, eventually producing 740 hp. For the 1929 and 1931 Supermarine Schneider Trophy seaplane racers the supercharged R-type engine was developed. It was a 2,780-hp V-12 which led to the Merlin (the PV.12 of 1933). Initially producing 1,045 hp, this V-12 unit achieved fame as the power plant of the SPITFIRE, the MUSTANG, the LANCASTER and the MOSQUITO, its later versions producing almost 2,000 hp. From the Merlin came the V-12 Griffon, which eventually developed 2,450 hp and equipped later Spitfires.

Rolls-Royce's first jet engine was the Welland (1943), a centrifugal engine of 1,700 lb. (770 kg) thrust which was installed in early METEORS. This was followed by the Derwent (1944), eventually developing over 4,000 lb. (1,612 kg) thrust, and the Nene (1944; 5,000 lb., 2,665 kg thrust). Rolls-Royce subsequently turned to axial-flow turbojets and produced the Avon, which powered the HUNTER. In its 300 series (RB.146) form, with a 16-stage compressor and cannular combustion chambers, it developed up to 17,000 lb. (7,700 kg) thrust.

Later civil turbojet designs have included the ultra-lightweight RB.162 (4,500 lb., 2,040 kg thrust), the problem-ridden advanced-technology RB.211 three-shaft turbofan of 53,000 lb. (24,000 kg) used in the Lockheed TRISTAR, and the Spey (1960), which powers the TRIDENT, BAC 1-11, Fokker Fellowship, and Grumman Gulfstream. The military version of the Spey generates over 12,000 lb. (5,436 kg) thrust and is used in the Nimrod and RAF PHANTOMS.

In 1971 Rolls-Royce went into bankruptcy; since that time Rolls-Royce's aircraft engine interests have been operated as government-owned industry. It is associated with a number of international projects, involving Allison, SNECMA, Turboméca (the Adour two-shaft turbofan for the JAGUAR), and Turbo-Union (the RB.199 three-shaft augmented turbofan for the TORNADO). In the late 1960s Rolls-Royce absorbed Bristol-Siddeley Engines, thus taking over development of the Olympus (the first British two-spool turbojet, which delivers 20,000 lb., 9,000 kg, thrust and

The 50-hp Gnôme, ancestor of the rotary engines of World War I

powers the VULCAN), the three-stage axial-flow Pegasus turbofan (used in the HARRIER), the single-spool Orpheus (employed in the Gnat), and the Viper, which powers the Aermacchi MB 326 trainer.

Rotary engine

A rotary engine is a type of radial engine in which the crankshaft remains stationary and the crankcase carrying the cylinders rotates around it to drive the propeller. Engines of this type were widely used in fighters during World War I. They included the Oberursel employed by Fokker EINDECKERS and TRIPLANES, and the Le Rhône, Clerget, and Bentley engines of Britain's Sopwith CAMELS, PUPS, and TRIPLANES, and France's NIEUPORTS.

The very large mass of the rotating engine tended to produce a marked swing to one side when the throttle was opened. It also made a fighter sluggish in turning in one direction but very quick in turning in the other.

Royal Air Force

Early British military flying was by the Royal Flying Corps, formed in May 1912, which had an Army and a Navy wing. A separate Royal Naval Air Service was established in November 1913, but on April 1, 1918, the RFC and the RNAS became the Royal Air Force, with Sir Hugh Trenchard as the first Chief of the Air Staff. By the end of World War I the new independent British air arm had 99 operational squadrons on the western front, but the postwar demobilization rapidly reduced this to a mere 20 or so operating in Europe, mostly undermanned as personnel returned to civilian life. In Britain, the Middle East, and India, however, there were another 144 squadrons.

In the 1920s a modest expansion took place, although naval flying became the responsibility of the Fleet Air Arm (1924). New aircraft were introduced (Siskin, Grebe, Gamecock fighters; Fairey IIIF, Wapiti, Virginia and Hyderabad bombers) and an RAF College was established together with an apprentices' college.

The RAF was involved in the suppression of terrorist insurrections in Iraq during the 1920s, using principally DH9A aircraft, but still found time to complete many exploratory flights over then unknown territory. An air mail service was maintained between Baghdad and Cairo for several years, four Southampton flying boats flew from England to Australia and then back to Singapore in 1927–1928, and flights were made to Africa with the FAIREY Long Range Monoplane; a High Speed Flight (unit) was established (primarily for the Schneider Trophy team) and there was a Long-Range Development Flight at Upper Heyford.

In the 1930s biplanes were gradually replaced by monoplanes (HURRICANE and SPITFIRE fighters; Hampden, Whitley and WELLINGTON bombers), but the Hawker biplanes that had been such a feature of the RAF's annual Hendon air display in the early 1930s played an important policing role on the turbulent Northwest Frontier of India.

During World War II the RAF expanded its fighter force after the Battle of BRITAIN to fulfil an offensive role, using the very fast Typhoons and Tempests, while Bomber Command concentrated on night raids and area-bombing using STIRLINGS, HALIFAXES and LANCASTERS. Coastal Command's Sunderlands and LIBERATORS fulfilled a maritime patrol function with Beaufighters and MOSQUITOES for anti-shipping strikes, and overseas commitments included Burma and North Africa (the Desert Air Force, which eventually followed the advancing allies through Sicily and into Italy).

Postwar contraction coincided with the replacement of piston-engined combat aircraft with jets (METEOR, VAMPIRE and HUNTER fighters; CANBERRA, Valiant, VULCAN and Victor bombers).

A strong force was maintained in Germany as a Nato contribution in the 1950s, but the RAF had become a much smaller, highly efficient force with 500 first-line aircraft by 1977, equipped principally with PHANTOMS, LIGHTNINGS, Vulcans, JAGUARS, HARRIERS, Nimrod maritime reconnaissance aircraft, and Buccaneer strike planes. The RAF will shortly be taking delivery of the new multirole Panavia TORNADO.

R-planes

The largest German bombers of World War I were the R-planes (Riesenflugzeug) or "giants," built by a number of manufacturers. These enormous machines had from three to six engines and a crew of as many as seven. Over a short distance they could carry 4,400 lb. (2,000 kg) of bombs, including 2,200-lb. (1,000-kg) bombs—the largest to be dropped from an airplane during World War I. The maximum range of the R-planes was about 500 mi. (800 km) with a reduced bomb load.

The only R-plane to achieve production status was the four-engined Zeppelin-Staaken R.VI, with a wingspan of 138 ft. 5 in. (42.19 m) and a length of 72 ft. 6 in. (22.10 m). Eighteen of these aircraft, which first saw action in 1917, were built. The Zeppelin-Staaken series originated in the three-engined V.G.O. I and V.G.O. II of 1915, which were sent to the eastern front but proved to be underpowered. The V.G.O. III and R.IV were both six-engined; the five-engined R.V led to the R.VII, which crashed on its delivery flight. The last Zeppelin-Staaken R-planes to see action were the three R.XIVs (1918), one of which was shot down by the RAF over the western front. The R.XVs were completed too late to fly operationally.

The AEG R.I, with four engines buried in the fuselage, never entered operational service. Only two were built, and one crashed when a propeller disintegrated. The DFW R-plane design also had four buried engines, driving propellers by bevel gears. The sole DFW R.I flew on the eastern front from April 1917, but only two examples of the DFW R.II had been completed by the end of the war.

The ungainly Linke-Hofmann R.I of 1917 crashed when its wings collapsed. The four buried engines of the R.II drove a huge single propeller, 22 ft. 8 in. (6.90 m) in diameter, the largest ever used on an aircraft. Two R.IIs were built, but only one was ever flown (after the Armistice).

The eight Siemens-Schuckert three-engined R-planes were notable for their fork-tails, intended to give the rear gunner a better field of fire. Four of these aircraft saw service in Russia, but the six-engined R.VIII, with a wingspan of 157 ft. 6 in. (48.00 m) and an orthodox fuselage, was never flown. It would have been the heaviest German aircraft of the war.

R-planes operated on both fronts. From bases in the Ghent area they attacked England and targets in France as far away as Le Havre. None of the Zeppelin-Staaken R.VIs was brought down over England, and they succeeded

Zeppelin-Staaken R.VI, most successful of all the German R-planes

in dropping 59,926 lb. (27,150 kg) of bombs in the course of raids on Britain between September 1917 and May 1918.

Russian Air Force

Although a Military Aeronautical Cadre with balloons was formed at St. Petersburg in 1885, the Russian Air Fleet dates from February 6, 1910. The first army and naval officers were sent to France to learn to fly a month later. The air force was initially only a small section of the War Department or Admiralty administration, but on September 24, 1916, the Administration of the Military Air Fleet (UVVF) was created as part of the War Department.

At the beginning of World War I there were 39 Army Aviation Detachments equipped with 263 airplanes, as well as some naval aviation units. The famous bomber unit which was equipped with SIKORSKY's giant four-engined ILYA MOUROMETZ was formed on December 10, 1914, and called the Squadron of Flying Ships.

The Soviet air force, the VVS (Voenno-novozdushnye sily) came into being on November 10, 1917, when the Commissariat for Aviation and Aeronautics was set up. During the next two and a half months, six Socialist Aviation Detachments (each consisting of 12 aircraft) were created on the basis of the former Csarist air force units. This air force took part in the Civil War although there is little evidence of air battles between the two sides and in 1919, of 79 Aviation Detachments (usually 2–6 aircraft) only 12 were fighters and even these were used mostly against ground forces.

The Chief Administration of the Workers' and Peasants' Red Military Air Fleet (GU RKKVVF) which took over the role of the Csarist Government's UVVF, was created on May 24, 1918. On April 15, 1924, it became the Administration of the Military Air Forces (UVVS). During these early years while the air force was being built

up there were several reorganizations, but there were relatively few changes in commanders-in-chief despite a fairly rapid turnover during the pre-war purges. In January 1938 the control of naval aviation was given back to the Navy (VMF).

The untried VVS gained combat experience in Spain from 1936 to 1939, and in 1938 Russian "volunteer" airmen backed up the Chinese Communists in operations against the Japanese; in 1939 the VVS was officially in action against the Japanese at Khalkin Goll, and later that year against Finland.

The German invasion of the USSR in June 1941 caught the VVS on the ground and on the first day about 1,200 aircraft were lost. The VVS fought desperately in the early days of the war with obsolescent equipment, with the aircraft industry in a state of changeover to the five basic types (Yak-1/9/3, LaGG-3/La-5/La-7, Pe-2, Tu-2, Il-2) used during the war. The first "Heroes of the Soviet Union" rammed their opponents and much publicity was given to this method of destroying enemy aircraft. Apart from token raids against Germany, almost all bomber operations were in support of Russian ground forces. The SHTURMOVIK had been specially designed for this and the Pe-2 bomber was often used as a dive-bomber against tactical targets. Soviet fighters were also used extensively in the ground attack role and almost half of all wartime combat sorties were of a tactical nature.

A Soviet innovation was the use of old (1928) U-2 (Po-2) training aircraft for night bombing or nuisance raids, drifting engine-off over the lines and dropping darts and small caliber bombs.

From the beginning of 1943 the VVS began to take the initiative with the newly introduced La-5, Yak-9, and Tu-2. The Russians eventually claimed 57,000 German aircraft destroyed on the ground and in the air between 1941 and 1945.

Since the war the VVS has not been

officially in action, although Soviet pilots have served as "volunteers" in Korea and as advisers in many other countries, particularly in the Middle East.

The VVS today is the largest air force in the world, with a front line strength of about 10,000 aircraft. There are five commands: Frontal Aviation (FA) is by far the largest and flies principally the MiG-21, -23 ("Foxbat") and -25, Su-7 ("Fitter"), Tu-28, Yak-28 ("Firebar"), and Mi-6, -8 and -24 helicopters; Air Defense Fighter Aviation (IA-PVO) has the MiG-19, -23 and -25, Su-9 ("Fishpot"), -15 and -17, Tu-28P and Yak-28P; Long Range Aviation (ADD) uses the Mya-4 ("Bison"), Tu-16 ("Badger"), -22 ("Blinder"), -95 and VG (the "Backfire"); Naval Aviation (A-VMF) operates the Be-12 ("Mail"), Il-38, Mya-4, Tu-16, -22 and -95, the "Forger" V/STOL fighter, and Ka-25 and Mi-8 helicopters; Air Transport Aviation (VTA), with the An-12 ("Cub"), -22 ("Cock"), -24 ("Coke") and -26, and the Il-14 ("Coot") and -18, is backed up by the Aeroflot fleet.

Ryan

Ryan Aircraft (originally Ryan Airlines) was founded in 1922, and its earliest staple product was the M-1 high-wing cabin monoplane, generally powered by a 300-hp Wright Whirlwind engine. From this aircraft was developed the special long-range NYP (New York/Paris) SPIRIT OF ST. LOUIS, built for Charles LINDBERGH, and the six-seat Brougham monoplane. During the Depression the firm went out of business, but Ryan himself had severed his connections with the company in 1927, devoting his energies to the Ryan School of Aeronautics at San Diego. In 1933 he reentered manufacturing, establishing Ryan Aeronautical at Lindbergh Field, San Diego, and building the S-T series of low-wing trainers, with spatted wheels and wire-braced wings. Thousands of aircraft in the series were built as standard primary trainers during the years 1939–45; the principal types were the PT-16, PT-20, PT-21, PT-22, powered by in-line Ranger or radial Kinner engines.

In 1944 Ryan produced an advanced combat aircraft, the FR-1 Fireball carrier-based fighter, powered by a nose-mounted piston engine and a jet in the tail, and with features that included nosewheel landing gear. Although technically successful, it was made in only modest numbers immediately after the end of the war. The XF2R jet/turboprop fighter remained no more than a prototype, and Ryan's last association with

production aircraft was to build the Ryan Navion lightplane designed by NORTH AMERICAN. Ryan did, however, also design several important VTOL aircraft. It participated in the LTV-Hiller-Ryan XC-142 (1964), one of the most successful CONVERTIPLANE designs, and it produced the propeller-driven VZ-3RY (1958) and the XV-5B (1963), the latter using wing-mounted lift fans powered off the conventional jet engines and covered during normal flight. In 1969 Ryan became part of Teledyne; today Teledyne Ryan is a leading producer of target drones and RPVs (remotely piloted vehicles), many of original Ryan design.

S

Saab

Svenska Aeroplan Aktie Bolag (Saab), founded in 1937, is Sweden's principal aerospace manufacturer. Among its first products were the Saab-18 single-engined reconnaissance-bomber, produced as the B17A (with a Swedish-built Pratt & Whitney Twin Wasp engine), B17B (Swedish-built Mercury 30 engine), and B17C (imported Piaggio P.XII*bis*), and corresponding reconnaissance versions including seaplanes. The Saab-18A twin-

The Saab-37 Viggen, which features an unusual canard layout, photographed during a looping maneuver

engined bomber made its maiden flight in 1942, and reached 289 mph (465 km/h) on two Pratt & Whitney engines; the B18B attained 357 mph (575 km/h) on two Daimler-Benz DB 605Bs imported from Germany. A total of 182 were delivered between 1944 and 1948; the series included 62 torpedo bombers armed with a 57-mm cannon. In the mid-1940s Saab also produced 299 single-engined twin-boom J21 fighters, powered by a pusher propeller and with armament that could include a cannon gun and nine 13.2-mm heavy machine guns. A Goblin turbojet was later substituted for the original piston engine to produce the highly successful J21R, 60 of which served as fighter and attack aircraft (the only example of a direct conversion from piston to jet engine in otherwise similar production aircraft).

Production of the Scandia twin-engined airliner was cut short in the late 1940s to release capacity for the Saab-29 Tunnan ("Flying Barrel"), the first transonic swept-wing fighter made in Western Europe. The Saab-29 made its maiden flight in 1948; 661 were produced in six main fighter (J29), attack (A29), and photo-reconnaissance (S29) versions and served with the Swedish air force until 1976. In the same period the trim Saab-91 lightplane, with retractable nosewheel landing gear, was beginning production: it has proved to be a popular trainer with many of the world's air forces. At the end of 1952 Saab flew the first prototype of the tandem-seat Saab-32 Lansen, initially intended as an attack aircraft, and powered by a license-built Avon turbojet with afterburner. Eventually some 450 were delivered; these included a night and all-weather fighter version and a multisensor reconnaissance aircraft. Variants with more powerful engines served until 1976.

The next major aircraft to enter production was the Saab-35 DRAKEN fighter, which made its maiden flight in 1955 and has seen considerable success in the air forces of several Scandinavian countries. The latest Saab aircraft is the delta-wing Saab-37 Viggen multirole combat aircraft. Powered by an afterburning RM8 turbofan of 26,000 lb. (11,780 kg) thrust, the Viggen has a canard configuration with small foreplanes with flaps and is capable of Mach-2 speeds. First flown in 1967, Viggen versions already in production include the AJ37A attack (about 100 delivered), SF37 multisensor reconnaissance, SH37 maritime reconnaissance/ attack, and SK37 dual trainer. Deliveries of the Ja37 fighter began in 1977. Powered by an RM8B engine of 28,100 lb. (12,730 kg) thrust, it employs

multimode pulse-doppler radar of medium pulse-frequency and is armed with a 30-mm Oerlikon KCA high-velocity cannon and missiles that include British Skyflash and Swedish Rb 72 air-to-air weapons.

Saab also produces two smaller families of aircraft. The Saab 105 twin-jet trainer began its career with Turboméca Aubisque turbofans, first flying in 1963, and is now produced with General Electric J85 turbojets; 150 of the original model were delivered, and 40 of the 105Ö type for Austria; the 105G can carry a weapon load of 5,150 lb. in attack missions. The MFI-15, taken over in 1968 with Malmö Flygindustri, is now produced as the Saab Safari civil trainer and utility aircraft. The Supporter has been developed from the Safari to serve as a multirole light military machine.

Sabena

Société Anonyme Belge d'Exploitation de la Navigation Aérienne (SABENA) was formed in 1923, as the successor to SNETA, which had established routes to London, Paris, and Amsterdam in 1919. Sabena opened passenger services to Basel in 1924, and a three-engined Handley Page airliner was acquired as a prototype for 10 aircraft constructed in Belgium between 1925 and 1927 for European and projected African routes. Belgian-built, 17-passenger Savoia Marchetti SM.73s arrived in 1935, followed by SM.83s in 1938, by which time the route network totaled 10,565 mi. (17,000 km).

The European network reopened in 1946 with DC-3s and DC-4s, and the first Brussels–New York service was flown on June 4, 1947, employing DC-6s. Short-haul DC-3s were replaced by Convair 440s and, eventually, CARAVELLES, while DC-4s and DC-6s gave way to DC-7s in 1955. Boeing 707s inaugurated a North Atlantic jet service on January 23, 1960. Modernization began in 1966 with Boeing 727s, followed by 737s and 747s. Present routes serve Europe, the Middle East, South Africa, and the Americas. Some 1,680,000 passengers are carried annually.

F-86 Sabre

F-100 Super Sabre

Sabre

The prototype of the North American F-86 Sabre, first flown in October 1947, was the first U.S. swept-wing fighter. Early production F-86A Sabres were sent to Korea late in 1950 to counter the MiG-15. Although heavier and inferior in climb and ceiling, the Sabre was still able to gain superiority over the Russian aircraft, particularly when the more advanced F-86F was introduced in 1952. Wingspan of the F-86F was 37 ft. 1 in. (11.30 m) and length 37 ft. 6 in. (11.43 m). The F-86D was an all-weather version with a radar nose and an armament of 24 air-to-air rockets instead of the interceptor's six .50 nose guns. Early Sabres used the General Electric J47 turbojet and had maximum speeds of between 650 and 690 mph (1,045 and 1,110 km/h). The F-86H fighter-bomber had the more powerful J73 and could exceed 700 mph (1,126 km/h). A total of 8,664 Sabres were built, including some in Canada and Australia, and were used by large numbers of the world's air forces.

The North American F-100 Super Sabre was the first true supersonic fighter. Flight testing began in 1953, and although aerodynamic and control problems delayed service acceptance the F-100 eventually became the major U.S. Air Force tactical fighter in the mid-1950s. A total of 2,294 Super Sabres were built, and they saw extensive service in Southeast Asia between 1964 and 1971. Powered by a Pratt & Whitney J57 engine, the principal model, the F-100D, had a maximum speed of 860 mph (1,380 km/h). Armed with four 20-mm cannon, the Super Sabre could carry an ordnance load of 7,500 lb. (3,400 kg).

Sailplane

The earliest successful flights with heavier-than-air machines were made with gliders (LILIENTHAL, the WRIGHT BROTHERS, CHANUTE, PILCHER).

As the use of thermals was explored and developed, gliders evolved from the very light, sensitive designs of the 1920s intended for hill soaring, to sailplanes specifically conceived for turning in thermals at maximum rates of climb and flying across country at speeds of 150 mph (241 km/h) (see GLIDING).

Instrumentation for cloud flying was installed and robust methods of construction were adopted to withstand turbulence, although dive brakes had to be used to limit diving speeds. With the emphasis on soaring, wing loadings rose and the minimum sinking speed ceased to be a primary consideration. Sailplanes, or gliders, are unpowered

The long, narrow wings of a typical sailplane: the Eon Olympia

heavier-than-air planes. They are usually towed up into the air by a small powered airplane and released at an altitude of 2,000 to 3,000 ft. (610 to 915 m), but can also be launched like a kite by an automobile or a winch; more primitive methods include pushing or catapulting the glider off a hilltop, while modern powered sailplanes or "motor gliders" have an auxiliary engine for takeoff and for emergency use in the air. Once aloft, sailplanes can simply glide back to earth, or they can make use of rising currents of air for soaring, and by spiraling up such upcurrents or "thermals" they may ascend to heights of over 10,000 ft. (3,050 m) and remain airborne for many hours. Modern sailplanes have made flights of over 900 mi. (1,448 km), climbed to over 40,000 ft. (12,192 m), and reached speeds of 150 mph (241 km/h). Although flown mainly for sport, sailplanes have been employed as troop carriers, and are used in meteorological and aeronautical research.

Any plane can glide, but a normal powered craft will descend rapidly as it is relatively heavy and its wings are designed to provide high lift only at fairly high airspeeds. A sailplane designer seeks to achieve maximum lift with minimum weight and drag. Long, narrow wings are used, sometimes with an aspect (length to width) ratio as high as 20:1 to produce the greatest efficiency

with minimum induced drag, and any taper is usually straight rather than curved to simplify construction. For maximum performance the wing's AIRFOIL shape is a thin laminar flow cross-section (see AERODYNAMICS; WING), and as such is very sensitive to any surface irregularities; even dust and raindrops need to be removed before flight. Construction materials include plywood, aluminum, fiberglass, and fabric, for minimum weight.

With high aerodynamic efficiency and low weight, a sailplane descends at a very shallow angle, and performance is a measure of this angle expressed as the ratio between distance traveled horizontally and loss of altitude. Thus a medium performance sailplane with a glide ratio of 25:1 sinks 1 ft. (0.3 m) for every 25 ft. (7.5 m) of forward travel, and has a "sink speed" of 2½ ft. (0.8 m) per second. The glide ratio on a high-performance sailplane may exceed 40:1.

Controls and instruments. Cockpit controls comprise a column for the elevator and ailerons, pedals for the rudder, and a hand-operated spoiler lever. This raises hinged wing panels to reduce lift and is mainly used to increase diving angle during a landing approach. It is also sometimes needed to reduce excessive speed in a dive, or to control excessive lift in a particularly strong upcurrent. Long, narrow ailerons are usual, while elevators may be attached to a fixed tailplane in the normal way or an "all flying" tail may be used. Some sailplanes have V or T shaped tails (the latter design increases the effect of the fin), but a conventional inverted T is usual.

The instruments include altimeter, airspeed indicator, compass, turn-and-slip indicator, artificial horizon, and a variometer which responds to changes in pressure and indicates rate of climb or sink. In addition sailplanes designed for aerobatics may have an accelerometer as a precaution against excessive acceleration (which can damage the structure), while those designed for long distance and high altitude flights normally carry two-way radio and oxygen.

History. Gliders were the earliest successful man-carrying heavier-than-air flying machines. The first to fly (1853) was built by Sir George CAYLEY, and this was followed during the 1890s by the hang gliders of Otto Lilienthal (Germany), Percy Pilcher (Britain), and Octave CHANUTE (United States). Their experience and successes encouraged and assisted the work of the Wright Brothers, and in recent years has led to the popular sport of hang gliding.

Hang gliders are controlled by body movements, and the early types were not designed for soaring. The first soaring flight was achieved in a Weiss glider in England (1909), but most pioneering work in sailplane design and the use of air currents took place in Germany in the 1920s and 1930s, during which period the variometer was evolved, sailplanes were equipped with instruments for cloud flying, and more robust construction enabled the craft to withstand turbulence. After World War II synthetic waterproof adhesives became available, together with the results of wartime research on airfoils, and during the 1950s low-drag wing designs developed by NACA came in use, as did fiberglass construction. A further reduction in drag was achieved in 1960 by adopting a reclining position for the pilot, and this was followed by the development of metal construction, retractable undercarriages, and various other refinements.

Saint Exupéry, Antoine de (1900–1944)
Pioneer French airline pilot, essayist and novelist. He learned to fly privately during military service at Istres (1921), qualified as an air force pilot at Casablanca and was posted to the 34th Air Regiment, Le Bourget (1922).

St. Exupéry joined Lignes Aériennes Latécoère on the Toulouse–Casablanca mail services (extended to Dakar in 1925), flying Breguet 14 biplanes over the desert. He later opened the Aéropostale Buenos Aires–Patagonia route in a Latécoère 25 (November 1929) and developed night flying on the Rio de Janeiro–Buenos Aires–Natal route.

He became a test pilot to Latécoère (1930–31), flew Mediterranean float-plane services (1932), was a news correspondent in Russia (1935) and Spain (1936), and crashed in Egypt during a Paris–Saigon record attempt in a Caudron Simoun (December 1935). With Henri Guillaumet he flew a six-engined Latécoère 521 flying boat *Lieutenant de Vaisseau de Paris* to New York and back in July 1939, but was recalled to the air force in September 1939 and sent to the United States in 1940.

On July 31, 1944, he failed to return from a reconnaissance mission over the Mediterranean in a Lightning. Although an important pioneer of civil aviation, St. Exupéry is best known as the poet of flying. His books celebrate the heroism and adventure of aviation, which for him provided an ideal and a purpose. He records in simple but moving language his belief in civilization, and in the mystical experience of the pilot facing danger on behalf of mankind. Books such as *Terre des Hommes* (1939—*Wind, Sand and Stars*) and *Pilot de Guerre* (1942—*Flight to Arras*) describe his own experience, while his novels include *Courrier-Sud* (1929—*Southern Mail*) and *Vol de Nuit* (1931—*Night Flight*).

Sakai, Saburo (1916–)
Japan's best known fighter ace of World War II, Sakai scored at least 62 victories in China and the Pacific and finished the war as his country's third-ranking and senior surviving pilot. Sakai first saw combat outside China on December 8, 1941, when he flew with his ZERO Squadron 450 mi. (725 km) from Formosa to the Philippines to raid Clark Field. In February 1942 he fought over Java, and then saw service in New Guinea. In August, on a long-range mission from Rabaul to Guadalcanal, he was badly wounded by the rear gunner of an AVENGER, losing the sight of one eye and only just succeeding in bringing his damaged Zero back to base.

After leaving hospital in January 1943 he remained in Japan as an instructor until he was assigned to Iwo Jima in the summer of 1944. He was eventually assigned to the home island of Shikoku, to intercept B-29 Superfortress raids, moving with his unit to Kyushu in the spring of 1945. His last victory was the shared destruction of a Superfortress on the night of August 13/14.

Sally
Probably the best known Japanese bomber of World War II, the twin-engined MITSUBISHI Ki.21 Type 97 first saw service over China in 1938, and by the time of Pearl Harbor it was established as the Army's standard bomber.

With 850-hp engines, a top speed of 270 mph (435 km/h) and a 1,650-lb. (750-kg) bomb load, the Ki.21 was a formidable bomber of its day, although its defensive armament of three machine guns proved inadequate. A new version, therefore, had five or six machine guns, some armor protection for the crew, and partially leak-proof tanks—the original Ki.21 had proved to be readily ignitable when hit.

For the advance through Malaya to New Guinea in 1941–42 an improved Ki.21, code-named "Sally" by the Allies, became available. The Ki.21-II, with 1,490 hp engines, still remained vulnerable, however, and a pedal-operated dorsal turret mounting a .50 machine gun was fitted to the Ki.21-IIb, together with still more armor protection for the crew. Its wingspan was 72 ft. 10 in. (22.20 m) and length 52 ft. 6 in. (16.00 m).

Not until September 1944 did "Sally" go out of production, by which time a total of over 1,800 were built.

Santa Cruz, Battle of
A large-scale air battle in the fighting for Guadalcanal and Papua during World War II. By late September 1942, the U.S. Navy had only a single aircraft carrier (the *Hornet*) operational in the Solomon Islands area. Taking advantage of this weakness, the Japanese attempted to retake Henderson Field, the vital airfield on Guadalcanal, and troops were landed on October 15 after sea bombardment. A Japanese task force that included the carriers *Shokaku*, *Zuikaku*, *Zuiho*, *Junyo*, and *Hiyo* stood by to find and destroy the *Hornet* as soon as Henderson Field had been seized, but American counterattacks drove the invaders back.

By October 25 the *Enterprise*, damaged by bombs two months earlier, had reached the Solomons from Pearl Harbor, and early the following day the opposing carrier forces were each located by their opponent's patrol planes. The two sides launched massive air strikes, with 133 Japanese aircraft heading southeast meeting 74 American planes on a northwest track. While

WILDCATS and ZEROS fought each other off, the Japanese "Val" dive-bombers and "Kate" torpedo-bombers broke through to cripple the *Hornet*, while U.S. DAUNTLESSES set fire to the *Shokaku*. In addition, two patrolling Dauntlesses had found the *Zuiho* and severely damaged her flight deck with bombs. Just before midday Japanese strike planes located the *Enterprise*, hitherto hidden beneath a squall, and bombed her flight deck, making it impossible for her to launch aircraft, while in the afternoon the "Vals" and "Kates" hit the *Hornet* again to leave her in a sinking condition.

With air supremacy now virtually theirs, the Japanese forces were nonetheless forced to withdraw with almost empty fuel tanks. The delay while the American defenders of Henderson Field repulsed attacking Japanese troops had forced the Japanese carriers to use up their fuel reserves while waiting in vain for the seizure of the vital airfield.

Santos-Dumont, Alberto (1873–1932)

Brazilian aviation pioneer, one of the most colorful and ingenious experimenters with balloons, dirigibles and powered, controlled flight in the earliest days of flying. Santos-Dumont, who worked in France, first visited Europe in 1891 with his father, a successful coffee planter. He learned how to fly a balloon from Henri Lachambre and Alexis Machuron, who then built him a small lightweight balloon to his own design. In 1898 he flew a powered non-rigid airship, and by 1899 had developed the design sufficiently to steer it around the Eiffel Tower. His airship crashed on the Trocadero Hotel during a flight over Paris in July 1901, but in September of the same year (after some dispute) he won the Deutsch Prize for a flight from St. Cloud to the Eiffel Tower and back, within half an hour.

In 1902, Dumont moved to Monaco, where Prince Albert I had offered to build him an airship hangar. The dapper Brazilian made various successful flights off the coast until his airship was wrecked on February 14, 1902, as a result of incorrect inflation.

By 1903 he was experimenting with two large airships (No. 7 and No. 10) while using his small No. 9 to fly all over Paris, occasionally astonishing his friends—and strangers on the street—by dropping in from above. His well-timed appearance at the July 14 Review at Longchamp racecourse resulted in an agreement to cooperate with the French military authorities.

Rumors of heavier-than-air flights by the WRIGHT BROTHERS reached Santos-Dumont in 1904. He experimented on aerodynamics, drew up designs for helicopters, and in 1905 embarked on the construction of an airplane with the assistance of Gabriel VOISIN.

This machine had dihedral box-kite wings spanning 33 ft. (10.05 m), a boxed canard elevator-and-rudder unit, and a 24-hp engine driving a pusher propeller. Initial trials were carried out with the airplane secured below airship No. 14, and the heavier-than-air device was named the No. 14*bis*. A 50-hp engine was installed, the lateral control function of the canard structure was dispensed with, and on October 23, 1906, Santos-Dumont succeeded in flying about 200 ft. (65 m) at a height of 6–10 ft. (1.8–3.0 m), thus winning the Archdeacon Prize for the first sustained flight of over 25 m (82 ft.). After modification to give a measure of lateral control, the No. 14*bis* secured the Aero Club Prize on November 12, 1906, by flying 722 ft. (220 m) in 21¼ seconds. This was the first accredited sustained flight made by an airplane in Europe.

In 1909 Santos-Dumont flew his *Demoiselle*, a small monoplane with a wingspan of only 17 ft. 9 in. (5.41 m) and a length of 26 ft. (7.92 m), powered by a flat-twin 35-hp engine driving a tractor propeller (later developments of this design had engines of greater power).

In the spring of 1910 Santos-Dumont developed symptoms of multiple sclerosis. He never designed or piloted another aircraft and eventually committed suicide.

SAS

Scandinavian Airlines System (SAS) was formed in 1946 by Swedish Airlines (ABA), Danish Airlines (DDL), and Norwegian Airlines (DNL), with the purpose of linking the Scandinavian capitals with the United States and South America. The initial fleet of seven DC-4s was augmented by DC-6s in 1948, the year that the European services of the parent companies were also merged. The route network was subsequently expanded to include East and South Africa, the Middle East and Far East. A transpolar route, bringing Tokyo to within only 30 hours of Copenhagen, was opened with DC-7Cs on February 24, 1957. Caravelle jetliners were introduced on the European routes in 1959, and the DC-7Cs on the Copenhagen–New York route were replaced by DC-8s on May 1, 1960. DC-8s gradually took over all intercontinental routes, flying their first service over the Pole on October 11, 1960, and their first South American route, to Rio de Janeiro, Buenos Aires, and Santiago, on December 5, 1961. The present fleet includes 2 Boeing 747s, 3 DC-10s, 12 DC-8s, and 47 DC-9s. Some 6,700,000 passengers are carried annually.

Saunders-Roe

The British manufacturer Saunders-Roe Ltd. was formed in 1928 when A. V. ROE, formerly founder of AVRO, joined the boatbuilders S. E. Saunders Ltd. and began production with a series of three flying boats. These were the twin-engined Cutty Sark amphibian, the three-engined Windhover, and the Cloud amphibian. In the years before the war Saunders-Roe also built the A-10 landplane fighter and the London and Lerwick flying boats, which were delivered to the RAF between 1935 and 1940. Probably the company's most famous product was the huge 200-passenger Princess flying boat. Powered by 10 3,780-hp turboprops, the prototype made its first flight in 1952, but only three of these last of the giant flying boats were ever built, and they never entered service. The SR-A1 fighter, the world's first jet flying boat, made its initial flight in 1947 and was a successful design, but the project was eventually abandoned.

Saro's Helicopter Division, formed 1951, took over Cierva Autogiro Co. Ltd. and built the twin-seat Skeeter scout

Saunders-Roe Princess, last of the great flying boats

helicopter for the RAF, the British Army, and West Germany during the 1950s, as well as the XROE rotorcycle, the Black Knight research rocket, and the long series of SR-N hovercraft. Saro became part of WESTLAND Aircraft in 1960.

Savoia-Marchetti

The Italian aircraft manufacturer Savoia(or SIAI)-Marchetti gained considerable fame in the interwar years, primarily as a manufacturer of flying boats. The early 1920s witnessed the appearance of the S.23 dual-control flying boat trainer, the 300-hp S.51 racer (which finished second in the 1922 Schneider Trophy race), the single-engined S.57 fighter/reconnaissance flying boat with a 250-hp V-6 Isotta-Fraschini engine driving a pusher propeller (developed into the 300-hp S.58), and the famous twin-hulled S.55, which originally had two 300-hp Fiat power units but was later fitted with 800-hp Isotta-Fraschinis and cleaned-up aerodynamically to become the S.55X, which Italo BALBO used on his famous international formation flights.

Later designs included the single-seat all-metal S.52 pursuit landplane with twin Vickers guns, powered by a 300-hp Hispano Suiza giving a top speed of 175 mph (280 km/h), followed by the S.56 biplane training amphibian with a side-by-side two-seat cockpit, the S.62 three-seat biplane reconnaissance flying boat, and the S.67 ship-launched single-seat flying boat fighter, which had twin fuselage-mounted Vickers guns and a 400-hp Fiat engine providing a maximum speed of 162 mph (260 km/h).

During the 1930s a further development of the S.55 type appeared, the S.66 twin-hulled flying boat powered by three 750-hp engines. Savoia-Marchetti was also becoming a leading champion of trimotor layouts. These included the 8-passenger SM.71 airliner and its bomber derivative (the SM.72), the SM.79 SPARVIERO bomber (descended from the SM.73P), the SM.81 211-mph (340-km/h) Pipistrello bomber (also an SM.73 development) with six machine guns and a 4,400-lb. (2,000-kg) bomb load, the SM.75 24-passenger airliner, the 10-passenger SM.83, and the SM.82 Canguro transport and bomber. The SM.78 was a three-seat reconnaissance flying boat, the SM.80 an amphibian with either two seats (with a single engine) or four (with twin engines), and the SM.74 a high-wing 27-passenger airliner with a fixed undercarriage and four 700-hp engines.

After World War II SIAI-Marchetti resumed production. Its first major

Savoia-Marchetti S.55X twin-hull flying boat

project was the SM.95 30-passenger four-engined airliner, actually a wartime design. Today the main group of products is the SF.260 family of piston-engined military trainers and private aircraft. The turboprop SM.1019 observation/utility aircraft is also in production, and the company participates in construction of license-built Boeing-Vertol helicopters.

Schneider Trophy. See RACING

SE5

Rivaled only by the Sopwith CAMEL among British fighters of World War I, the SE5 was designed by the Royal Aircraft Factory and made its maiden flight in November 1916. The SE5 was powered by a 160-hp Hispano-Suiza engine and had a Vickers machine gun on the forward fuselage with a Lewis gun rail-mounted above the upper wing center section. It began active duty with the Royal Flying Corps in France in April 1917.

A slightly modified, more powerful version, the SE5A, soon replaced the early production machines. This aircraft had an engine of up to 240 hp and could reach a top speed of 132 mph (212 km/h). Wingspan was 26 ft. 7 in.

(8.10 m) and length 20 ft. 11 in. (6.38 m). Eventually 24 British squadrons, together with 2 American and 1 Australian, flew the SE5A, and it became the favorite mount of many leading fighter pilots, including J. T. B. MCCUDDEN and Edward MANNOCK. In all, 5,025 SE5As were built.

Seaplane

Any airplane designed to operate off water is technically a seaplane. FLYING BOATS have a planing hull with either wing-mounted stabilizing floats or fuselage-attached sponsons; FLOATPLANES employ pontoons; and AMPHIBIANS have wheeled landing gear incorporated in their structure so that they can also operate from dry land.

Seaplanes were for many years in the forefront of aircraft design, at least in part because PIONEER FLIGHTS over wide stretches of ocean were obviously marginally safer in a machine that was designed to float if it came down in the sea, although the ability of even a flying boat to ride out rough seas in a severe storm is limited.

High-speed flying was also at one time almost a prerogative of seaplanes. This was partly due to the fact that 50 years ago seaplanes could be landed at higher speeds than the undercarriages

SE5A

of contemporary landplanes permitted, so that they used smaller-span, lower-drag wings. It was also due to the incentive provided by the Schneider Trophy races, which were open only to seaplanes. In fact, for a period from the mid-1920s to the mid-1930s, a series of British and Italian floatplane racers held the world speed record for aircraft of any type.

A modern landplane with a retractable undercarriage is inherently more efficient in terms of aerodynamics than either a floatplane or a flying boat, both of which inevitably incur a degree of drag from their special adaptations for operating off water. One notable attempt to produce a high-speed seaplane set out to tackle this problem. This was the twin-jet Convair Sea Dart XF2Y-1 of 1953, which employed a pair of hydroskis that were extended for takeoff and landing, providing a faster, shorter takeoff run than could a conventional flying boat hull. The third prototype of the Sea Dart (the first YF2Y-1) had only a single hydroski, which threw up less spray but was not quite as stable as the paired arrangement. The YF2Y-1 exceeded Mach 1 in a shallow dive at 34,000 ft. (10,350 m) on August 3, 1954, but three months later it exploded, probably because of a fuel leak, during a high-speed low-level pass at a public demonstration, and the project was canceled.

Floatplanes still have a use in wilderness areas where lakes and rivers remain the only possible landing places, but large military flying boats are now only built by Japan (SHIN MEIWA) and Russia (BERIEV). Airfreighting is the only likely future use for civil flying boats, and seaplanes in general would appear to have very limited prospects.

Sesquiplane

A sesquiplane is an airplane of fundamentally BIPLANE configuration, but with its lower wings considerably smaller than its upper ones—"sesqui" is Latin for "one and a half." The upper wing is frequently of parasol construction, attached to the fuselage by supporting struts, while the lower wing is joined to the lower fuselage longerons. A sesquiplane incurs less drag than would be the case with a biplane, and less adverse effect on the airflow around the upper wing caused by the presence of the lower wing.

Shenandoah

Built at Lakehurst, N.J., between 1919 and 1923, the airship ZR 1, commissioned as the U.S. Navy *Shenan-*

A sesquiplane fighter of the 1920s: the RAF's Armstrong Whitworth Siskin IIIA

doah, was the first rigid airship to be inflated with helium. She was 680 ft. long and had a capacity of 2,115,000 cu. ft. (59,220 m³) of gas. Five 300-hp Packard engines gave her a speed of 60 mph (96 km/h) and a range of 2,770 mi. (4,450 km), carrying a useful load of 33 tons.

The *Shenandoah* was intended to make the first flight to the North Pole in 1924, but broke away from her mooring mast in a gust, damaging the bow. A skeleton crew left on board returned the ship to her base, and on October 7, 1924, the *Shenandoah* flew from Lakehurst to San Diego, making intermediate stops en route.

Shortage of helium kept the *Shenandoah* inactive until June 1925, when she took part in fleet exercises, and on September 2 of that year she began her 57th flight—a five- or six-day cruise round Midwest cities and state fairs. Caught in a rising air current over Ohio, she was swept up to over 6,000 ft. (1,830 m), then fell to 3,000 ft. (915 m) before again catching a powerful upcurrent. The ship broke her back and disintegrated. Only 29 of the 43 crew survived (they were in the severed nose section, which free-ballooned to safety).

Shiden

An outstanding combat fighter by any standards, and probably the finest production interceptor flown by the Japanese Navy during World War II, the KAWANISHI N1K2-J Shiden-Kai ("George 21") owed its origin to the 15-Shi Kyofu ("Mighty Wind") floatplane fighter of 1942. A landplane derivative of this machine, the N1K1-J Shiden or "Violet Lightning," (codenamed "George" by the Allies), was first flown in the summer of 1943 and proved to possess outstanding combat maneuverability—due largely to the use of combat flaps that deployed automatically.

During 1944 this four-cannon fighter was encountered by American combat flyers in increasing numbers. It proved far superior to the HELLCAT, despite the unreliability of its 1,990-hp Homare 18-cylinder radial engine, which gave a maximum speed of 362 mph (582 km/h) at 17,715 ft. (5,400 m), and handling qualities that demanded an experienced pilot.

As the Pacific war drew toward its close, a handful of N1K2-J ("George 21") fighters reached Japanese Navy units. This redesigned, simplified aircraft used only 43,000 parts compared to the 66,000 in an N1K1-J. It was capable of 370 mph (595 km/h) at 18,370 ft. (5,600 m) with the same Homare engine, and could climb to 35,300 ft. (10,760 m). Wingspan was 39 ft. 4 in. (12.00 m) and length 30 ft. 8 in. (9.34 m).

Shin Meiwa

The Japanese four-turboprop STOL SS-2 (PS-1) and SS-2A (PS-1 Mod) flying boats are the products of Shin Meiwa Industry, formerly the KAWANISHI Aircraft Company. Shin Meiwa had rebuilt a Grumman Albatros as the UF-XS flying scale model in the mid-1960s and then went on to the SS-2 (1967), with four 3,060-ehp Ishikawajima-built General Electric T64 turboprops, a boundary layer control system, and extensive flaps for propeller slipstream deflection. Wingspan is 108 ft. 8¾ in. (33.14 m) and length 109 ft. 11 in. (33.50 m); maximum speed is 340 mph (547 km/h). After two prototypes were built, a short production run of aircraft was undertaken to equip the 31st Squadron of the Japanese Maritime Self-Defense Force for antisubmarine work. These machines were equipped with MAD (magnetic anomaly detection) radar, sonobuoys, depth charges, torpedoes, and air-to-surface rockets.

The SS-2A air/sea rescue amphibian first flew in 1974. It incorporates a hydraulically retractable tricycle land-

F-80 Shooting Stars

Short-Mayo composite aircraft: the Mercury on a test-launching from the Maia

Korean War, but in the first phase of the conflict bore much of the brunt of aerial combat. On November 8, 1950, an F-80C brought down a Chinese MiG-15, in the first victory gained by one jet pilot over another.

A reconnaissance version of the Shooting Star, the RF-80, was produced, but the major variant was the T-33 two-seat advanced trainer, production of which ran to 5,691 aircraft in the United States, with an additional 656 manufactured in Canada and 210 in Japan (1948–60).

Short Brothers

The Englishmen Horace, Eustace, and Oswald Short formed the aircraft manufacturer Short Brothers in 1908. During World War I a series of Short floatplanes was built for the Royal Naval Air Service. Among them was the famous Short 184, a reconnaissance and antisubmarine machine of which nearly 1,000 were built, and which was the first airplane to sink a ship with a torpedo. These wartime designs were followed by the first British all-metal airplane, the Silver Streak biplane of 1920. The tiny Cockle flying boat (1924) and the Mussel floatplane were among Short's designs in the field of light aircraft, but the company's efforts were soon largely concentrated on the design and production of large military and civil flying boats. Among the earliest of these was the Singapore I of 1926, leading to three-engined 15-passenger Calcuttas (1927) for Imperial Airways, the four-engined RAF Singapore II, and the six-engined Sarafand (1932). Sixteen-passenger Kent flying boats, produced for Imperial Airways in the early 1930s, were so successful that a fleet of 28 four-engined C-class "Empire boats" were ordered in 1935. With a maximum speed of 200 mph (322 km/h)—as fast as many fighters—these aircraft carried 24 passengers and two tons of mail on routes from London to South Africa and Australia and on Bermuda–New York services. Also in the late 1930s came the famous record-breaking Mayo composite aircraft. This unusual design, actually a four-engined S.20 Mercury floatplane mounted on a larger four-engined S.21 Maia flying boat and released in midair to complete its flight, was intended for regular transatlantic services, but only a small number of experimental flights took place.

During World War II the Sunderland, a military development of the C-class flying boat, played a major role in maritime and antisubmarine reconnaissance duties, and the STIRLING became Britain's first modern four-engined

ing gear and has a crew of 9 with accommodation for 12 survivors.

Shooting Star

The first U.S. jet aircraft to enter service, the Lockheed P-80 (later F-80) Shooting Star first flew in prototype form on January 9, 1944. The first prototype was powered by a de Havilland Goblin engine, but later prototypes and production models used the General Electric/Allison J33. Two pre-production aircraft were tested in Italy in 1945 but saw no combat, and with the end of the production plans were cut back. A total of only 1,731 Shooting Star fighters were built between 1944 and 1950, the major model being the 580-mph F-80C, armed with six .50 nose guns and capable of carrying 10 air-to-surface rockets or two 1,000-lb. (450-kg) bombs. Wingspan of the F-80C was 39 ft. 11 in. (12.17 m) and length 34 ft. 6 in. (10.57 m).

Shooting Stars served as ground-attack aircraft during most of the heavy bomber. In the immediate post-war years Sandringham and Solent flying boats (derivatives of the Sunderland) were used by BOAC and other airlines, but Short Brothers and Harland (as it became known after 1947) began to broaden its range of products. It was responsible for the SC-1, Sherpa, and SC-5 research aircraft; naval Sturgeons and Seamews; Sealand amphibians; and the enormous Belfast military transport. It also did major subcontracting work on the Canberra and Britannia. Skyvan freighters and SD3-30 transports are still in production.

Shturmovik

"Shturmovik," the Russian term for any ground-attack aircraft, was the name by which the ILYUSHIN Il-2 became universally known in World War II. The Il-2 was the first Russian ground-attack aircraft to go into production, and 36,163 were eventually produced between 1941 and 1944. (It was replaced by the Il-10, essentially a modified Il-2.) The Shturmovik made its first flight on December 30, 1939. Late production Il-2s, which were two-seaters, were powered by a 1,750-hp AM-38F engine and had a top speed of 251 mph (404 km/h) at 4,920 ft. (1,500 m). They were armed with two 23-mm cannon, one .50 hand-operated machine gun and two .30s, and carried up to 1,320 lb. (600 kg) of bombs or eight rockets. Wingspan was 47 ft. 10 in. (14.60 m) and length 38 ft. 1 in. (11.60 m).

Sideslip

The basic feature of a sideslip is that the relative wind (the airflow past the airplane) is not from nose to tail but obliquely from one side. Simple yaw will produce a sideslip, and this can be caused either by failure of an outboard engine on a large aircraft or strong application of rudder in otherwise symmetric flight.

Short Sunderland II flying boat of World War II

The sustained sideslip requires both lateral control, to roll the aircraft to a banked attitude, and application of rudder. In a correctly harmonized turn the rudder helps the aircraft around the turn. Application of rudder in the opposite sense will cause a sideslip, and in most aircraft the pilot can set up a stable and safe condition in which the aircraft sideslips in order to lose height rapidly without significantly changing either heading or airspeed. Successive reversals of sideslip can produce the aerobatic maneuver called a falling leaf. Sometimes severe sideslip may be encountered inadvertently, for example by inertia coupling in long and heavy (especially supersonic) aircraft when rolled rapidly.

Sikorsky, Igor (1889–1972)

Russian born aircraft designer, Igor Sikorsky constructed two unsuccessful helicopter-like machines in 1909–1910, before turning his attention to conventional airplanes. He built the first successful four-engined plane, *Le Grand* (1913), and its development, the ILYA MOUROMETZ bomber series, which equipped Russia's "Squadron of Flying Ships" during World War I.

After the Russian Revolution, Sikorsky worked in Paris and then, in 1919, moved to the United States, where he established the Sikorsky Aero-Engineering Corporation in 1923 and soon turned his attention to flying boats. His S-38 twin-engined amphibian of 1928 was used by the U.S. Army and Navy and by Pan American Airways. The S-40 was a four-engined flying boat supplied to Pan American, which christened the aircraft the "American Clipper," while the S-42 flying boat of 1934 was designed for transoceanic travel, and in use by Pan American pioneered Pacific air routes and participated in the first commercial North Atlantic services.

Sikorsky renewed his interest in helicopters in the late 1930s, producing pioneering models that included the VS-300 of 1939–40 and the R-4 (1944), the first helicopter to be used in large numbers (130 before the end of World War II) by the military (see HELICOPTERS). Sikorsky went on to produce the S-51, S-55, and S-58, all highly successful piston-engined helicopters that found numerous military and civilian customers at home and abroad. These were followed by turbine-powered designs that include the S-61 (the U.S.

Sikorsky S-42 flying boat, a pioneer of Pan American's transoceanic services

Sikorsky S-64 "Skycrane" (U.S. Army CH-54A)

D-558-2 Skyrocket

Navy's Sea King, known to soldiers in Vietnam as the "Jolly Green Giant"), S-62 (a successor to the S-55), S-64 (the "Skycrane," with a backbone fuselage below which containers or bulky cargo can be carried), S-65 (a transport using the S-64's rotor system), and S-67 (the Blackhawk, which reached a speed of 220.885 mph, 355.485 km/h).

Skyraider

The Douglas Skyraider was one of the longest-serving and most versatile warplanes of the postwar years. First flown in March 1945, and originally intended as a shipboard strike aircraft, the Skyraider entered U.S. Navy service in 1947 and a total of 3,180 were built before production ceased in 1957. Used in numerous roles in the Korean War, when it flew from both carriers and land bases, the Skyraider went on to serve in counterinsurgency and close-support operations in Vietnam.

Some 30 variants of the Skyraider were built, including antisubmarine and AEW versions. The aircraft is powered by a Wright R-3350 radial engine and armed with two or four 20-mm cannon. Capable of carrying some 3,000 lb. (1,360 kg) of bombs or napalm tanks, the Skyhawk (in its two-seat A-1E version) has a maximum speed of 310 mph (500 km/h). Wingspan is 50 ft. 9 in. (15.47 m) and length 40 ft. 1 in. (12.22 m).

Skyrocket

The swept-wing Douglas D-558-2 Skyrocket experimental aircraft was the descendant of the D-558-1 Skystreak, a straight-winged jet-powered research aircraft which set a world speed record of 650.92 mph (1,047.33 km/h) in 1947. Three Skyrockets were built; the first made its maiden flight in February 1948. It was initially powered by both a jet and a rocket engine, but with the jet

removed and the rocket fuel capacity doubled, a Skyrocket reached a speed of 1,327 mph (2,135 km/h) in November 1953 after release from a modified Superfortress carrier aircraft. In August 1953 an altitude record of 83,235 ft. (25,370 m) had already been established after a similar air launch. Wingspan of the Skyrocket was 25 ft. (7.62 m) and length 45 ft. 3 in. (13.79 m).

Slipstream

The airflow generated by an aircraft's power unit(s) is known as a slipstream. In jets, the hot exhaust gases are always discharged clear of the airframe, and the aerodynamic effects are nil. Conventional tractor propellers, however, project air backward, and in single-engined aircraft a rotating column of fast-moving air is pushed back down the fuselage. This slipstream is more pronounced on one side than on the other (due to the direction of the propeller's rotation) and causes yaw, particularly when the throttle is opened.

Smith, Ross Macpherson (1889–1922)

The Australian aviator Ross Smith was pilot of the first aircraft to fly from England to Australia (November 12–December 10, 1919). Smith qualified as a pilot in 1916 and joined the Australian Flying Corps serving in Palestine, where he flew a Handley Page O/400 against the Turks in the last months of World War I. Near the end of 1918 he flew as copilot on the first flight ever made between Egypt and India.

In 1919 the Australian government offered a £10,000 prize for the first flight

Douglas Skyraider

from England to Australia completed within a time limit of a month. Six teams entered. Ross Smith, with his brother Keith as copilot and navigator, and mechanics J. M. Bennett and W. H. Shiers, was first to leave England, flying a Vickers VIMY. They arrived at Darwin, Australia, with only 52 hours to spare, having covered a distance of 11,294 mi. (18,157 km). Ross and Keith Smith managed to coax the battered Vimy southward across Australia to arrive at their birthplace of Adelaide on March 23, 1920.

The brothers were knighted for their achievement; Ross Smith was killed in April 1922, when the Vickers Viking amphibian he had ordered for a round-the-world flight crashed in England.

SPAD

Following the conviction of Armand DEPERDUSSIN for embezzlement, his aircraft company was taken over in August 1914 by a group of industrialists (led by Louis Blériot) and renamed Société anonyme pour L'Aviation et ses Dérivés (SPAD).

The new company's first aircraft (designed by Louis Bechereau) was the A-1 military reconnaissance plane (1915), with a nacelle for the observer rigged in front of the propeller to give a good field of fire. Developments of this machine were the A-2 (which served with the French and Russian air forces), A-3, and A-4. In 1916 a new Bechereau design, the Type V appeared—a conventional but very strong single-seat biplane scout with an eight-cylinder 150-hp Hispano-Suiza engine. This became the Spad-7, which had only a single machine gun and lacked the rate of climb possessed by contemporary German fighters, although it was later equipped with a new engine of 180 hp. The reworked Spad-12 of 1916–17 had a 200-hp engine with a 37-mm cannon firing through the propeller hub, but only a few saw operational service: the reduction gear of the Hispano engine caused excessive vibration and the engine was itself unreliable.

The Spad-13 combined the 200-hp Hispano-Suiza engine with an armament of twin Vickers machine guns. A slightly larger plane than the Spad-7, the Spad-13 began to reach operational units in the summer of 1917. Nearly 800 were in service with the French air force by November 1918, and final production exceeded 7,000. The aircraft still lacked sufficient visibility for the pilot, a competitive rate of climb, or adequate maneuverability, but it was nonetheless flown by British, American, Italian, and Belgian units during the war,

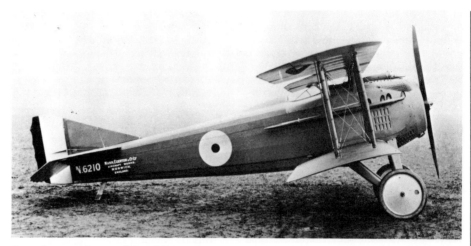

British-manufactured Spad-7

and saw later use in Czechoslovakia, Poland, Spain, and Japan. Other Spad developments were the Spad-17 (300-hp engine), Spad-18 (cannon gun, 300-hp engine), Spad 21 (a Spad-17 with modified wings), Spad-22 (sweptback upper wing, swept-forward lower wing), and Spad 24 (a landplane version of the Spad-14 floatplane, itself based on the Spad-12). Of these, only the Spad-17 reached service units, and then in very small numbers.

Spanish Civil War

Fought between General Franco's Nationalists and Republicans defending the left-wing government, the Spanish Civil War (1936–39) provided the opportunity for several nations to test in combat the aircraft and tactics they were soon to use in World War II. Almost immediately after the outbreak of the civil war Germany and Italy organized military assistance—especially aircraft with pilots and ground crew—to help the Nationalists, while the Soviet Union sent forces to assist the Republicans. The chief aircraft types dispatched to the war by the Germans were the HEINKEL 51, 70 and 111; the Henschel 123; the JUNKERS 52/3m and Ju-87 STUKA; the Do-17 FLYING PENCIL; and the BF 109. These aircraft equipped the Spanish Nationalist air force and the Condor Legion, commanded by Hugo Sperrle. From Italy came the FIAT C.R.32, the SM.79 SPARVIERO and SM.81, and (in 1939) the B.R.20 and the Fiat G.50. The Soviet Union sent the TUPOLEV SB-2 bomber and POLIKARPOV I-16 (RATA) and I-15 fighters, while the Republicans also bought an assortment of combat types from other sources.

Command of the air shifted back and forth, but the presence of the Rata did give the Republicans some measure of

air superiority before the summer of 1938, while the Bf 109 subsequently swung the battle in favor of the Nationalists until their final victory in 1939. By the last phase of the war foreign forces had returned to their native countries, taking with them very definite views on how air power should be deployed and what it could accomplish. The Soviet Union maintained rigid political control over its personnel and tactics, even dictating individual behavior in a dogfight. The result was that Soviet pilots learned very little and consequently suffered heavy casualties against the Luftwaffe in 1941. The Italians were confirmed in their views that fighter design should sacrifice performance and firepower in favor of pilot visibility and maneuverability. The Germans continued to subordinate the role of the Luftwaffe as little more than a support weapon for the army: strategic bombing in any real sense was ignored, although a raid was launched against the Basque city of Guernica on April 26, 1937, when He 111 bombers

Italian aircraft in action in the Spanish Civil War: SM.79 Sparviero bombers with an escort of C.R.32 fighters

and He 52 fighters left 1,654 dead and 889 injured. With the lack of sophisticated fighter resistance few conclusions could be drawn about how a modern bomber should be armed, how bombers or dive-bombers would fare unescorted against tough opposition, or how a carefully designed strategic bombing program might affect the course of a war.

Sparviero

The trimotor SAVOIA-MARCHETTI SM.79 was the principal Italian bomber at the outbreak of World War II. The prototype of the Sparviero ("Hawk") first flew in November 1934, and two years later a bomber version entered production. It could carry a bomb load of up to 2,756 lb. (1,250 kg) and was armed with three to five .30 or .50 machine guns. Wingspan was 66 ft. 3 in. (26.79 m) and length 53 ft. 2 in. (16.20 m). The bombardier occupied a ventral gondola, and there was generally an additional .50 gun fixed above the flight deck, firing ahead. The Sparviero first saw action in the Spanish Civil War, and transports and reconnaissance versions as well as bombers served in the Mediterranean and the Balkans during World War II. Some 1,200 SM.79s were built, together with about 100 export models.

Spinning

An airplane will enter a spin if one wing drops and becomes fully stalled, probably as a result of coarse use of the rudder that has caused yawing when the air speed is close to the stalling point. The dropped wing is subject to an upward air flow which imparts to it a high angle of attack, leading to the stalled condition. The opposite wing will only be partially stalled as the relative airflow strikes it from above when its counterpart drops, thus lessening the effective angle of attack.

In a fully developed spin (autorotation) there is a combination of pitching-up, yawing, and rolling, with the wing on the outside of the spin traveling faster than the stalled inner wing and retaining some lift. Air speed remains near the stall, and the ailerons are ineffective. To recover, the yaw has to be corrected by applying opposite rudder to the direction of spin, with a forward movement of the control column to restore airspeed.

Spirit of St. Louis

The single-seater RYAN high-wing monoplane in which Charles LINDBERGH made his historic solo crossing of the Atlantic in 1927 was specially designed for the purpose and was called the Ryan NYP (New York–Paris). Lindbergh named the aircraft *Spirit of St. Louis* in honor of the city whose businessmen had put up the money to make the flight possible. It was powered by a 237-hp Wright J-5C radial engine and had a wingspan of 46 ft. (14.02 m) and length of 27 ft. 5 in. (8.36 m), with a top speed (when fully loaded) of 124 mph (200 km/h). It carried 450 gallons (1,705 liters) of fuel, which gave it a range of over 4,000 mi. (6,400 km).

The *Spirit of St. Louis* was based on the modified Ryan Brougham (a five seater), and had a fabric-covered steel-tube fuselage and spruce wings. It was designed and built in 60 days for Lindbergh's solo flight—and was flown only by Lindbergh himself. Most of the standard Brougham cabin space was filled with fuel tanks, and Lindbergh even left out a radio and parachute in order to save space and weight. The windshield was replaced by streamlined cowling and the only view ahead was through a periscope.

First flown on April 28, 1927, the *Spirit of St. Louis* covered the 1,500 mi. (2,214 km) to St. Louis in 14½ hours (May 10), reached Roosevelt Field, New York, in 7½ hours (May 12), and then flew the 3,609 mi. (5,809 km) to Paris nonstop in 33½ hours (May 20–21).

After touring the United States and Latin America between June 1927 and April 1928, the aircraft was donated to the Smithsonian Institution.

Spitfire

Probably the best-known British aircraft of World War II, the SUPERMARINE Spitfire was the most advanced single-seat fighter in RAF service at the outbreak of the war, and distinguished itself above all in the Battle of Britain, in which it proved a match for the BF 109. Designed by R. J. MITCHELL, and evolved from a long line of racing floatplanes, the Spitfire made its first flight on March 5, 1936. By September 1939, 400 365-mph (587-km/h) Mark I aircraft were in RAF service. Their length was 29 ft. 11 in. (9.12 m) and wingspan 36 ft. 10 in. (11.22 m).

Altogether 24 basic versions of the Spitfire were designed, and the aircraft remained in production until after the war, with over 20,000 eventually manufactured. The Marks V (1,470-hp Merlin engine), IX (1,720-hp Merlin, giving a speed of 404 mph, 650 km/h), and XVI (Packard-Merlin) were built in large numbers. The XII, XIV, and XVIII were instead equipped with Griffon engines. The XI and XIX were unarmed photo-reconnaissance aircraft. Eight Seafire variants were produced for the Fleet Air Arm (I, II, and III with Merlin engines, XV, XVII, 45, 46, and 47 with Griffons).

The original armament of eight machine guns was reduced to four guns supplemented by two 20-mm cannon on the Vc and IXc; 21, 22, and 24 had four cannon only. The Mark 24 of 1946 had a 2,050-hp Griffon engine, and a top speed of 450 mph (724 km/h).

Spoiler

When an airplane makes a banked turn, the aileron on the outer wing is lowered to increase the lift generated by that wing and so initiate banking.

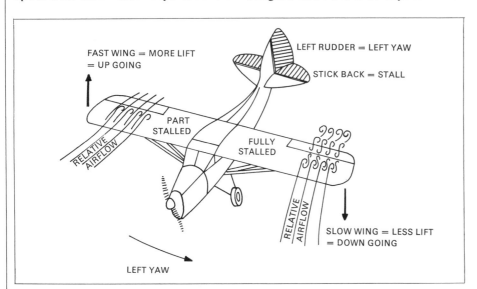

Spinning: Coarse use of the rudder, making an aircraft develop yaw when it is near its stalling speed, may bring about a spin. The wing on the outside of the spin, only partly stalled, retains some lift; the opposite wing, fully stalled, drops

This effect can be enhanced by correspondingly reducing the lift and augmenting the drag on the inside wing. Raising the aileron will achieve this, but a retractable spoiler can also be extended upward to "spoil" the airflow by increasing drag and partly destroying the lift. A spoiler may be interconnected to the aileron and is often installed behind a leading-edge slot.

SR-71A

One of the world's fastest military airplanes, the Lockheed SR-71A long-range high-altitude reconnaissance aircraft has a maximum speed of Mach 3 at an altitude of 100,000 ft. (30,000 m). A demonstration of the airplane's remarkable capabilities was given in 1974 when it flew from New York to London in 1 hour 55 minutes. The SR-71A is a long-nosed twin-engined delta-wing aircraft with a thin flattened fuselage and twin fins; wingspan is 55 ft. 7 in. (16.94 m) and length 107 ft. 5 in. (32.74 m). Three prototype YF-12A interceptors were designed in parallel with the SR-71A but failed to enter production.

SST

An SST (supersonic transport) is an airliner designed to fly at supersonic speeds. The two first generation SSTs in service today—the Anglo-French CONCORDE and the Soviet Tu 144, both fly at around 1,400 mph (2,253 km/h)—a little over Mach 2—and are constructed primarily of aluminum alloy.

American plans for a Mach 3 variable-geometry ("swing wing") SST, which would use heat-resistant metals such as titanium alloys and stainless steel, were abandoned in 1971 largely because of expense and environmental questions.

Although several hypersonic (faster than Mach 5) transports have been proposed, we are unlikely to see even a second generation of SSTs until environmental problems (particularly those of noise, and damage to the Earth's protective ozone layer) have been resolved. Most experts agree that a small fleet of SSTs would have little effect on the ozone layer, but the U.S. National Academy of Sciences concluded that 100 operational SSTs could raise skin cancer incidence by 1.4 percent—and that a 400 strong fleet might raise that figure to 20 percent. (See also SUPERSONIC FLIGHT.)

Stalling

As long as the airflow over a wing's airfoil is maintained, LIFT will be de-

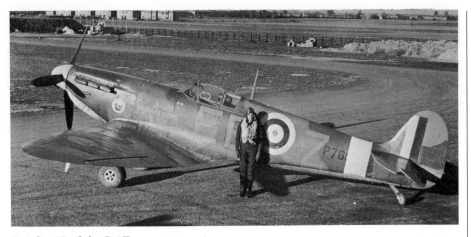

Spitfire II of the RAF

The SR-71A touching down at Britain's Farnborough Air Show after its record-breaking transatlantic flight, September 1974

veloped. If speed is reduced, lift can be increased by adopting a greater angle of attack; this augments both the pressure below the wing and the backward downwash of air.

Once the angle of attack exceeds 15–20° (depending on the aircraft type), the drag becomes too great for forward speed to be maintained. TURBULENCE spills over the trailing edge and around the wing tips as air moves from the high pressure region below the wing to the low pressure area above it, and the airflow breaks down. The wing can no

longer generate lift and becomes stalled.

If one wing drops when an aircraft stalls, SPINNING may develop unless corrective action is quickly taken. Shock stalls occur at sonic speeds and are due to a breakdown of the airflow as a result of shock-wave formation. (See SUPERSONIC FLIGHT.)

Starfighter

A widely-used and somewhat controversial Mach-2.2 interceptor, fighter-bomber, and multirole aircraft, designed

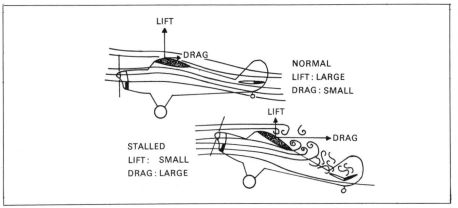

Stalling: Airflow during normal flight, and at the stalling angle, where the smooth airflow breaks down into eddies and lift decreases

179

F-104G Starfighter of the Royal Norwegian Air Force

by Lockheed and used by a number of Nato countries. The F-104 Starfighter was first flown in 1954, and possessed a top speed of nearly 1,400 mph (2,252 km/h). Its relatively small dimensions (wingspan 21 ft. 11 in., 6.68 m, and length 54 ft. 9 in., 16.69 m) gave it the popular name of "the missile with a man in it."

The U.S. Air Force acquired only a small number of F-104s because of the aircraft's poor endurance and its lack of all-weather capability, but a tactical fighter version was sold in large numbers to Nato countries, notably Germany, the Netherlands, Belgium, and Italy. An exceptionally high accident rate—approximately 25 percent of the Luftwaffe force—brought this aircraft a considerable amount of unfavorable publicity. (It became known in Germany as the "widowmaker.")

Powered by a General Electric J79 afterburning engine of 15,800 lb. (7,167 kg) thrust, the F-104G, produced in several European nations and Japan, is armed with a six-barrel 20-mm cannon and a variety of externally slung weapons of up to 4,000 lb. (1,812 kg) total weight. The F-104S, used by Italy, employs a more powerful engine; the TF-104G is a trainer variant.

Static tube

The operation of an airspeed indicator (see INSTRUMENTS) depends on the difference between two air pressure readings: that of the air driven by the aircraft's forward movement into the PITOT TUBE, and that of the ambient air around the airplane registered by the static tube.

A static tube has a sealed forward-facing end. A series of small holes located along its length provides access for air that is uninfluenced by the aircraft's forward movement and thus capable of providing a measure of the actual atmospheric pressure. The Pitot tube and static tube are frequently in-

corporated in a single unit, but the static vent may need to be located elsewhere on the airframe if the pressure in the vicinity of the Pitot tube varies from a true atmospheric reading because of the effect of the airflow.

Stinson

U.S. aircraft manufacturer, established by E. A. Stinson in 1925, which produced mainly light aircraft. Stinson began flying in a Wright biplane in 1913 with his sisters Katherine and Marjorie. Together, they established a flying school at San Antonio in 1915. Katherine made a flying tour of Japan and China (1916–1917), and flew nonstop from San Diego to San Francisco in 9 hours 10 minutes (December 1917).

E. A. Stinson was a test pilot and army flying instructor during World War I. When the war ended he barnstormed for a while and eventually became a Detroit charter operator (1922) before forming the Stinson Airplane Syndicate in 1925. He built the Detroiter cabin biplane (with brakes, heater, and starter) and the Detroiter monoplane (1927), which won the Ford Tour and flew from Harbour Grace, Newfoundland, to London in $23\frac{1}{2}$ hours.

After merging with the Cord Corporation and moving to a new factory in Wayne, Michigan, in 1929, the firm built eight-passenger trimotor airliners,

eventually producing low-wing versions for American, Australian, and Chinese airlines. Single-engined Detroiters led to a family of Juniors, Reliants, Voyagers, Sentinels, and Vigilants, which provided liaison and observation for British and American forces throughout World War II. Stinson became a subsidiary of VULTEE in 1940 and was eventually absorbed into Convair.

Stirling

The first four-engined monoplane bomber to enter RAF service, was the SHORT Stirling. The prototype flew in May 1939, and the first squadron received the heavy bomber in August 1940. Approximately 2,500 machines were produced, in four major variants.

Powered by four 1,650-hp Bristol Hercules XVI engines, the Stirling suffered from performance limitations due largely to the short-span (99 ft., 30.17 m) wings specified by the British Air Ministry to permit the use of existing hangars. A total of 23 squadrons were equipped with Stirlings of various types during the war, modified aircraft being used as glider tugs, supply aircraft, and paratroop carriers. The final Stirling bombing operation was flown on September 8, 1944.

Maximum speed was 270 mph (435 km/h) and ceiling only 20,500 ft. (6,350 m). Armament consisted of .303

Stirling I bomber, the RAF's first four-engined "heavies" of World War II

machine guns in nose, tail, and dorsal turrets; the bomb load 14,000 lb. (6,350 kg).

STOL

Unlike VTOL aircraft STOL (Short Take-Off and Landing) craft resemble conventional aircraft in everything but their performance. Common means of producing STOL performance may involve increasing the wing area, fitting more powerful engines, and, particularly in the case of larger aircraft, using complex high-lift devices.

In multiengined airplanes, the power units may be mounted so that their thrust is directed backward through high-lift FLAPS. This artificially speeds up the airflow and augments the lift. Another method of increasing lift is to bleed compressed air from the engine(s) and expel it from slits located along the leading edges of the wings and tail. The injection into the BOUNDARY LAYER of this extra air represents a source of additional energy and inhibits the tendency of the airflow to breakdown in TURBULENCE as the speed drops. Air bled from the engine compressor can also be liberated from slots located immediately in front of the flaps and ailerons.

Vortex generators are small vanes 2–3 in. (5–7.5 cm) high located on the wing surface at an angle to the local airflow. As the stalling speed is approached they create vortices which introduce energy from the smooth-flowing outside airstream into the sluggish boundary layer and delay the onset of turbulence.

Stratocruiser

The civilian version of the Model 367 C-97 Stratofreighter military transport (1944), the Model 377 Stratocruiser was Boeing's first postwar airliner. It employed a pressurized double-deck fuselage married to B-29 SUPERFORTRESS wings and tail assembly.

An airliner seating between 55 and 117 passengers, the Stratocruiser was luxuriously comfortable. The 56 that were eventually built served initially with Pan Am, SAS, American Overseas, Northwest, BOAC, and United during the period from 1948 until the advent of jets in the late 1950s. Powered by four 3,500-hp engines, the Stratocruiser had a maximum speed of 340 mph (547 km/h) and a range of up to 4,000 mi. (6,400 km). Its wingspan was 141 ft. 3 in. (43.06 m) and length 110 ft. 4 in. (33.63 m).

A number of Stratocruisers and C-97s have been rebuilt as GUPPY freighters with extra-capacious fuselages

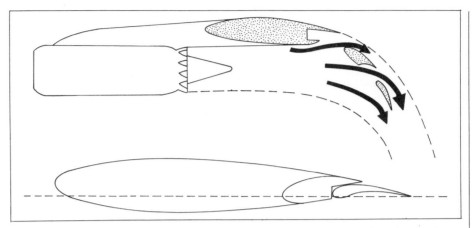

STOL: Mounting the jet engines of the YC-15 STOL aircraft so that their exhaust is directed through special flaps augments lifts on takeoff and landing and reduces the stalling speed

Boeing Stratocruiser

to carry such outsize loads as Saturn rockets. Production of the C-97 transport and the KC-97 tanker totaled 888.

Stratofortress

Mainstay of the U.S. Strategic Air Command for two decades, and probably the world's last giant heavy bomber, the Boeing Model 464 B-52 Stratofortress made its first flight in 1952. The initial models, 3 B-52A and 50 B-52B/RB-52B aircraft were powered by eight J57 turbojets. The B-52C, like the B variant, had reconnaissance as well as bomber versions; but the B-52D was a bomber only; the E and F variants had improved targeting and navigation systems.

Major changes embodied in the B-52G included reduction in the size of the underwing fuel tanks, automatic operation of the multibarrel 20-mm cannon in the tail, and a shorter fin. The warload included two Hound Dog missiles carried beneath the wings.

TF-33 turbofans were introduced on the B-52H, which could carry Skybolt

B-52G Stratofortress

missiles. Total bomb load of this final variant was 50,000 lb. (15,240 kg). It had an unrefueled range of 12,500 mi. (20,000 km), and a maximum speed of 665 mph (1,070 km/h). Length was 157 ft. 6¾ in. (48.00 m) and wingspan 185 ft. (56.39 m).

Stratofortress production concluded in 1962 after delivery of the 744th aircraft.

Stratojet

First flown in December 1947, the Boeing Model 450 B-47 Stratojet was a very advanced bomber design powered by six General Electric J47 turbojets mounted on pylons below its swept

B-47A Stratojet

wing, which flexed in flight. Tandem main undercarriage units retracted into the fuselage, and outrigger wheels into each inner engine pylon. Maximum bomb load was 20,000 lb. (9,072 kg), range 4,000 mi. (6,436 km), and maximum speed over 600 mph (965 km/h); the only defense point was the remote-controlled tail-gun cone. Wingspan was 116 ft. (35.36 m) and length 109 ft. 10 in. (33.47 m). Some 2,000 B-47s were built between 1947 and 1957, and at one period over 1,500 were operational in Strategic Air Command squadrons; B-47 bomber types were finally withdrawn from service between 1966 and 1969. There were also reconnaissance (RB-47) and electronic countermeasures variants.

Stratoliner

The first fully pressurized airliner to enter service anywhere in the world, Boeing's 33-seat Model 307 Stratoliner of 1938 employed the wings and tail surfaces of the B-17C FLYING FORTRESS. With four 1,100 hp Cyclone engines, the aircraft's maximum speed was 246 mph (396 km/h) and range 2,390 mi. (3,850 km). The wingspan was 107 ft. 3 in. (32.69 m) and length 74 ft. 4 in. (22.66 m).

Three Stratoliners flew on Pan Am's South American routes; five served with TWA, and a ninth machine was supplied to Howard Hughes.

During World War II Stratoliners were employed as military transports (C-75s), flying principally to South America and across the Atlantic. In 1951 the ex-TWA machines, fitted with 1,200-hp engines and B-17G wings, were sold to Aigle Azur in France. These aircraft, as well as the Pan Am machines and the ex-Howard Hughes plane, were later sold to other operators and some were still flying in the 1960s.

Stratosphere

Layer of the atmosphere extending from an altitude of 6–8 mi. (10–13 km) to about 19 mi. (30 km). The temperature of the Earth's atmosphere falls as the altitude increases, until a region called the tropopause, the lower boundary of the stratosphere, is reached. The tropopause lies at about 50,000 ft. (15,000 m) over the poles and 100,000 ft. (30,000 m) over the Equator. The stratosphere is the layer of the atmosphere above the tropopause, in which the temperature remains constant. Unlike the troposphere (the layer of air up to the tropopause) conditions in the stratosphere are generally stable, so that its "weather" is virtually non-existent. Some high cirrus cloud rises well into the stratosphere, but the air is almost always clear, the sky dark violet-blue (becoming darker with increasing height, owing to the lessening dust content), and wind very slight except in JET STREAMS. Humidity in the stratosphere is negligible, but above about 11 mi. (18 km) the concentration of ozone increases considerably, making it inadvisable to use outside air for pressurizing the cabins of aircraft without prior treatment.

The pressure of the atmosphere also decreases with height, though unlike the temperature this decrease continues through the stratosphere. For jet aircraft this has great significance, because it enables them to cruise at high altitude, above the uncomfortable turbulence of the lower levels. Because the density is much lower, there is less resistance to motion, and so for the same amount of power an aircraft can travel much faster; in fact if the power remained the same, the speed would build up excessively. But the thrust of a jet engine decreases with height, because there is less oxygen entering the engine, and so less fuel is burned. Thus jet engines become more economical with increase in altitude.

Another advantage of high-altitude flight is that aircraft are able to benefit from the jet stream, whose speeds may reach from 70 to over 300 mph (110 to over 480 km/h), provided they are traveling in the same direction.

Stuka

In the 2½ years that followed its appearance in the Spanish Civil War at the end of 1937, the JUNKERS 87 Stuka (*Sturzkampfflugzeug*) DIVE-BOMBER established a deadly reputation. Development of the two-seat Stuka began in 1934, and the first production aircraft (the Ju 87A-1) reached the Luftwaffe in the spring of 1937. Both this model and the 1,100-hp Ju 87B saw extensive use in the Spanish Civil War; by September 1939 there were 335 Stukas in service. The Luftwaffe's air supremacy over Poland (1939) and the Low Countries and France (1940) enabled the Stuka to operate with great success, but Ju 87 units suffered heavy losses early in the Battle of BRITAIN, when they met the determined resistance of strong fighter defenses (see WORLD WAR II).

Ju 87B Stukas of the Condor Legion in action over Spain

The Ju 87D (1940) had a 1,400-hp Jumo 211 engine, giving a maximum clean speed of 255 mph (410 km/h) and enabling bombs of 1,110 lb. (500 kg) to be carried beneath the fuselage (smaller bombs could be fitted instead beneath the wings). Length of the Stuka was 37 ft. 9 in. (11.50 m) and wingspan 45 ft. 3 in. (13.79 m). Armament was increased to twin machine guns in the rear cockpit and one in each wing.

A measure of success still attended Stuka operations against shipping and in North Africa and Russia (including the use of special cannon-armed ground-attack versions), and production continued until 1944. A total of over 5,700 Ju-87s were constructed, and they served with Italy and Axis satellites, as well as the Luftwaffe.

Sukhoi, Pavel Osipovich (1895–1975)

Soviet aircraft designer, responsible for several of the best known Russian fighter planes. He studied at the Central Aero and Hydrodynamic Institute (TsAGI) and later became chief designer of a team working there under A. N. TUPOLEV's direction. His first design was the I-4 all-metal single-seat sesquiplane fighter, of which 370 were produced between 1928 and 1933. Powered by a 480-hp M-22 engine, the I-4 had a maximum speed of 143 mph (231 km/h) at 16,400 ft. (5,000 m); a monoplane version had a maximum speed of 167 mph (268 km/h).

In 1933 Sukhoi designed the first Soviet fighter to have an enclosed cockpit and a retractable undercarriage, the Wright Cyclone-powered I-14. This had a maximum speed of 280 mph (449 km/h) at 11,150 ft. (3,400 m).

He was well known before World War II for his record-breaking long-distance aircraft. The single-engined ANT-25 (RD-1) took a number of distance records, including the world record for distance in a straight line with a 6,306-mi. (10,090 km) flight from Moscow to San Jacinto, Cal., across the North Pole. The ANT-37 Rodina (Motherland) DB-2 was a twin-engined bomber development, which was abandoned as a combat design but broke the women's distance record with a 3,670-mi. (5,909-km) flight across Siberia in the hands of V. S. Grizodubova.

Sukhoi set up his own design bureau in 1936. His first aircraft, the Su-2, was a single-engined light bomber produced in small numbers immediately before World War II. He went on to design the Su-1 and Su-3 high-altitude fighters, the Su-6 and Su-8 ground-attack aircraft, and the Su-5 mixed power-plant fighter, but none went into quantity

Sukhoi Su-7B fighters of the Soviet Air Force

production.

In 1946/7 the Sukhoi bureau originated the Su-9 and Su-11 twin-engined jet fighters, and the four-engined jet Su-10 bomber (which was not completed). The design bureau was closed down in 1946 but reestablished about 1953 to produce the Su-7 swept-wing fighter-bomber and Su-9 delta-wing interceptor, both of which were tested in 1956. The Su-7B "Fitter" (maximum speed about Mach 1.7) was subsequently widely used by the Soviet Union and also exported.

In the mid-1970s the Sukhoi bureau was working on the Su-15 delta-wing interceptor, two variable-geometry fighters and a delta-wing bomber.

Supercharger

Piston engines normally develop power in proportion to the rate at which they can burn fuel, and this in turn depends on the volume of air drawn into the cylinder on each induction stroke. A supercharger is a blower or compressor, driven either by step-up gearing from the crankshaft or by a turbine spun by the exhaust gas (turbo-supercharger or turbocharger), which forces additional air into the engine cylinders. In aircraft, a supercharger's purpose may be to maintain sea level power as the aircraft climbs into the thinner air of higher altitudes, in which case the supercharger is either turned at lower speed or not used at all at sea level, cutting in only at a predetermined height. Alternatively, it may be used from takeoff to increase the power of the engine by supercharging it to a "boost pressure" above that of the atmosphere.

In World War II some engines, such as the Rolls-Royce Merlin, could operate for short periods with boost pressure about twice that at sea level. Many American aircraft, notably the LIGHTNING, THUNDERBOLT, FLYING FORTRESS, LIBERATOR and SUPERFORTRESS

used General Electric turbochargers, giving extremely good performance at heights of 30,000 ft. (9,144 m) and above. The Merlin and Griffon had two-stage superchargers (two blowers in succession, with an intercooler to cool the air between them), while the Germans tried complex schemes involving several blowers feeding one engine.

The term supercharger is sometimes also applied to a mechanical cabin blower feeding a pressurized cabin (not normally used in jet aircraft).

Superfortress

Conceived as a replacement for the B-17 FLYING FORTRESS and B-24 LIBERATOR, the Boeing Model 345 B-29 Superfortress made its maiden flight on September 21, 1942, and became the largest bomber used in World War II. The first operational squadrons began moving to China in the spring of 1944 for raids against Japan under 20th Air Force direction, but operations shifted to the Marianas in November 1944. High-altitude precision bombing proved ineffective and costly, and in March 1945 a gamble was taken in initiating low-altitude night raids. The 10 turret-mounted remote-controlled .50 machine guns were removed (only the tail armament was retained) in order to carry a greater load, and subsequent fire-bomb raids over Tokyo succeeded in killing at least 80,000 and rendering 1,000,000 homeless in a single night. (March 9/10). Using both incendiaries and conventional high explosives, B-29s continued their devastating raids over most of Japan's major cities. Powered by four 2,300-hp Cyclone engines, the B-29 could reach a top speed of 367 mph (590 km/h) and carry a 16,000-lb. (7,250-kg) bomb load over 2,600 mi. (4,200 km). Wingspan was 141 ft. 3 in. (43.05 m) and length 99 ft. (30.18 m).

On August 6 and 9, 1945, Superfortresses dropped the two atomic bombs on Japan that brought the war to an end. B-29s were later used in the Korean War. A modified version (the B-50 of 1947) had 3,000-hp engines and was distinguishable by its taller tail fin. The B-50 was used to launch the Bell series of supersonic research aircraft (see x-1), and later KB-50 tanker versions were equipped with two additional jet engines to boost performance.

Supermarine

British aircraft manufacturer founded just prior to World War I. Almost all its early products, with the exception of the P.B.29E Night Hawk and P.B.31 anti-Zeppelin fighters, were marine aircraft or amphibians. The Supermarine P.B.7 of 1914 was a single-engined pusher biplane flying boat that could unlatch its wings, rear fuselage, and tail, and operate as a seaworthy motorboat with marine propeller. A pusher biplane configuration was used for many Supermarine flying boats over the next 30 years. Such aircraft included the Seaking fighter of 1920, the Schneider Trophy-winning Sea Lion II racer of 1922, the six-passenger Sea Eagle of 1923, and the mass-produced Walrus search and rescue amphibian of World War II.

From the Swan flying boat of 1924, chief designer Reginald MITCHELL developed the Southampton patrol flying boat, which entered RAF service in September 1925. Most of the 78 eventually built were of Duralumin, instead of wood, construction. Powered by two 500-hp Napier Lion engines, Southamptons served in the RAF for over 10 years. A further development was the Scapa (with two Rolls-Royce Kestrels) of 1935, which led to the larger and faster Stranraer (Bristol Pegasus engines) of 1936. The last of the company's marine types to enter service was the Sea Otter amphibian, first flown in 1938 but not delivered until 1943; 290 of these tractor (Bristol Mercury-engined) biplane amphibians were produced as replacements for the pusher Walrus.

During the 1920s Mitchell's energies were largely devoted to producing high-performance seaplanes for the Schneider Trophy races. The S4 was an extremely clean monoplane design but crashed in 1925. The S5 won the 1927 race at 281.65 mph (453.25 km/h); the bigger S6 won the 1929 event at 328.63 mph (528.86 km/h); and the S6b took the 1931 race at 340.08 mph (547.23 km/h). All three types also gained world speed records: the S5 (unofficially) at 284.21 mph (457.39 km/h), the S6 at 357.75 mph (575.62 km/h), the S6b at 407.02 mph (654.90 km/h). The experience of designing these racers helped Mitchell and Rolls-Royce create the famed Merlin-engined SPITFIRE fighter of World War II.

After the war Supermarine produced the Swift fighter (1952). Powered by a Rolls-Royce Avon engine, the Swift had an armament of two or four 30-mm Aden guns, but suffered from a long succession of faults, mainly aerodynamic in origin. The FR5 Swift served in a tactical reconnaissance role and the F7 as an indoctrination trainer. The last Supermarine production type was the Scimitar naval fighter, powered by two Avon turbojets that also provided air for flap-blowing. The Fleet Air Arm received 76, which served between 1958 and 1965. Supermarine, together with its parent company VICKERS, became part of the BRITISH AIRCRAFT CORPORATION (BAC) at the beginning of the 1960s.

Supersonic flight

In warm air at sea level the speed of sound is about 1,100 ft. (335 m) per sec. (760 mph or 1,216 km/h), while at a height of about 37,000 ft. (11,000 m) its velocity is only 660 mph (1,056 km/h).

At subsonic speeds, pressure waves preceding the aircraft cause the air ahead of it to begin moving out of the way before the plane actually arrives, so that the air is not subject to any compression. When the speed of sound is reached, the aircraft is traveling as fast as the pressure waves preceding it, and the air does not begin to move aside ahead of the machine. As the wings meet this wall of air a shock wave (representing an increase in pressure due to compression of the air) is generated from the point on the upper surface at which the airflow is accelerated over the airfoil's curvature to reach the speed of sound. Behind this shock wave there is an increase in air pressure and density, accompanied by a fall in the speed of the airflow, resulting in TURBULENCE, loss of LIFT, and increased drag (shock stall). When a shock stall occurs the pilot experiences similar sensations to those associated with a normal stall: buffeting, shaking, change of trim, and loss of control. Instead of becoming sloppy, however, the controls may be too stiff to move, except with the aid of the trimmer, but while an ordinary stall occurs at a high angle of attack, a shock stall occurs when diving or in level flight, and the aircraft's nose must be pulled up to overcome it. Two basic solutions to the problem have been evolved: sweepback and laminar-flow airfoils. If a wing is swept back, the speed of the airflow across the chord is less (relative to the speed of the aircraft) than it would be over a straight wing because a proportion of the airflow's speed becomes directed along the wing instead of across it. As a result, the chordwise airflow has a lower MACH NUMBER than would be the case with a straight wing at the same air speed, and the formation of sonic shock waves is deferred.

Sweepback also has the effect of making the chord effectively longer if it is measured parallel to the fuselage (in the direction of airflow); hence the wing section is proportionately thinner and produces less drag.

A straight wing begins to incur a sharp rise in drag at Mach 0.85, and the maximum drag is experienced at Mach 1.1. The higher the Mach number, the greater must be the sweepback if compressibility effects are to be delayed. A sweep of 45° will delay the onset of maximum drag until Mach 1.4, but 70° is necessary to achieve a further delay to Mach 1.5, and at Mach 1.7 the drag of a swept wing will be comparable to that of a straight wing (the drag coefficient of which begins to fall once the speed of sound has been exceeded). Straight wings are consequently often used on missiles and have also appeared on a number of high-speed aircraft (for example the F-104).

A laminar-flow airfoil has an approximately equal convexity (often angular) both above and below, with a sharp leading edge. Sonic shock waves form above and below it as the speed of sound is approached, and just beyond Mach 1 this shock wave becomes established at the trailing edge, while a second shock wave begins to form slightly ahead of the leading edge.

Once above Mach 1, the second shock wave gradually moves back until it touches the leading edge, and the angles at which both shock waves leave the wing become progressively more acute. Beyond the speed of sound, the shock wave pattern stabilizes and flight becomes smoother. Another device to facilitate the attainment of supersonic flight is AREA RULE.

As shock waves from an airplane traveling faster than sound reach an observer on the ground, they give rise to sonic booms or bangs. Although the strength of these shock waves is attenuated by distance, the sudden increase in pressure (compressibility) which they represent is capable of doing considerable damage. The magnitude of sonic bangs will be greater if a large, heavy aircraft is involved and the faster it flies, the more severe will be the shock waves generated.

Compressibility first became a problem during World War II, when fighters such as the P-38 and P-47 began to approach the speed of sound in power dives. The first man credited with reaching Mach 1 was Major Charles Yeager of the U.S. Air Force, who attained Mach 1.45 at 60,000 ft. (18,000 m) in 1947, flying the rocket-propelled Bell x-1 research plane. In 1949 the Douglas SKYROCKET became the first supersonic jet-powered aircraft when Gene May flew it to Mach 1.03 at 26,000 ft. (7,900 m).

Swing-wing design

In an attempt to combine the low-speed virtues of straight wings with the high-speed advantages of swept wings, variable-geometry airplanes have been designed with "swing-wings"—wings that can be rotated to a straight position for landing and taking off, or swept back for maximum speed, with an intermediate cruising configuration.

Although the mechanical complexity of swing-wings results in a greater structural weight, the fact that drag is less and that maximum aerodynamic efficiency can be obtained at any given speed enables a lighter, less powerful engine to be used. A swing-wing aircraft will also use less fuel than a comparable fixed-wing machine because of its enhanced aerodynamic efficiency, and so any additional structural weight may be offset by reducing the fuel capacity.

The first swing-wing aircraft to enter military service was the General Dynamics F-111 tactical strike fighter with a maximum speed of Mach 2.5.

Variable-geometry wings have also been incorporated in Grumman F-14 TOMCAT, the TORNADO, the MiG-23, and the Russian BACKFIRE strategic bomber.

Swissair

Swissair, the Swiss national airline, was formed on March 26, 1931, by the merger of Balair (which opened services from Basel to the German Rhineland and Lyon in 1926, using Fokker F.VII/3m aircraft) and Ad Astra Aero (the first Swiss international airline, which had operated a Junkers F-13 to Nuremburg from Geneva via Zurich as early as 1922). Swissair (Schweizerische Luftverkehr) began with 13 aircraft and 10 pilots inherited from its parent companies, but in 1932 the acquisition of two ORIONS enabled the airline to establish a Zurich–Munich–Vienna service with aircraft 60 mph (100 km/h) faster than those of their competitors.

Swissair followed KLM's lead in buying the new DC-2 (1935) and, subsequently, the DC-3, eight of which were in service by 1939. After World War II began, Swissair attempted to maintain some international services, but when Italy entered the war the routes to Rome and Barcelona were halted, and in 1943 flights to Berlin came to an end. The line closed down completely until July 30, 1945.

DC-4s were added to the fleet in 1946, and the competing Alpar Bern company was taken over in February 1947. Convair 440 Metropolitans appeared in service in 1956, and an international network was progressively built up (DC-6B airliners in 1951, DC-7Cs in 1956). North Atlantic services from Zurich to New York via Geneva, Shannon, and Gander, had already begun on July 4 1949, and a route to Brazil via Lisbon opened in May 1954. The first scheduled flights to the Far East came in 1957. Jet services to North America started in May 1960 with DC-8s; Convair 990 Coronados were acquired in 1961, followed by Caravelles the following year.

DC-9s began replacing the Metropolitans in 1966, and the airline's first Boeing 747 was delivered in 1971, with DC-10s arriving in 1972. Swissair carries some 5,600,000 passengers annually. Its present fleet includes 2 Boeing 747s, 8 DC-10s, 6 DC-8s, and 30 DC-9s.

T

Tail

To endow an airplane with directional stability it is equipped with a stabilizer or tailplane (incorporating the ELEVATORS) and a fin (with the RUDDER). Without a stabilizer, a conventional (nondelta-wing) aircraft would tend to somersault forward or backward. If the tail drops during level flight, however, the tailplane acquires a greater angle of attack and its lift is therefore increased, raising the tail to the normal position. Should the nose fall, the tailplane acquires a negative angle of attack, its lift decreases and the tail drops to level the airplane again. The elevator controls diving and climbing maneuvers, and overrides the stabilizing effect of the tailplane.

Stability in the yawing plane is provided by the fin, with its hinged rudder giving directional control. Both the elevators and the rudder may be provided with trim tabs, which enable the flight attitude to be maintained.

If the wings of an aircraft are positioned sufficiently far back, the wing tips themselves act to stabilize the aircraft and the machine may not need a tailplane. CANARDS, or "tail-first" aircraft, have their small stabilizers located near the nose, with the wings moved to a position well aft.

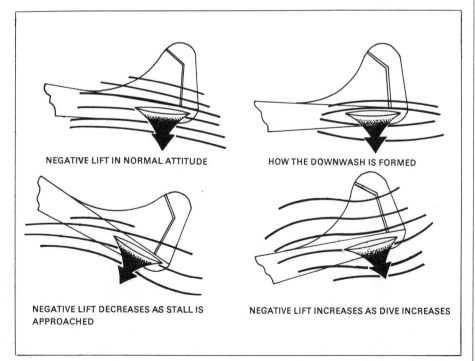

NEGATIVE LIFT IN NORMAL ATTITUDE

HOW THE DOWNWASH IS FORMED

NEGATIVE LIFT DECREASES AS STALL IS APPROACHED

NEGATIVE LIFT INCREASES AS DIVE INCREASES

Tail: Forces on the stabilizer (tailplane) of an aircraft in various flight attitudes. The "negative lift" the stabilizer exerts in normal flight to keep the aircraft from pitching alters in magnitude to achieve equilibrium in other attitudes

Takeoff

Helicopters and VTOL (vertical take-off and landing) aircraft like the HARRIER need no runway. But conventional aircraft must use the normal force of LIFT (which keeps them airborne in flight) to take-off, and since this depends on air flowing over the airfoil wing surface, they must accelerate along the ground—usually under maximum engine power—until adequate lift is generated. Wing flaps are extended to increase wing surface and therefore lifting power, and whenever possible the takeoff is made into the wind to increase airspeed (and therefore lift).

In practice, the moment of takeoff is normally delayed until more lift is developed than is needed simply to get off the ground—the aircraft must reach a speed higher than the minimum at which the wings can support the weight at stalling angle of attack.

The takeoff is made along a runway, which is followed by a stopway (on which an aircraft can be brought to rest after a "rejected" takeoff) and a clearway (a length of ground which though unsafe for deceleration is clear of obstructions to an aircraft already airborne). The actual takeoff stretch of the runway is known as the TORa (takeoff run available) while the overall length of all three sections is the TODa (takeoff distance available).

At the TODa point is an imaginary screen 35 ft. (10.67 m) high. All aircraft are designed to clear this screen, although at high-altitude or hot-climate airfields their weight must be reduced according to published WAT (weight/altitude/temperature) calculations.

During takeoff the pilot notes three crucial speeds: V1, the decision speed which marks the point at which takeoff can no longer be aborted; VR, rotation, in effect the speed at which the aircraft can be tilted nose-up to leave the ground; and V2, safety speed, reached at screen height (after traveling a distance less than TODa), which gives enough margin above VS (stall at the WAT condition) for the climb-out to be safely started.

NOISE SUPPRESSION takeoffs demand a very steep rate of climb and are therefore more dangerous.

Tank, Kurt (1898–   )

The German aircraft designer Kurt Tank, best known for his association with the manufacturer Focke-Wulf during the 1930s and early 1940s, began his career in 1924 by joining Rohrbach established in 1922 and specializing in seaplanes. Tank worked on a series of twin-engined flying boats, notably the Ro IIA and IIIA Rodra (sold to Turkey), the Robbe I and II, the Rostra, and the Rocco. He was also involved in design of the Rofix monoplane fighter (in which aerobatic ace Paul Bäumer was killed in 1927), the successful 10-passenger Roland trimotor airliner, and the huge Romar flying boat. (The Roland II, with an enclosed cockpit, was one of Rohrbach's last products before it went out of business in 1932.)

By now a competent test pilot, Tank foresaw a greater future in landplanes and moved briefly to Messerschmitt in 1930, but in November 1931 joined Focke-Wulf, founded in 1923 by Heinrich Focke (a scientist) and Georg Wulf (an ex-combat pilot).

The manufacturer Albatros had amalgamated with Focke-Wulf in 1931, and Tank worked on both the 220-hp Albatros L.102 trainer and the three-seat Fw 43 Falke, before developing the outstanding Fw 44 Stieglitz two-seat biplane trainer (1932), which was widely exported and license-built, and the high-wing Fw 56 Stosser monoplane in which Ernst UDET made dive-bombing experiments, and in which Luftwaffe pilots later received intermediate training. The twin-engined Fw 57 turret-fighter was not ordered by the Luftwaffe, and the parasol-wing Fw 159 fighter, which had a retractable undercarriage, was a failure. The twin-engined Fw 58 Weihe, a popular trainer and general utility plane, was a successful Tank design.

In the late 1930s Tank achieved great success with the Fw 200 KONDOR transport and reconnaissance aircraft and the FOCKE-WULF 190 fighter (with its Ta 152 derivative). The Fw 187 Falke twin-engined Zerstörer fighter did not go into production despite its top speed of 326 mph (525 km/h), and the twin-boom Fw 189 Euler reconnaissance plane found only limited application during World War II (principally on the eastern front). Among Tank's wartime designs were the Fw 191 twin-engined medium bomber with all-electric controls, and the Ta 154 Moskito all-wood twin-engined night fighter, designed to catch the fast-flying British Mosquitoes. Neither of these received Luftwaffe orders.

At the end of World War II Tank was designing a jet fighter (Ta 183). He went to Argentina, where he produced this machine as the Pulqui, powered by a Nene engine.

Taylorcraft

The original Taylor CUB monoplane, one of the most important light aircraft of all time, was designed by C. G. Taylor in 1928–30. The Taylor Aircraft Company, which was set up at Lock Haven, Pa., built 211 Cubs in 1935, and the total of 550 sold in 1936 represented over 35% of that year's sales of American civil aircraft. The company was reorganized by W. T. Piper in 1937 as the PIPER Aircraft Corporation.

Taylor himself established Taylorcraft Aviation Corporation of America at Alliance, Ohio, where a similar airplane known as the Taylorcraft Model A was built with side-by-side instead of tandem seating. This was followed by the Model B two-seater trainer, with a choice of engines; the 65-hp Model C tandem trainer, which reverted to Cub seating; and the Taylorcraft Model D, which was also manufactured in England under an agreement with Fairchild and the Taylor Young Corporation, forming the basis of the well-known Auster series.

A military version of the Model D (the L-2A Grasshopper) was built in quantity during World War II, with variants that concluded with the L-2M of 1944. The 65-hp BC-12D was produced in 1945, but the company ceased operation the following year. It was subsequently reorganized by C. G. Taylor as Taylorcraft Inc., with a factory at Conway-Pittsburgh Airport, Pa., where the fiberglass-covered 145-hp Tourist, 85-hp Sportsman, 225-hp Ranch Wagon, and 225-hp Zephyr 400 were produced.

Tempest

The Hawker Tempest was an advanced British fighter of World War II that entered service with the RAF in 1944. It owed its origins to proposals put forward in 1937 for an aircraft that would succeed the HURRICANE, and would be powered by the 24-cylinder horizontal-H Napier Sabre engine. Prototypes of this new machine with both the Sabre engine (Typhoon) and the unreliable Rolls-Royce Vulture (Tornado) were produced, but only the Typhoon entered production and began to equip service units in the autumn of 1941. Mounting either 12 .303 Browning machine guns (Typhoon IA), or four 20-mm cannon (Typhoon IB), the Typhoon had a poor rate of climb, an indifferent altitude performance, and tended to buffet severely in dives. The Sabre engines also proved unreliable, but by the end of 1942 this 400-mph (640-km/h) fighter had established itself as an effective ground attack machine, armed with either two 1,000-lb. (450-kg) bombs or eight rocket projectiles. The

Tempest was essentially a development of the Typhoon with a thinner wing section. Only a single Sabre IV-engined Tempest I was built, and the first version to enter service was the 435-mph (700-km/h) Tempest V, with a Sabre IIB engine and four 20-mm cannon. It proved itself to be an excellent low-level fighter over Europe in 1944–45. Wingspan of the Tempest V was 41 ft. (12.50 m) and length 33 ft. 8 in. (10.26 m). The Tempest II had a Bristol Centaurus radial engine (tested in a Tornado in 1941) and could attain 440 mph (708 km/h). It did not reach RAF squadrons until November 1945. The Tempest VI was a "tropicalized" Mark V; the Hawker series of piston-engined fighters was continued by the FURY.

Test pilot

Testing aircraft calls for the most skilled of pilots. The work falls into two categories: testing the prototypes, and production-testing before acceptance by the customer. Aircraft are tested not only to ensure that they are safe and that all systems function satisfactorily, but also to determine that they meet the design specifications the airline or military service has set. All aircraft must also meet stringent operation regulations made, in the case of the United States, by the FAA. In military or airline service, new types of aircraft may pass through the hands of test pilots who supervise the writing of operating manuals and participate in setting any operating restrictions. After each major overhaul an aircraft will be test flown, but for such flights an ordinary pilot may suffice.

Texan

An advanced trainer used extensively by Allied air forces during World War II and after, the Texan (as it eventually became known in U.S. service) was first produced as the BC-1 in 1938. It was a development of the BT-9 basic trainer with which NORTH AMERICAN AVIATION had obtained its first Army Air Corps order in 1936. A low-wing monoplane powered by a 600-hp Pratt & Whitney R-1340 radial engine, the Texan had a top speed of 208 mph (335 km/h). The main undercarriage retracted into the wing, and the two pilots were seated in tandem. Wingspan was 42 ft. (12.80 m) and length 29 ft. (8.84 m).

A refined version of the aircraft was ordered by Britain in 1938 and became known as the Harvard, the post-1940 American designation being AT-6 (T-6 after 1948). The aircraft stayed in production into the 1950s, and saw combat service as an artillery spotter in the Korean War. A total of over 15,000 were produced, and many are still in service as trainers with some of the world's smaller air forces.

Thrust

Force generated by a jet or rocket engine, or by a propeller driven from a piston engine or turboprop, corresponding to mass multiplied by velocity. A jet develops thrust by displacing a relatively small volume of air at high velocity; a propeller displaces a large volume of air at a low velocity. See JET ENGINE.

Thunderbolt

When the Republic P-47 Thunderbolt made its first flight on May 6, 1941, it was the largest and heaviest single-seat fighter ever built. Designed around the massive Pratt & Whitney R-2800 two-row radial engine and its associated turbosupercharger ducts, the XP-47B weighed over 12,000 lb. (5,440 kg). Production P-47Bs, with eight .50 machine guns, began to reach U.S. Army Air Force units in the summer of 1942 and flew operationally from England the following spring. Speed was 430 mph (692 km/h) and ceiling 42,000 ft. (12,800 m), but the Thunderbolt had a range restricted to only 550 mi. (885 km) because of its thirsty engine. Its rate of climb and maneuverability were also poor.

The P-47C had a lengthened fuselage and could carry drop tanks to extend its range; the F-47D, the most numerous variant, was powered by a 2,500-hp engine and had a range stretched to 1,800 mi. (2,900 km). Wingspan of the P-47D was 40 ft. 9 in. (12.42 m) and length 36 ft. 1 in. (11.00 m). The P-47D was used in the Pacific as well as Europe. The 2,800-hp P-47N was, however, specifically designed for escort-ing B-29 SUPERFORTRESSES over Japan (it had redesigned wings and a range of 2,350 mi., 3,780 km). The P-47M, with a speed of 453 mph (729 km/h) and an improved rate of climb, was a specially boosted version used only by the 56th Fighter Group in Europe. Over 15,000 Thunderbolts of all types were built.

Thunderjet

The second operational American jet fighter, the single-seat Republic F-84 Thunderjet was first flown on February 28, 1946, and entered service two years later. The principal Air Force tactical fighter of the early 1950s, the F-84E saw extensive action in the Korean War. The later F-84G of 1951, with a 2,000-mi. (3,200-km) range and in-flight refueling capability, became the first Air Force fighter able to deliver tactical nuclear weapons. Wingspan of the F-84G was 36 ft. 5 in. (11.10 m) and length 38 ft. 1 in. (11.61 m).

All variants of the Thunderjet were powered by the General Electric/Allison J35 turbojet and had top speeds in the 580–620 mph (933–999 km/h) range. A development of the Thunderjet, the swept-wing F-84F Thunderstreak (1950), used the Wright J65 engine (a British license-built Sapphire), and could reach 700 mph (1,126 km/h). Thunderstreaks served as long-range escorts as well as attack aircraft. A reconnaissance version of the Thunderjet, the RF-84F Thunderflash, was distinguished by its wing-root air intakes for the engine, introduced so that camera equipment could be housed in the nose. A total of 4,457 Thunderjets, 2,711 Thunderstreaks, and 715 Thunderflashes were built in the 12 years that the F-84 and RF-84 were in production.

Tomcat

The Grumman F-14 Tomcat is a U.S. Navy variable-geometry ("swing-wing") multipurpose fighter evolved to replace the canceled F-111B. A two-

P-47N Thunderbolt

F-14 Tomcat

seater, the F-14A has two TF30 turbofans of 20,600 lb. (9,330 kg) thrust, which give it a maximum speed of Mach 2.34 at 40,000 ft. (12,200 m). It is 62 ft. (18.89 m) in length, with a wingspan of 33 ft. 2½ in. (10.12 m) with the wings swept (64 ft. 1½ in., 19.54 m, with the wings forward). The Tomcat is equipped with small retractable foreplanes housed in the fixed parts of the wings. These are extended to improve stability in the transition period while the main wings are being swept back.

First flown in 1970, the Tomcat is armed with a 20-mm rotary cannon and air-to-air missiles. Variants include the F-14B, powered by F40 turbofans, and the F-14D, employing simplified avionics and weapons systems.

Tornado

Modern European multirole combat aircraft with variable-geometry wings and twin RB-199 turbofans of 14,500 lb. (6,577 kg) thrust. Produced by an Anglo–Italian–German agency (Panavia, consisting of BAC, Aeritalia, Messerschmitt-Bölkow-Blohm), the Tornado is designed to fulfil requirements for roles that include close air support/battlefield interdiction, interdictor strike, air superiority, interception, reconnaissance, and naval flying.

First flown in 1974, it represents Europe's largest aircraft program to date, and is intended to replace the RAF's VULČANS, PHANTOMS, and Buccaneers, and the STARFIGHTERS and G91s of the LUFTWAFFE and the Italian air force.

The Tornado's length is 54 ft. 9½ in. (16.70 m) and its wingspan 45 ft. 7¼ in. (13.90 m). Its basic armament is a pair of 27-mm Mauser cannon, other weaponry varying according to designated function. Its weight of about 45,000 lb. (20,385 kg) is some 30 percent less than that of an aircraft with a fixed wing designed to perform the same functions. Radar signals from a scanner in the nose pass direct to a highly advanced autopilot, which steers the plane automatically at a height of some 200 ft. (61 m). Navigation is also automatic, relying on an inertial guidance system.

Trident

A short/medium haul British airliner with its three jets mounted at the rear, the Trident began as a de Havilland design (the DH 121), but after aircraft industry mergers was manufactured and marketed by Hawker Siddeley. The prototype flew for the first time on January 9, 1962, and the Trident 1 (103 passengers) entered service with British European Airways on April 1, 1964. The 1E, with engines uprated to 11,400 lb. (5,160 kg) thrust seats 115; the 2E (1967), with wingspan of 98 ft. (29.87 m), length of 114 ft. 9 in. (34.98 m), and still more powerful engines, carries up to 149 passengers. The 3B (1969) is a 179-passenger short-haul development of the 1E with a fuselage stretched to 131 ft. 2 in. (39.98 m); the Super 3B, which possesses extra fuel capacity, was ordered by China in 1972.

Trim

Airplanes, airships, and most rotary-wing aircraft must be trimmed to fly in the correct attitude, tending neither to pitch nor to roll. An airplane is trimmed when there is no rolling moment (tendency) and when the nose-down pitching moment is exactly balanced by an equal and opposite nose-up moment caused by the CG (center of gravity) lying just behind the CP (center of pressure of the LIFT on the wing). The aircraft will then fly with the pilot's hands off the controls. Ideally, it should remain trimmed after being disturbed by flying through a sharp gust and, though momentarily rotated in pitch and/or roll, will recover straight and level flight. Trimming in simple airplanes may be achieved either by adjusting springs or rubber cords in the flying-control system or by bending trim tabs fixed to the trailing edge of the control surfaces. Most airplanes have hinged trim tabs whose incidence is

Panavia Tornado

Trident 1E

controlled from the cockpit, either by handwheels or by electric switches.

Triplane

An aircraft with three wings or sets of wings possesses even greater maneuverability than a biplane can provide, but at a cost of reduced speed due to the extra weight and drag of the additional wing.

The first powered triplane to fly was the Ambroise Goupy-designed machine built by the Voisin Brothers in 1908; the Englishman A. V. ROE flew a series of fully successful triplanes at almost the same time. Interest lapsed until the appearance in 1916 of the Sopwith Triplane fighter, designed to combat the formidable German ALBATROS scouts over the western front. Although underpowered with its 110-hp Clerget rotary engine it could attain 117 mph (188 km/h) at 5,000 ft. (1,500 m) and equipped with only a single Vickers gun, the Sopwith Triplane became so successful in the hands of Royal Naval Air Service pilots that a captured example was shown to Anthony FOKKER with a request that he produce a comparable fighter.

The result was the Fokker Dr.I (Dr. stood for "Dreidecker," or triplane). It had a 110-hp Le Rhône engine and twin Maxim ("Spandau") machine guns. A series of crashes caused by structural failure—the result of faulty manufacture—grounded all Dr.Is shortly after their introduction in the fall of 1917, but the Fokker Triplane went on to achieve eventual fame as the mount of the legendary RICHTHOFEN. It was also flown by VOSS, UDET and GOERING.

Although immensely maneuverable the Dr.I lacked speed—at 14,000 ft. (4,300 m) it could only make 86 mph (138 km/h). By the spring of 1918 the triplane had become obsolescent, replaced by the Sopwith CAMEL in Allied service and by the Fokker D.VII in the Luftwaffe. The Dr.I had virtually disappeared from the front by the Armistice, and altogether only about 175 saw operational service.

At the height of enthusiasm for triplanes (1917–18) there were, however, over 30 triplane fighters under development in Germany alone, and in Italy CAPRONI produced a series of three-engined twin-boom triplane bombers, culminating in the Ca 42, which had a wingspan of nearly 100 ft. (30.5 m) and could carry 3,910 lb. (1,770 kg) of bombs.

A few giant triplanes were built after the war, none of them successful. Among them was the British-designed six-engine Tarrant Tabor (1919), which was intended to bomb Berlin had the

Fokker Dr.I Triplane

Lockheed L-1011 TriStar

war lasted another six months. On its first attempted takeoff, the aircraft dug its nose into the ground and its two pilots were fatally injured. The American Barling XNBL-1 (1923), an experimental night bomber with wingspan of 120 ft. (36.58 m), did manage to fly successfully, but its range with full payload was found to be only 170 mi. (275 km) and the project was abandoned. Probably most spectacular of all postwar triplanes was the huge Caproni Ca 60 Transaero eight-engine flying boat (1919). With three sets of triplane wings the giant aircraft was intended to carry 100 passengers and 8 crew members on transoceanic flights. The engines of the time were not, however, powerful enough even to lift the plane off the water, and after the first attempted flight, in which serious damage was incurred, this project too was abandoned.

TriStar

Second (after the DC-10) of the new generation of wide-bodied AIRBUSES, the Lockheed L-1011 TriStar first flew on November 16, 1970. Length of the TriStar is 178 ft. 8 in. (54.35 m) and wingspan 155 ft. 4 in. (47.35 m). Powered by three Rolls-Royce RB-211 turbofans of 42,000 lb. (19,025 kg) thrust (one of them tail-mounted), the original L-1011-1 TriStar's cruising speed was 562 mph (904 km/h). The aircraft entered service with Eastern Air Lines in April 1972 and by the end of 1973 it was being flown by six other commercial operators, including TWA and Air Canada.

The original L-1011-1 had a range of only 2,500 mi. (4,000 km). Its successor, the L-1011-100, could fly 4,000 mi. (6,400 km); the L-1011-200, with more powerful engines, had improved takeoff and climb performance for hot climates and high-altitude airports. The L-1011-250 was given extra fuel capacity to extend its range to nearly 5,000 mi. (8,000 km). All versions of the TriStar are capable of carrying up to 400 passengers.

Tu-16

The TUPOLEV Tu-16 medium-range

Tu-20 "Bear" heavy bomber

bomber (Nato code name "Badger") entered service in 1954, providing the Long-Range Air Arm of the Soviet Air Force with a strike power broadly comparable to that of the RAF's Valiant and the Boeing B-47 STRATOJET of the U.S. Strategic Air Command. The Tu-16, with its two 20,950 lb. (9,503 kg) thrust Mikulin engines, was an important front-line aircraft until the 1960s, when the growing effectiveness of surface-to-air missiles rendered it too vulnerable. It was gradually transferred to maritime patrol and antishipping duties. Today most of the total production of 2,000 are used in maritime roles, though some were supplied as bombers to countries such as Indonesia and Egypt. The Tu-16 has a wingspan of 109 ft. 11 in. (33.50 m) and a length of 120 ft. (36.58 m).

For antishipping missions the Badgers carry air-to-surface missiles under their wings. Some versions have a bulbous nose that holds electronic sensing equipment to detect and measure radio and radar transmissions from Western ships and aircraft.

Tu-20

Russia's Tu-20 bomber (originally known as the Tu-95), code-named "Bear" by Nato, is the world's only turboprop bomber and unique among aircraft in combining propellers and swept wings. Propellers are still the most efficient propulsive devices at low to moderate speeds (up to about Mach 0.65), while swept wings become necessary above about Mach 0.7; normally the two are not well matched. But the TUPOLEV design bureau in the mid-1950s decided to combine them in order to achieve long range and high speed. The four 14,795-hp Kuznetsov turboprop engines were geared down to turn the huge contra-rotating propellers quite slowly, so that they retained their efficiency at higher speeds than usual.

The Tu-20 was the final outcome of a line of propeller-driven bombers that began with the Tu-4 of 1946. This was a close copy of the U.S. B-29 SUPER-FORTRESS, three of which had force-landed in Russia after bombing Japan in 1945. The turboprop Tu-20 bomber was introduced into the Long-Range Air Arm of the Soviet Air Force in 1956, and for the next 10 years was the backbone of the strategic bomber force. A civilian version, the Tu-114 (code-named "Cleat"), was the world's largest commercial transport, with a wingspan of 163 ft. (49.68 m) and a length of 150 ft. (45.72 m), until the introduction of the Boeing 747 in 1970.

Initial versions of the bomber could carry a payload of some 25,000 lb. (11,325 kg) over a range of about 7,800 mi. (12,480 km). Defensive armament was cannon controlled from the flight deck. Later versions carried the Kangaroo strategic missile, with a range of about 400 mi. (643 km) after launch.

Eventually, the Tu-20 was gradually transferred to maritime patrol duties for which it was particularly suitable because of its great range and endurance. Some aircraft have been adapted as electronic "ferrets," carrying complex equipment to measure the characteristics of Nato's early warning radar and other systems.

Tu-144

The Tupolev Tu-144 supersonic airliner (Nato code name "Charger") made its first flight on December 31, 1968—more than two months before that of the rival BAC/Aérospatiale CONCORDE. The Russian aircraft seats up to 140 passengers, and with a length of 215 ft. 6 in. (65.68 m) and a wingspan of 94 ft. 6 in. (28.80 m) it is somewhat larger than the Anglo-French design. The configuration is that of a fully cambered delta with four Kuznetsov NK-144 engines (44,090 lb., 20,000 kg, thrust with afterburner) mounted in two separate ventral ducts (originally they occupied a single housing). Cruising speed is Mach 2.4.

The second production aircraft was destroyed in a crash at the 1973 Paris Air Show. Later modifications to the design during the Tu-144 test program include the repositioning of the engines and the incorporation of retractable foreplanes immediately behind the cockpit.

The Tu-144 became the first supersonic airliner to operate a scheduled service when a twice-weekly mail and freight run was inaugurated on December 26, 1975, between Moscow and Alma Ata, Kazakhstan (a distance of 2,190 mi., 3,510 km).

Tu-144 "Charger" supersonic transport

Tuck, Roland Robert Stanford
(1916–)

British World War II fighter ace. Tuck joined the RAF in 1935 and survived a midair collision in 1938. By May 1940 he was flying in combat over Dunkirk with one of the first Spitfire units; he then commanded a Hurricane squadron in the Battle of Britain, in which he was shot down twice, and flew fighter sweeps over Europe in 1941 before going to the United States with an RAF liaison mission. Returning to operational flying in December 1941, Tuck assumed command of the Biggin Hill Spitfire Wing, but in January 1942 his Spitfire was shot down by ground fire over France and he was taken prisoner by the Germans.

While being marched east in 1945 as Anglo-American forces advanced across Europe, Tuck managed to escape and reached the Russian lines. He fought as an infantryman with the Red Army, but eventually slipped away from the front line and made his way to Moscow, where the British Embassy arranged for his return to England.

Tupolev, Andrei Nikolaevich
(1888–1972)

Possibly the best known Soviet aircraft designer, responsible for many of Russia's heavy bombers both before and after World War II. In 1908 he joined Professor ZHUKOVSKI's Aeronautical Course and took part in early gliding experiments. In 1920 Tupolev was put in charge of the aircraft design department at the Central Aero- and Hydrodynamic Institute (TsAGI), and in 1922 was made chairman of a commission to design and build all-metal aircraft. The Russians had been inspired by the work of Dr. Hugo JUNKERS, and in 1926 the Soviet authorities took over the Junkers factory in Moscow and built Tupolev designs there. Tupolev's early aircraft were very reminiscent of Junkers' models, and the twin-engined ANT-4, which became the TB-1 heavy bomber, owed much to the Junkers G.24. Over 200 TB-1s were built, and the design was used in experiments that included carrying a single-seat fighter on each wing. The wings and tail unit of the TB-1 were joined to a new fuselage to produce the ANT-9 passenger aircraft; it was the ancestor of a line of giant landplanes that were the largest in the world at the time they were built. The first of these was the four-engined ANT-6 or TB-3 heavy bomber (1930), which had a wingspan of 132 ft. 9 in. (40.46 m). This was followed in 1933 by the experimental six-engined ANT-16 (TB-6), on which the eight-engined

Tu-114 "Cleat" turboprop airliner

ANT-20 MAXIM GORKI propaganda airplane was based.

Tupolev was arrested and imprisoned during the Purges of the 1930s and was released in 1942 when he was awarded a Stalin Prize for designing the Tu-2 medium bomber. He was also responsible for Soviet production of the Tu-4 copy of the Boeing B-29 SUPERFORTRESS, which was followed by more giants that included the Tu-85, powered by four 4,300-hp VD-4K 28-cylinder radial engines, and the turboprop TU-20 heavy bomber and Tu-114 (1955) airliner, which used the same wings and tail unit. These two aircraft were also the world's fastest propeller-driven aircraft and the only such with swept wings. The Tu-104 (1955), also developed from a bomber, was the world's second production turbojet airliner, and between 1956 and 1958 the only one in scheduled airline service.

The TU-144 supersonic transport flew for the first time on December 31, 1968, making its maiden flight several months before that of CONCORDE. It cruises at about Mach 2.4 and seats 140 passengers. Also of Tupolev origin is the variable-geometry BACKFIRE bomber, capable of delivering nuclear weapons to North America.

Turbofan

A turbofan is a jet engine in which part of the air drawn into the engine by a huge fan bypasses the combustion chambers and gives additional thrust. It is also called a bypass turbojet, or a fanjet.
See JET ENGINE.

Turboméca

Established in 1938 to develop aeronautical compressors, superchargers and turbines, Société Turboméca is the leading European manufacturer of gas turbines for small aircraft. Among its engines are the lightweight Astafan, the Marboré turbojet (used on the Morane-Saulnier Paris, Super-Magister), and the Astazou turboprop with two-stage axial and single stage centrifugal compressors

(Skyvan, Pilatus Turbo-Porter, Jetstream, Nord 260A) and Astazou turboshaft (Alouette). The Adour turbofan has been developed in conjunction with Rolls-Royce for the JAGUAR and Hawk.

In its first 25 years of operation. Turboméca built 12,500 engines, with another 12,000 license-manufactured in Britain (Rolls-Royce), the United States (Teledyne CAE), Spain, India, and Yugoslavia.

Turboprop

The turboprop engine consists of a propeller coupled to a turbine. Air passes through a compressor into a combustion chamber; the exhaust gases are then discharged through the turbine.

A small amount of thrust is generated by the exhaust gases, but propulsion is primarily derived from the propeller, and the speed of turboprop-engined airplanes is limited to the subsonic region (up to about 550 mph, 885 km/h), where they generally possess better performance than a pure-jet engine. Compared to PISTON ENGINES, turboprops are lighter, have a lower frontal area, are more efficient at the top end of the speed range, and are easier to maintain.

The general layout of a turboprop closely resembles that of a JET ENGINE. Compressors may be either centrifugal (single- or two-stage), or axial (one- or two-spool), or even a combination of both (the Proteus has a 12-stage axial compressor feeding forward into a single-stage centrifugal compressor, the overall pressure ratio being 7.2:1). A free-turbine turboprop has a separate additional turbine to drive the propeller ("free" because it is not mechanically coupled to the compressor). This enables the compressor turbine and the propeller turbine to be operated independently at their most efficient speeds, and permits a wide range of propeller speeds while maintaining a constant compressor speed.

Turboprop engines are now used principally in smaller transport aircraft, particularly those intended for STOL operations. Compared with jet and piston engines, turboprops are quieter.

Turbulence

Molecules of a fluid in random motion are said to be turbulent, the opposite being smooth laminar flow. The BOUNDARY LAYER, the air closely surrounding an aircraft, can be kept laminar by making the aircraft surface regular and very smooth, but the slightest roughness makes the flow turbulent, causing increased drag. Turbulence is caused by the motion of unstreamlined bodies, and was a serious problem when improperly designed aircraft tried to achieve SUPERSONIC FLIGHT.

The atmosphere is always turbulent, especially around sharp-edged gusts (vertical or horizontal winds in otherwise almost still air) or where surface wind encounters trees and buildings. At greater heights this type of turbulence disappears, but certain conditions of humidity and uneven cooling of the surface produce strong turbulence for aircraft. In thunderstorms, and particularly in the interior of thunderclouds, turbulence increases greatly, and so-called clear air turbulence can be experienced above 30,000 ft. (9,144 m) in the vicinity of the jetstream. The danger of turbulence increases with the airspeed.

TWA

The initials TWA originally stood for Transcontinental and Western Air, Inc., an airline formed on July 16, 1930, by the amalgamation of Western Air Express (founded 1925) and Transcontinental Air Transport.

On October 25, 1930, TWA inaugurated U.S. coast-to-coast services, four years later it introduced the DC-2 on its Columbus–Pittsburgh–Newark route (May 18, 1934). Five pressurized Boeing Stratoliners joined the airline in July 1940 and cut the transcontinental time to 14½ hours. During the war they were operated on military flights across the Atlantic (TWA's first international operations). On February 5, 1946, TWA opened its commercial transatlantic flights with a Lockheed L-049 Constellation service between New York and London via Gander and Shannon.

By the time the airline's name was changed to Trans World in 1950, its routes extended throughout Europe to North Africa and India. In 1953 the network reached Ceylon, with services to Thailand in 1958 and Hong Kong in 1966; the first direct flights between New York and Rome were begun in 1957, a nonstop schedule from Chicago to Paris starting the following year. Boeing 707 jets were introduced on the North Atlantic route in November 1959 (TWA's domestic services having used them since the previous March).

Flights to East Africa began in 1967, followed two years later by a transpacific route to Hong Kong. By 1961 all TWA's international services were operated by jets; in 1967 the last propeller-driven aircraft left U.S. domestic operations.

TWA's present fleet consists of some 200 jet airliners (Boeing 747s, 727s, and 707s; TriStars; DC-9s). The airline carries 15,000,000 passengers annually.

U

Udet, Ernst (1896–1941)

Credited with 62 victories, Udet was Germany's second highest scoring World War I ace. He gained his first aerial victory on March 18, 1916, and was then assigned to *Jagdstaffel* 15, with which he claimed five more victims. Appointed commander of *Jagdstaffel* 37, he had scored 14 victories by March 1918, and was invited by RICHTHOFEN to join *Jagdgeschwader* 1. As a member of Richthofen's Flying Circus (serving in *Jagdstaffel* 11), he scored a further 42 victories, although twice shot down (successfully bailing out on both occasions). After the war Udet became a stunt pilot, and eventually joined the new Luftwaffe as a high-ranking officer. He continued to fly, testing many new aircraft designs, but became the scapegoat for GOERING's many errors in directing the Luftwaffe's wartime operations and, unwilling to accept the continuing disgrace, he shot himself on November 17, 1941.

Undercarriage

Landing gear can assume many forms in landplanes and amphibians. The classic arrangement had two mainwheels ahead of the CG (center of gravity) and a tailskid or tailwheel; adoption of a tailwheel came with the general introduction of mainwheel brakes. Today the usual arrangement is the so-called tricycle, with a nosewheel and two mainwheels behind the CG. Tricycle undercarriages provide the pilot with better visibility during take-offs and landings, give a level fuselage position which facilitates loading, and keep jet exhausts from damaging runway surfaces. Aircraft with very low aspect ratio wings cannot be stalled on landing and have short nose wheels to give a negative angle of attack at touchdown so that they are prevented from ballooning.

Many aircraft have their weight supported by one, two, or more wheels

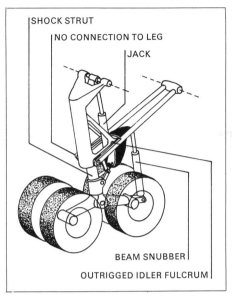

The four-wheel main undercarriage assembly of the Russian Tu-114, the world's largest and fastest propeller-driven airliner. The growth in size and weight of aircraft over the past decades has led to the introduction of such multiwheel units and complex shock-absorber systems in order to spread weight safely

along the longitudinal centerline (the B-52 has four steerable twin-wheel trucks) with outriggers to keep the wing tips off the ground. Helicopters often have skids, while RPVs, drones, and some research aircraft have no proper gear at all (some have small retractable skids). Heavy aircraft spread the load over large areas by using numerous wheels, commonly in pairs or in four-wheel or six-wheel assemblies. Military freighters need "high flotation" gear to enable them to use unpaved surfaces; the C-5A GALAXY has four six-wheel assemblies and a four-tire nose gear. In contrast, carrier-based aircraft can use tires with much higher inflation pressure, so that heavy machines can use fewer and smaller tires, which occupy less space when retracted.

United Air Lines

The world's largest privately owned airline, United Air Lines came into being as United Aircraft and Transport in 1931, the result of the merger of four pioneer American airlines—Boeing Air Transport, National Air Transport, Pacific Air Transport, and Varney Air Lines. United began operating under its present name in 1933. Today it serves 113 cities, flying from Vancouver in the north to Mexico City in the south, and westward as far as Hawaii.

DC-8s were introduced on the New York–San Francisco/Los Angeles route and services to Hawaii in 1959. On July 5, 1960, United was the first airline to put the Boeing 720 into service. Capital Airlines was acquired by United in June 1961. All cargo DC-8Fs came into use in 1964, and by July 1967 United was operating the world's largest DC-8 fleet. Boeing 727s took over short-haul routes in 1964, phasing out DC-6s, Convair 340s, Caravelles, and Viscounts. In 1970 Boeing 747s came into use on services to Hawaii and transcontinental routes.

United Air Lines covers a total of 19,300 route miles (31,000 km); some 29,000,000 passengers are carried annually. Its present fleet consists of 18 Boeing 747s, 37 DC-10s, 100 DC-8s, 150 Boeing 727s, and 57 Boeing 737s.

U.S. Air Force

On August 1, 1907, the U.S. Army established an Aeronautical Division in the Office of the Signal Corps to take charge of all matters pertaining to military ballooning and air machines. The following year a contract was signed for the purchase of a Wright biplane but not until 1911 did the Army possess more than one airplane and pilot. During the uncommitted years of World War I both the U.S. Army and Navy developed small aviation services, based largely on Curtiss biplanes. With America's entry an expansion was begun, particularly in the Army where, influenced by British, French, and Italian experience, units were formed with distinct missions—pursuit (fighter), bombardment, and observation. Participation in the European war zone involved some 1,200 aircrew and 740 Army Air Service aircraft were in first line units at the time of the Armistice in November 1918, although all were types designed by their Allies; the U.S. ace Eddie RICKENBACKER, for example, flew a French-designed Spad.

In the immediate postwar period there was a drastic reduction in both manpower and equipment. Unlike their British counterparts, who had obtained autonomy as the Royal Air Force, the principal American military air service was still part of the U.S. Army. The fighters were largely Boeing PW-9s and, later, P-12 biplanes, the bombers Martin MB-1s (also biplanes). A number of officers strongly promoted the cause of an independent air force, basing their case primarily on the potential of the bomber as an offensive weapon for strategic warfare. The voluble General Billy MITCHELL arranged some highly successful bombing trials against redundant battleships, but these sinkings failed to convince the tradi-

The B-52 bomber, still a mainstay of the Air Force's Strategic Air Command

tionalists of the Army and Navy that a separate air force was either necessary or desirable. Eventually Mitchell provoked his own court martial and resigned from active service. But he was by no means the only advocate of a powerful and independent air force—only the most vocal.

In 1926 the Air Service became the U.S. Army Air Corps and plans were made for limited expansion and re-equipment. A body of opinion within the Air Corps saw the potential of strategic bombing as the key to the long-delayed recognition of air power. At the beginning of the 1930s the first all-metal production fighter appeared, the Boeing P-26 PEASHOOTER. Various other technical advances in the 1930s fueled the effort to establish a force of bombers. One of the first of these was the technically advanced Martin B-10, faster than most fighters of its day. To appease the military traditionalists, long range bombers such as the Douglas B-18 were at first ordered ostensibly for sea patrol.

The outbreak of wars in Europe and China brought a rapid change in the status of the Army Air Corps, with huge funds being made available for its expansion. Four-engined bombers such as the Boeing B-17 rapidly made their appearance. In June 1941 the U.S. Air Corps became the Army Air Forces and achieved a semiautonomous status. When the United States was brought into the war, strategic

bombing became a major objective of the AAF, which now flew not only the B-17 FLYING FORTRESS but also the larger consolidated B-24 LIBERATOR. In Europe it was only partly successful, but against Japan, using the high-altitude B-29 SUPERFORTRESS developed from the B-17, it was wholly so, shattering the economy of the country. As enemy interceptors began to take a heavy toll of the giant bombers, the escort fighters were developed. These were principally P-47 THUNDERBOLTS and P-51 MUSTANGS. The strategic value reached its zenith with the advent of the atomic weapon which at first could only be delivered by very heavy bombers. With peace, the huge USAAF, largest air force in the world, was quickly and drastically reduced, from 2,400,000 personnel and 80,000 aircraft in mid-1944 to 306,000 personnel and 25,000 aircraft by 1947, of which less than 10,000 were available as airworthy.

The independent USAF was established by an Act of Congress on September 18, 1947, around three major combat commands—Strategic, Tactical, and Air Defense. Deteriorating relations with the Soviet Union gave the Strategic Air Command a major deterrent role in the two decades following World War II. From the single B-29 "atom bomb" group of 1946, SAC grew by the mid-1950s to embrace 45 heavy bomber wings, armed with the six-engined Boeing B-47 Stratojets, which had nuclear capability. In the following decade intercontinental missiles took over a major share of the deterrent role. Nevertheless, air power was a principal factor in the Korean War and again in Southeast Asia during the late 1960s. The first conflict was notable for the introduction to combat of the new U.S. jet fighters, the Lockheed F-80 Shooting Star and the North American F-86 Sabre. These limited wars led to the development of substantial tactical air forces.

From the post-World War II peak of 997,000 personnel and 28,000 aircraft following the cessation of hostilities in Korea, strength has gradually declined, particularly in the 1970s. In 1976 the Air Force had 9,300 aircraft, 1,000 intercontinental ballistic missiles, and 571,000 men and women.

SAC's share of responsibility for delivering the nuclear deterrent is borne by some 450 B-52 and FB-111 manned bombers, plus 1,050 ICBM, all based in the United States. Tactical Air Command and its allied commands in Europe and the Pacific have the largest aircraft forces, predominantly fighter-bombers and reconnaissance types. It is also charged with the support of the Army, by tactical assault aircraft such as the Lockheed C-130 Hercules. Home defense fighter units have been gradually phased out as this role has been shifted to the part-time Air National Guard squadrons. A major organization in the USAF is Military Airlift Command controlling the largest fleet of military air transports in the world during the 1960s and 1970s. The two mainstays here are the Lockheed C-141 Star Lifter and the more recent Lockheed C-5A GALAXY.

U.S. Marine Corps

Marine Corps officers were among the early American naval pilots, the first of whom flew solo in August 1912. The early mainstay of U.S. maritime aviation was the Curtiss A-series flying boat. Since the airplane appeared equally suited to supporting land operations, a body of opinion within the Marine Corps favored a separate aviation section to meet the requirements of their ground forces. This idea was approved by the Navy in 1915, and two distinct air services under Navy jurisdiction were allowed to develop. Marine Corps pilots flew most types of Navy aircraft during these formative years, and with America's entry into World War I units were sent to Europe to form part of a combined Navy/Marine Corps bombing force to carry out a campaign against German North Sea submarine bases. Flying de Havil-

The A4M Skyhawk, a versatile attack aircraft of the U.S. Marine Corps

HRP Marine Corps rescue helicopters

land DH 4s, the Marine Corps operated by day, carrying out its first major operation only a few weeks before the Armistice. Marine fighter pilots who had previously been sent to France (without aircraft) gained combat experience with British fighter squadrons.

Troubles in the Caribbean, where the United States was involved in policing activities, brought active service for Marine flying units for 15 years from 1919. Marine-operated Curtiss flying boats, together with Curtiss Jenny and DH 4 landplanes, were operational in the Dominican Republic and Haiti that year. During these Caribbean actions, pilots developed the technique of dive-bombing for greater accuracy, and a specialized aircraft, the Curtiss HELL-DIVER, was built. In 1927 activities were extended to Nicaragua and China, and it was in these areas that an organized form of air transportation was regularly employed for personnel.

While Marine aviation originally

extended to most activities undertaken by Navy aviation, in the 1930s its forces were concentrated largely into fighter, attack, and observation roles. Most of its equipment continued to be developed from that used by the Navy. During this period Marine fighter squadrons received some of the first monoplane fighters, Brewster F2As, Buffaloes and Grumman F4F Wildcats. When the Japanese struck in the Pacific on December 7, 1941, the Marines had a squadron of 12 F4Fs on Wake Island and another with 10 in Hawaii, as well as units with some 38 other types, chiefly SBD Dauntless and SB2U Vindicator dive-bombers. Most of these aircraft were eliminated in the early action but Marine fighter squadrons were to play a significant role in most of the island battles that followed during the next three and a half years of Pacific war. They were particularly successful with the F4U CORSAIR, a deck-landing fighter that at first proved unsatisfactory for carrier operation, but showed itself an outstanding land-based airplane. The Corsair was again employed by Marine squadrons in the Korean War, where the Marine Corps effort amounted to some two-thirds of that of the U.S. Navy.

In the 1960s and 1970s Marine combat strength varied between a quarter and a third of that of the U.S. Navy. In 1975 there were 600 first-line aircraft. Twelve fighter squadrons with F-4 PHANTOMS and ten attack squadrons with A-4 Skyhawks and A-6 Intruders plus three close support squadrons with AV-8 HARRIERS, vertical takeoff fighters. Other units operated reconnaissance, observation, and tanker aircraft, and troop transportation was handled by 15 helicopter units.

U.S. Navy

American naval aviation dates from November 14, 1910, when a Curtiss biplane piloted by Eugene Ely successfully took off from a specially constructed platform on the U.S.S. *Birmingham*. On February 17 the following year Glenn CURTISS' floatplane was tested in San Diego bay and hoisted on and off the U.S.S. *Pennsylvania*. This met the first formal requirements for the use of airplanes for naval observation duties, leading the Navy Department, on March 4, 1911, to create the Naval Aviation Service and recommend the purchase of three aircraft and the setting up of the first Navy flying camp at Greenbury Point, Maryland. In 1912 the first flying boat was acquired and for the next six years naval aviation was developed predominantly around this type.

When America entered World War I, Navy squadrons were sent to Europe to fly sea patrols and attack enemy bases. Flying from bases in Britain and France, U.S. Navy flying boats were credited with sinking four and damaging twelve submarines in ten months. The Curtiss flying boats were the only American-built aircraft used in combat in Europe.

After the war naval aviation was cut back, although the Navy Bureau of Aeronautics was established in 1921 to allow this branch of the service centralized development. Many influential admirals attached little importance to aviation and deplored the funds diverted to it. This attitude changed appreciably after a series of trials instigated by Billy MITCHELL in the immediate postwar years when surplus warships were sunk by aerial bombing. Naval aviation was expanded in the early 1920s to embrace fighter, scout-observation, and torpedo bomber units. In

The Forrestal class carriers of the U.S. Navy are mobile and self-contained airfields. Seen here on the flight deck of the U.S.S. Ranger *are A-4 Skyhawks, A-7 Corsairs, F-8 Crusaders, E-1 Tracers, and a C-1 Trader*

view of British success with aircraft carriers a collier was converted for this purpose, becoming the U.S.S. *Langley*. The first takeoff from the *Langley* was on October 17, 1922, by a Vought VE-7. Two battle cruiser hulls were adapted to provide the carriers *Lexington* and *Saratoga*, which were commissioned late in 1927. A five-year program approved in 1926 called for 1,000 Navy aircraft by 1931, and new types for carrier use were to have air-

cooled engines because of their simplicity and greater reliability. The first true aircraft carrier design was the *Ranger* of 1935, followed by four others during the next six years.

At the time of Pearl Harbor, the U.S. Navy had seven first-line carriers equipped with fighter, dive and torpedo bomber units, largely F4F Wildcats, TBD Devastators, SBD Dauntlesses, and PBY Catalinas. In the battles of the CORAL SEA and MIDWAY the Japanese lost five and the U.S. Navy two of their carriers, largely through strikes and counter strikes by carrier-based aircraft. These and subsequent operations established the aircraft carrier as the capital ship of World War II. In the later stages of the Pacific war, carrier aircraft were used in numerous pre-invasion strikes against land objectives. At the end of hostilities the U.S. Navy had 28 large and 71 small carriers, 41,000 land and ship-based aircraft, 61,000 aircrew and some 400,000 support airmen. New airplanes had appeared too: the TBF Avenger (carrying two torpedoes) replaced the TBDs and SBDs, while the standard fighter was now the F6F Hellcat.

U.S. Navy carrier strike forces also operated off Korea and Southeast Asia during those conflicts. Offensive operations were principally against coastal or near coastal land targets. Low-level ground attack with bombs, rockets, and fixed guns was the usual form in the Korean War, and here AD Skyraiders came into their own. More sophisticated techniques were developed for Vietnam to deliver bombs against similar types of ground target with all-jet aircraft such as A-4 Skyhawks, F-4 Phantoms, A-7 Corsair IIs, and A-3 Skywarriors. Since 1945 only the U.S. Navy has maintained a significant conventional carrier fleet although the nuclear-armed submarine had superseded the carrier as the principal strategic weapon by the mid-1960s. In 1976 the U.S. Navy had 15 carriers in commission, usually with 70 to 90 aircraft each, including high-performance fighter, attack, reconnaissance, and antisubmarine types. Of the total 1,900 combat aircraft some 250 were land-based maritime reconnaissance types.

V-1

V

V-1

The pilotless flying bomb powered by a PULSE JET launched from ramps and carrier planes against England and liberated Europe by the Nazis during 1944–45 was officially the Fieseler Fi 103. V-1 stood for *Vergeltungswaffe-1* (vengeance weapon 1). Over 30,000 "buzz bombs" were produced, including a small number with cockpits that were intended for suicide attacks. The V-1 carried a 1-ton warhead, its wingspan was 17 ft. (5 m) and length 23 ft. (7 m). Range was 150 mi. (240 km) with a maximum speed of 400 mph (640 km/h).

V-2

In a period of seven months at the end of World War II, the Nazis launched 4,000 V-2 liquid-fueled rockets bearing 2,150-lb. (975-kg) warheads at targets in England, Belgium, the Netherlands, and France. The V-2 (*Vergeltungswaffe-2*, vengeance weapon 2) was the first true ballistic missile. It was

46 ft. (14 m) in length and weighed 13 tons; its range was 200 mi. (320 km), and it reached an altitude of over 50 mi. (80 km) during its flight, attaining 3,000 mph (4,800 km/h).

The V-2 design team was led by Wernher von Braun, who subsequently played a leading part in the American space program.

V-2 being prepared for launch

The U.S. Navy's latest swing-wing fighter, the F-14A Tomcat, being launched from the flight deck of the U.S.S. Enterprise

Vampire

The de Havilland DH 100 Vampire was the second jet aircraft in service with the RAF. Designed during World War II and possessing an unusual twin-boom layout, the Vampire fighter made its maiden flight in September 1943. The 540-mph (870-km/h) Mark 1, powered by a Goblin engine of 3,100 lb. (1,400 kg) thrust and armed with four 20-mm cannon, entered RAF service in 1946. Wingspan of the Mark 1 was 40 ft. (12.19 m) and length 30 ft. 9 in. (9.37 m). The Mark 3 had a longer range (1,390 mi., 2,235 km) and modified tail. In 1949 the Mark 5 fighter-bomber, carrying wing-mounted rockets or bombs, was introduced, and a special tropical version, the Mark 9, also saw service.

The DH 112 Venom, a new fighter-bomber development of the Vampire, served with the RAF from 1952 until 1962. It had swept-back wing leading edges and was powered by a Ghost engine of 4,850 lb. (2,100 kg) thrust. Capable of 640 mph (1,030 km/h), the Venom could carry up to 2,000 lb. (900 kg) of bombs or rockets; in two-seat form it was used by the Royal Navy as an all-weather fighter and by the RAF as a night fighter.

DH 115 Vampire T11 two-seat trainers were employed by the RAF from 1952 to 1967. The two-seat DH 113 Vampire NF10 night fighter saw service from 1951 to 1954, with a small number remaining as trainers until 1959.

VC-10

A highly successful long-range British-built jet airliner, the Vickers (later BAC) VC-10 had its four engines mounted at the rear, with a high "T" tailplane. First flown on June 29, 1962, the Type 1101 151-passenger VC-10 entered service with BOAC in April 1964. Similar machines were supplied to Ghana Airways (1102) and British United (1103); the Type 1106 was an RAF transport. With four Conway engines of 21,000 lb. (9,526 kg) thrust, the VC-10 had a cruising speed of 568 mph (914 km/h) and a range of 5,500–6,500 mi. (8,850–10,460 km). Its length was 158 ft. 8 in. (48.36 m) and wingspan 146 ft. 2 in. (44.55 m).

The Super VC-10 had a fuselage stretched to 171 ft. 8 in. (52.32 m) to accommodate 180 passengers, and Conways of 22,500 lb. (10,190 kg) thrust.

Vega

The original Lockheed "plywood bullet" was the Vega, a streamlined high-wing monoplane with a fixed

VC-10

undercarriage that became its maker's first highly successful airliner. The first Vega, George Hearst Jr.'s *Golden Eagle*, made its maiden flight on July 4, 1927, but was lost in the Dole race of that year. Twenty-eight four-passenger Series 1 Vegas with 220-hp Whirlwind engines were built, and there were 45 Series 5 aircraft (450-hp Wasp engines), of which two were six-seaters operated by Pan American–Grace Airways. These were followed by 6 Series 2 (300-hp) five-seaters and a single Series 2A carrying six passengers.

The 5B, with a 450-hp Wasp, had seating for six passengers and found employment with a number of airlines (among them Braniff, Alaska-Washington, Varney Speed Lines, and several foreign operators), and was also used by Wiley POST for his record flights. The 5C had a Wasp C1 engine and enlarged tail surfaces, but only 6 of these aircraft were new machines; the other 27 were converted from 5Bs.

A number of custom-built Vega specials were also built, as well as 11 DL-1 aircraft with Duralumin skins. Wingspan was 41 ft. (12.50 m) and length 27 ft. 6 in. (8.38 m).

Vickers

The British industrial giant and armaments manufacturer Vickers entered the field of aviation well before World War I. On November 19, 1912, it received an order from the British Admiralty for what became known as EFB.1 (Experimental Fighting Biplane), probably the world's first contract for a combat aircraft. A pusher biplane, it seated an observer, armed with a Lewis gun, in front of the pilot, and was the first of a long series of Fighting Biplanes. The best known of these was the F.B.5 "Gunbus," 210 of which (including the similar F.B.9) were built by Vickers and 99 by Darracq in France. Several other F.B. types did not see production, but the twin-engined F.B.27A VIMY bomber served with the RAF in the years

immediately following World War I, and made numerous pioneering flights, including the first nonstop crossing of the Atlantic (by ALCOCK AND BROWN).

The Vimy bomber was eventually replaced by the bigger Virginia, a biplane powered by two Napier Lion engines of 450–570 hp, which progressed through ten basic models during the period 1924–37, serving as an RAF heavy bomber. Vickers fighters of the 1920s saw limited production for South American air forces, and a few airliners were built. During these same years Vickers was also responsible for a series of RAF bomber-transports for colonial duties: the Vernon of 1921, Victoria of 1926, and Valentia of 1934. These were a family of fabric-covered biplanes with limited performance, although strong and reliable. Other important biplanes of the interwar years included the Vildebeest torpedo-bomber and Vincent general-purpose three-seater, sharing similar airframes and powered by the 660-hp Bristol Pegasus (except for the Vildebeest IV, which used an 825-hp Bristol Perseus sleeve-valve engine). These sturdy aircraft served in large numbers in many RAF units outside Britain between 1933 and 1943.

In October 1938 Vickers (Aviation) Ltd. and SUPERMARINE merged to form Vickers-Armstrong. The late 1930s saw the introduction of GEODETIC CONSTRUCTION with the Wellesley, a monoplane bomber of which 176 were delivered in 1937–38. Far more famous was its immediate successor, the WELLINGTON. Over 11,000 of these bombers were produced between 1938 and 1945. From the Wellington were derived the larger Warwick, an ocean patrol and transport, the Viking airliner, which had a conventional stressed-skin fuselage, the Valetta, a military transport, and the Varsity trainer, which served with the RAF until as late as 1976.

In the field of commercial aviation Vickers scored a striking success with the appearance of the VISCOUNT, the world's first turboprop airliner, which became one of the best-selling transports in the history of the British aircraft industry. In 1955 the RAF took delivery of the first of 108 four-jet Valiants, which served as tankers and reconnaissance and multirole aircraft as well as being Britain's first strategic nuclear bombers. The Valiant was withdrawn from service in 1965 following the discovery of metal fatigue cracks. After the commercial failure of the turboprop Vanguard, intended as a successor to the Viscount, Vickers developed the VC-10, the first long-range airliner to have its jet engines mounted at the tail. The VC-10 and Super VC-10 continued to be manufactured after the absorption of Vickers into the new BRITISH AIRCRAFT CORPORATION (BAC) in the early 1960s.

Vimy

Designed for the long-range strategic bombing of Germany during World War I, but too late to see action, the Vickers Vimy served as the RAF's standard heavy bomber until gradual replacement began in 1925. The principal production model, the Vimy IV, was powered by twin 360-hp Rolls-Royce Eagle VIII engines, giving it a top speed of 100 mph (161 km/h). Wingspan was 68 ft. 1 in. (20.75 m) and length 43 ft. 6 in. (13.25 m). The bomb load amounted to nearly 2,500 lb. (1,130 kg).

In service in the Middle East the

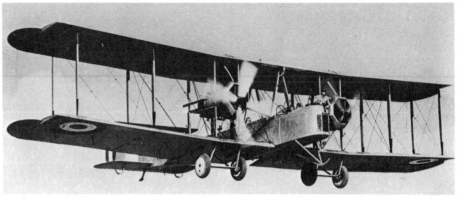

Vickers Vimy

Vimy saw additional use on airmail routes and, after its withdrawal from first-line duties, became the standard heavy trainer. The aircraft also distinguished itself in several pioneering long-distance flights. In 1919 a Vimy made the first nonstop crossing of the Atlantic, flown by ALCOCK AND BROWN, and the same year another Vimy made the first flight from England to Australia, piloted by Ross SMITH. Two attempts to reach Cape Town from London were made in 1920, also using Vimys (see PIONEER FLIGHTS).

Viscount

The world's first operational turboprop airliner, the VICKERS Viscount entered service with British European Airways on April 18, 1953. Initial series 700 production aircraft with

1,540-ehp Dart 506 engines had seating for 53 passengers. The 700D employed uprated (1,760-ehp) Dart 510s, and the series 800 (1956) had a fuselage stretched to seat 65. Final series 810 aircraft were powered by 1,990-ehp Dart 525 engines and cruised at 357 mph (575 km/h) with a range of 1,700 mi. (2,735 km). Wingspan of the series 810 was 93 ft. 8 in. (28.55 m) and length 85 ft. 8 in. (26.11 m). Among airlines to operate Viscounts were Aer Lingus, Air France, Trans-Canada, Capital, Continental, and Lufthansa. A total of 444 Viscounts were built before production ceased in 1964.

Voisin, Gabriel (1880–1919) and Charles (1882–1912)

French aviation pioneers, the Voisin Brothers built probably the earliest

Vickers Viscount

fully successful European aircraft. Gabriel Voisin joined Ernest Archdeacon in building a series of gliders derived from Archdeacon's first (April 1904) copy of the Wright Brothers' design, and the young Frenchman took over much of the test-flying. From the beginning the Voisins concentrated on the box-kite biplane design pioneered by Australian Lawrence Hargrave, and they also took an interest in float seaplanes, making the first manned flight from water while being towed behind a speed boat down the Seine at Paris on June 6, 1905. Of the two brothers, Gabriel was the designer and Charles the pilot. A Voisin biplane flew the first official circular kilometer, piloted by Henry FARMAN, friend and pupil of Gabriel, in January 1908. It was the first really successful airplane built by anyone other than the Wrights.

During the period 1907–12 the Voisin Brothers were among the world's leading aircraft builders. Their standard products were pushers with canard elevators, and the aircraft included the world's first successful amphibian, and by 1913 the thoroughly sound and robust military Type L. The latter became the basis for the LA.S, LB, LB.S, and later versions, of which a total of over 3,500 were built. Made of steel with fabric covering, these wartime Voisin pushers were slow but carried a useful bomb load and were often armed with a 37-mm or 47-mm quick-firing cannon. The only unsuccessful version was the Type XI, with a Fiat engine, which caught fire easily.

Volksjäger

The so-called German "people's fighter"—the Heinkel He 162 Salamander—was designed and built in 69 days, making its first flight on December 6, 1944. Powered by a single BMW 003 turbojet mounted above the fuselage, and intended for production by semi-skilled labor using nonstrategic materials, the Volksjäger was constructed largely of wood and Duralumin. The first prototype crashed at a public demonstration four days after its maiden flight, the result of wing failure; but production aircraft were being delivered to the Luftwaffe early in the new year. Armament consisted of two 20-mm cannon; maximum speed was 522 mph (840 km/h) and ceiling 39,370 ft. (12,000 m). Wingspan was 23 ft. 7 in. (7.20 m) and length 29 ft. 8 in. (9.05 m).

At the most, only 116 Volksjägers were produced (there is considerable skepticism on this point), and the aircraft saw little real operational use, since He 162 pilots avoided combat in

He 162 Volksjäger

their unproven aircraft. Plans to assign glider-trained boys from the Hitler Youth to Volksjäger squadrons never materialized.

Vortices

A vortex is any large mass of gas or other fluid having a rotary motion, for example, a whirlwind. A powerful vortex is generated at the tips of most wings in flight. The air under the wing is at higher than atmospheric pressure and spills outward around the tip, while the air above the wing is at reduced pressure and strengthens this rotary flow. The vortex extends behind the wing tips, and with large, highly loaded wings can be dangerous to other aircraft. When pulling out of a dive, or in a tight turn, the tip vortex can be so intense that moisture condenses at the center, leaving a visible white trail. Similar trails can be seen spiraling behind the tips of a propeller at full power. The complete system of left and right vortices from wing tips plus the vortex flow over the wing itself forms a U-shape in plan view called a horsehoe vortex.

In the slender-delta type of wing used on supersonic transport very large and powerful vortices are generated over the leading edge at high angles of attack, delaying separation of flow over most of the wing and greatly increasing the LIFT, thus almost eliminating the stall, even at extreme flight attitudes. The inboard leading edge is extended forward along the fuselage (often seen in modern fighters) to generate the vortices evenly on both sides. Vortices are also deliberately generated by turbulators or vortex generators, small vanes projecting upward in rows along wings or tail surfaces to re-energize the boundary layer (flow near the surface) and improve control or prevent stagnant boundary layer drifting outward toward the wing tips.

Voss, Werner (1897–1917)

German World War I ace, Voss joined a Hussar regiment in April 1914 as a private, and by May 1915 was a corporal; shortly after this he was accepted for pilot training and joined *Jagdstaffel* 2 on November 21, 1916. His first two aerial victories came six days later, and by the end of March 1917 his score had risen to 20, resulting in the award of a *Pour Le Mérite* on April 8. He was next given commands of *Jagdstaffeln* 5, 29, and 14 in quick succession, though by inclination he preferred to fight alone rather than lead a formation. In July 1917, RICHTHOFEN personally requested that he join *Jagdgeschwader* 1, and Voss succeeded to the command of *Jagdstaffel* 10 in that group. On September 3, flying a Fokker TRIPLANE, Voss scored his 39th victory, and in the next three weeks ran his tally up to 48, but on September 23, 1917, in what has come to be regarded as a classic of aerial combat, Voss in Fokker Triplane 103/17 became involved with SE5s from 60 Squadron of the Royal Flying Corps, and was then attacked by six SE5s from 56 Squadron, led by J. T. B. McCudden. Voss was finally shot down and killed within the British lines.

VTOL

The HELICOPTER is the most widely used VTOL (Vertical Takeoff and Landing) aircraft, but its limited speed has encouraged research into fixed-wing machines with VTOL capability.

Various methods of achieving VTOL without using rotors have been investigated. The most practical has proved to be vectored thrust—deflecting downward the exhaust from a jet engine. Other systems include using separate engines to provide lift only (Dornier Do 31, 1967), swiveling the engines themselves (Bell XV-3, 1955) or tilting the entire wing structure (Hiller X-18, 1959) to provide either vertical or horizontal thrust, and designing aircraft to take off and land on their tails (Convair XFY-1 Pogo, 1954). In addition, various experimental compound aircraft (such as the McDonnell XV-1) have been built, using helicopter-like rotors for takeoff but employing conventional wing surfaces and propellers or turbojets for level flight.

All these methods have their drawbacks. Considerably more power is required to effect vertical takeoff than is needed to maintain horizontal flight, even at high speed. Consequently, any VTOL aircraft using jet deflection must have an excessively large engine, representing substantial deadweight during normal flight. Separate lift engines have the same drawback, but special power units for this purpose can be built with a very high thrust-to-weight ratio. Fans driven off the main power plant to suck air in from above and discharge it downward have the virtue of a reduced noise level, but the installation (as in the Ryan XV-5A of 1963) is bulky. Rotating the engines, either by themselves or in concert with the entire wings, involves substantial mechanical problems (see CONVERTIPLANE), while airplanes that take-off and land on their tails have proved unacceptable because the pilot has difficulty in seeing the ground as he comes down to land.

The use of several small power units to provide lift was studied on the FLYING BEDSTEAD of 1953, and the later Short SC1 (1961) and Dassault Balzac (1963) both employed the same principle. These special lift-producing engines may be installed in a battery within the airframe with doors to seal off the intakes and nozzles during normal flight, or they can be housed in special streamlined containers located beneath the wings or at the sides of the fuselage.

The vectored-thrust HARRIER became the world's first operational VTOL fighter at the end of the 1960s, and in 1969 set a city center to city center New York–London record of 5 hours 57 seconds.

Vulcan

One of the world's first delta-wing bombers, and used exclusively by the RAF, the Hawker Siddeley (originally Avro) Vulcan first flew in 1952, and B1 aircraft entered operational service five years later. Initially fitted with four Olympus engines of 11,000 lb. (4,980 kg) thrust, the Vulcan had progressively more powerful turbojets installed, and the aerodynamically revised B2 of 1957 eventually mounted Olympus 301s of 20,000 lb. (9,000 kg) thrust.

Manned by a crew of five the B2 was

Ryan X-13 Vertijet experimental VTOL aircraft

Vulcan B2 strategic "V bomber" of the RAF, armed with Blue Steel H-bomb missiles and free-fall atomic bombs

99 ft. 11 in. (30.46 m) in length and had a wingspan of 111 ft. (33.83 m). Maximum speed was 645 mph (1,038 km/h) at 40,000 ft. (12,200 m), service ceiling was 65,000 ft. (19,800 m) and range 4,600 mi. (7,400 km).

Vultee

A U.S. aircraft manufacturer established by Gerald Vultee in 1932. Vultee had made his reputation with Douglas and Lockheed designing advanced stressed-skin monoplanes (he designed Lockheed's VEGA and Sirius). Vultee's very fast V-1 single-engined eight-passenger airliner of 1933 was bought by American Airlines, and in 1935 DOOLITTLE set a transcontinental record of 11 hours 59 minutes in a V-1A, while another V-1A flew from Los Angeles across the Arctic to Moscow. Vultee went into production at Glendale (Los Angeles) Airport with a V-1 military development, the V-11, which sold widely in many countries. In 1936 Vultee accomplished the feat of designing and building the BT-13 Valiant trainer in 88 days. In World War II over 11,000 of these monoplane basic trainers were constructed, but Vultee's design brilliance was lost just before the war when he was killed at the age of 38 in a crash.

In 1941, when the Vanguard fighter was in production, Vultee's parent company, Avco (Aviation Corporation) bought control of Consolidated to create Consolidated Vultee (CONVAIR), which was eventually acquired by GENERAL DYNAMICS. Other wartime Vultee aircraft included the Vigilant observation machine, the relatively mediocre Vengeance dive-bomber, the Vultee-Stinson L-5 Sentinel observation/liaison aircraft, and the XP-54 pusher fighter.

Wal

The DORNIER Do J Wal (Whale) was an eight-passenger, all-metal monoplane flying boat with its two engines (a pusher and a tractor) mounted in tandem above the wing. In the early 1920s Germany still was not permitted to build such relatively large aircraft, and so 150 of these Dornier-designed planes were constructed in Italy from 1923 onward. Wingspan of the Wal was 73 ft. 10 in. (22.50 m) and length 56 ft. 7 in. (17.24 m). Powered by Napier Lion, Rolls-Royce Eagle, or Hispano-Suiza engines, Wals were sup-

Warhawk: P-40F (Kittyhawk II) of the RAF

plied to the Brazilian airlines Condor and Varig, Scadta in Colombia, and Aero Lloyd in Germany. Italian airlines operated 26 Wals, and two were used on Roald Amundsen's abortive flight to the North Pole in 1925. Wals were also constructed in Japan, together with 40 in Spain (1928) and 40 in the Netherlands (1927–31).

Larger, German-built Wals with two 600-hp BMW engines made 328 Atlantic mail crossings for Deutsche Lufthansa in the 1930s, while Dornier also produced the Do R4 Super Wal four-engined flying boat, built as well in Italy and Spain. Another tandem twin-engined aircraft, the Do 18 (1935), was originally designed as a transatlantic mailplane based on the Wal design, but the entire production series of 100 served principally as Luftwaffe reconnaissance flying boats in World War II.

Wallis, Sir Barnes Neville (1887–   )

A British aeronautical engineer whose long career produced innovation in airplane and airship design. Trained as a marine engineer, Wallis joined the firm of Vickers as an airship designer. In 1929 he designed the *R-100*, an exceptionally stable airship incorporating many new features, which successfully crossed the Atlantic. Later, he headed the firm's research and development team, and during the 1930s he devised a new system of airframe design, GEODETIC CONSTRUCTION, in which compression loads in any member of a lattice structure are braced by tension loads in cross members. The famous WELLINGTON bomber of World War II was one of several aircraft constructed on these principles.

Wallis also did valuable pioneering research on SWING-WING design, draw-

ing up plans for aircraft that could change the sweep of their wings in flight. This early postwar work was little regarded in Britain but was followed closely by designers in the United States and elsewhere. Wallis is also remembered for designing the "bouncing bomb" used by the DAMBUSTERS to breach the Ruhr dams in Germany during World War II, and devised the large "Earthquake" bombs used in the last stages of the war. In his later years he continued to work on revolutionary ideas for hypersonic aircraft.

Warhawk

The Curtiss P-40 was one of the principal American-made fighters of the early years of World War II. Basically a Curtiss P-36 (HAWK) airframe with a liquid-cooled Allison engine in place of a Wright radial, its poor high-altitude performance initially restricted it largely to a tactical role. The RAF named the P-40B and C models Tomahawks and used them mainly in North Africa. An improved Allison powered the P-40E, and this and subsequent models were known as Kittyhawks in British and Dominion service; late models used by the United States were called Warhawks. The P-40F and L had Packard-built Merlin engines; other late models were powered by versions of the Allison V-1710. Wingspan was 37 ft. 4 in. (11.38 m) and length 33 ft. 4 in. (10.16 m). Top speed ranged from 340 mph (547 km/h) for the P-40C to 375 mph (603 km/h) for the P-40N (the experimental P-40Q achieved 422 mph, 678 km/h). Early P-40s had only two .50 synchronized nose guns; those sent to Britain also had two (later four) .30 guns in the wings. In the P-40D the nose guns were replaced by four wing-mounted .50s, which thereafter became standard armament until the advent of the P-40N, which had six wing guns.

The P-40 saw wide service in the Pacific as well as the Mediterranean. Although no match for the Japanese ZERO in dogfights, Fying Tiger pilots in China developed hit-and-run tactics with their P-40s that proved very successful in engagements. With their good handling qualities and sound construction, the P-40s were popular aircraft, and altogether more than 15,000 were built in 21 different models. P-40Cs were the first U.S. wartime aircraft to be used by the Soviet Union, and others were flown by the Chinese. Australian and New Zealand squadrons also made extensive use of the P-40.

Weapon systems

The concept of a weapon system originated in the United States in 1951, when it was recognized that the modern combat aircraft could no longer simply be regarded as a vehicle to carry equipment. With the emergence of automatic radar/autopilot fire control for all-weather interceptors, it was realized that providing an airframe and propulsion was only part of the problem of creating an effective warplane. Accordingly, the term weapon system was coined to describe the complete operative system, with the aircraft or missile (or other carrier or transport vehicle) being merely a part of it. Thus, MX-1164 was an extensive weapon system assigned entirely to Convair, which subcontracted nearly all parts to other companies except for the basic "air vehicle," which became the B-58 supersonic bomber. Another system, WS-107A-I, weighed more than 3,200 times as much as the unrefueled rocket vehicle which was the reason for all the rest (the vehicle portion became SM-65 Atlas, the first ICBM).

Later British usage blurred the concept of a weapon system by using the term to refer to the armament and mission electronics carried by a combat aircraft. This reversed the original concept by making the weapon system a part of the air vehicle instead of vice versa. In its original meaning a weapon system would include (for a fighter aircraft) the aircraft itself, its armament and electronics, ground handling equipment, servicing and maintenance equipment and tools, instructional equipment including simulators, all handbooks and manuals, and even such a minor point as the specification of paint applied to the wagons used to load the missiles. By thus defining a single system, the customer ensured that every part is compatible with every other part.

Wellington

The twin-engined VICKERS Wellington, designed by Sir Barnes WALLIS, was one of the most successful medium bombers of World War II. The prototype made its first flight on June 15, 1936, and the first production Wellington I flew in December 1937. By the outbreak of war 10 RAF squadrons were equipped with this sturdy midwing aircraft, whose GEODETIC CONSTRUCTION gave it an immense ability to withstand damage. Wellingtons were the backbone of the RAF's Bomber Command in the first three years of the war, and in the Mediterranean and Far East they continued to serve on bombing missions until 1945. British-based Wellingtons dropped a total of 42,440 tons of bombs on enemy targets; the final sortie by a Bomber Command aircraft took place on the night of October 8/9, 1943.

A total of 11,416 Wellingtons of all types left the factories. The principal bomber versions were the Mark IA and IC (1,000-hp Pegasus engines), Mark II (1,145-hp Merlins), Marks III and X (1,500-hp and 1,585-hp Hercules) and Mark IV (1,100-hp Twin Wasps). All these had .303 machine guns in nose and tail turrets; bomb loads were 4,500–6,000 lb. (2,040–2,700 kg). Maximum speed of the Wellington was 255 mph (410 km/h); wingspan (of the Mark X) was 86 ft. 2 in. (26.26 m) and length 64 ft. 7 in. (19.68 m).

Variants of the basic Wellington design included general reconnaissance and antisubmarine aircraft, high-altitude bombers, radar flying classrooms, and transports.

Westland

The British firm of Westland Aircraft was founded in April 1915, and began operations producing Short 225 and Canton-Unné seaplanes, Sopwith 1½-strutters, DH 4 and DH 9A bombers, and Vickers VIMYs. After the Armistice in 1918, many prototypes and short production runs sustained the firm until 1927, when the Wapiti was produced. This was a rugged, metal-constructed successor of the DH 9A in its role as the main RAF general-purpose aircraft in the Middle East and elsewhere overseas. A Jupiter-powered two-seat biplane, which was designed to use as many DH 9A components as possible, the Wapiti became one of Westland's staple products, with 512 built for the RAF and about the same number for export. More experimental aircraft followed, including the Westland-Hill Pterodactyl tailless machines and some CIERVA-derived autogiros. The high-wing Lysander, first flown in 1936, was a STOL observation and utility monoplane that was used to ferry secret agents to and from occupied Europe during World War II, but major wartime production centered on the Seafire (see SPITFIRE). The Whirlwind (1938) was the RAF's first twin-engined cannon-armed fighter, but only 112 were built.

In 1953–55 the Wyvern turboprop torpedo plane was delivered to the Royal Navy, but all subsequent Westland products have been helicopters. A SIKORSKY license led to production of the Dragonfly (based on the Sikorsky S-51), Whirlwind (S-55), Wessex (S-58), and Sea King (S-61), and partly through absorption of other companies Westland added the Sycamore, Scout, Wasp, and Belvedere to its list of helicopters.

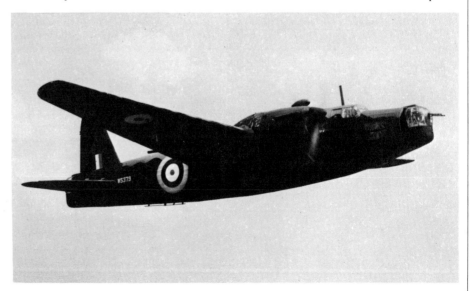

The Wellington II medium bomber, an almost indestructible aircraft due to its geodetic construction

Collaboration with the French manufacturer Aérospatiale brought Westland work on the Gazelle and Puma, and production of Westland's own Lynx high-performance multirole helicopter.

Whittle, Sir Frank (1907–   )

A British aeronautical engineer who was one of the pioneer developers of the JET ENGINE, Whittle served in the Royal Air Force from 1923 to 1948, initially becoming a test pilot and instructor. He patented his first turbojet design in 1930 and later joined the British government-backed Power Jets Ltd.

On April 12, 1937, he ran what is generally considered the world's first jet engine, the W.1U, which was eventually used to power Britain's first jet aircraft, the Gloster E.28/39 (May 15, 1941). From this airplane was eventually developed the twin-engined Gloster METEOR, the first and only operational Allied jet fighter of World War II. Engines based on Whittle's design and built by General Electric powered initial P-59 AIRACOMETS, the first American jet aircraft.

Wildcat

The GRUMMAN F-4F Wildcat was the U.S. Navy's standard carrier-borne fighter for the first two years of World War II. The original prototype flew in 1937 as the XF4F-2, Grumman's first monoplane design. Its initial performance with a 1,050-hp Wasp engine was disappointing, but modified as the 1,200-hp XF4F-3 it attained 335 mph (539 km/h) and entered service as the Wildcat at the end of 1940, armed with four (later six) .50 wing-mounted machine guns. Wingspan of the Wildcat was 38 ft. (11.58 m) and length 28 ft. 10 in. (8.79 m).

Production at Grumman was terminated in May 1943 in favor of the F6F HELLCAT, but continued at General Motors until September 1945. Altogether some 8,000 Wildcats were produced; the aircraft was also supplied to Britain's Royal Navy, with which it served as the Martlet.

Although the Wildcat had a performance inferior to that of most enemy fighters it met in combat, its firepower and the use of diving passes enabled the F4F to establish a creditable record, even against the Japanese ZERO.

Wind

Sustained horizontal motion of the atmosphere is caused basically by the tendency of air to flow from regions of

F4F-3 Wildcat

high pressure to regions of low pressure. Over a stationary Earth the wind would move perpendicular to the isobars (lines joining all places at the same pressure), but the Earth's rotation causes the wind to flow more nearly along the isobars, so that in the northern hemisphere the wind blows counter-clockwise around an area of low pressure and clockwise around a high (and the reverse in the southern hemisphere). Many local winds are caused, or modified, by the Earth's surface; thus, a catabatic wind flows down a cold mountainside at night, while a sea breeze crosses coasts warmed by sunshine. Wind varies with altitude, and wind shear (rapid change in wind speed with height) near the ground can be dangerous to aircraft preparing to land. The JETSTREAM is a well-defined powerful wind in the STRATOSPHERE.

Relative wind is the direction from which the airflow appears to come toward an aircraft, and though this is usually almost aligned with the longitudinal axis it can in some conditions (see SPINNING) be completely different.

In the early days of flying, aircraft performance was low, with speeds of perhaps 100 mph (160 km/h) and maximum altitudes of 5,000 ft. (1,500 m). Under these conditions they were largely at the mercy of the wind, which could either double their speed over the ground or prevent them from reaching their destination at all. Modern aircraft are so fast that wind exerts a much smaller influence. Wind patterns over the globe are mapped for every day of the year, and civil aviation can be routed so as to take advantage of tailwinds. Winds are not normally a factor determining takeoff or landing, since aircraft are directed to use the runway which most nearly points into the wind. But some airfields have only one runway, so if the effective cross-wind is strong aircraft are unable to land or takeoff because the pilot is unable to resist the natural tendency of his aircraft to weathercock, preventing him

from maintaining a straight course on the runway.

Strong cross-winds can cause an aircraft to drift, i.e., move over the ground in a direction different from that in which it points. To maintain its course under these conditions the aircraft must point slightly into the wind.

Wind tunnel

Wind tunnels are used to simulate air flow for aerodynamic measurement. They consist essentially of a closed tube through which air is circulated by powerful fans. Depending on the air speed expected to act upon the surfaces being tested, wind tunnels operate at low, high, supersonic, and hypersonic capacity. They operate continuously, intermittently, or, in impulse tunnels, for a split second only. In a closed system not only the velocity of the air, but also its pressure, temperature, and humidity can be controlled. These characteristics are especially important in simulating flight at very high speed.

The size of airplanes means that only models are normally tested in wind tunnels. In the continuous tunnel used for relatively low-speed flight a fan driven by an electric motor forces the air around a gradually widening tunnel fitted with guiding vanes until it passes through a honeycomb and out of a narrowing nozzle into the test chamber. The air is then sucked through the fan again and recycled. Balances to which the model is attached by fine wires or thin rods record the forces generated by the airflow.

The intermittent type of tunnel uses high-pressure air flowing out of a nozzle or induced airflow produced by a vacuum.

To achieve simulated speeds of above 120 mph (192 km/h), tunnels using compressed air were developed. When it became necessary to test supersonic designs, gas turbines were employed to produce a high pressure at the base of a convergent-divergent nozzle in which the test model is installed. Mach numbers of over 20 have been achieved in impulse tunnels by using shock waves lasting for only a thousandth of a second produced by a giant piston driven by pressure or explosion.

Because a model is only a fraction the size of the actual aircraft, wind tunnel tests must take into consideration the definitive criteria known as Reynolds numbers, which measure the characteristics of fluid flow at the interface with solid surfaces. These mathematically-calculated values are always of substantial numerical magnitude because of the units employed. A low Reynolds number (100,000) will be yielded by a low-speed test in a small wind tunnel; a high Reynolds number indicates that the scale of the test or the speeds simulated in the wind tunnel approach full scale (a large model being tested at very high speed might yield a Reynolds number of 20,000,000).

Apart from overall size differences between a model and the actual aircraft, allowances must also be made for details (protruding rivet heads, for example) that cannot be reproduced on a model. Furthermore, care must be taken to ensure that the tunnel's walls do not themselves influence the airflow in the vicinity of the model to give a false result.

Almost all aerodynamic characteristics are tested in wind tunnels, including the size, shape and location of airframe components and the effectiveness of all flight control elements. They are also able to measure the distribution of loads over the total surface of the aircraft.

Wings

Airplanes derive LIFT from their wings. The front of the wing is known as the leading edge; the back is the trailing edge; the breadth between the leading and trailing edges is known as the chord. Span is measured from wing tip to wing tip.

Wings with a low aspect ratio have a wide chord relative to their span. A high aspect ratio wing, on the other hand, has a wide span and relatively narrow chord. This arrangement is more efficient aerodynamically because the air flowing around the tips from the higher pressure below the wing to the lower pressure above it (causing induced drag) has relatively less influence on the overall airflow if the wing is of very wide span. Wide-span wings, however, pose structural problems, add weight, and restrict maneuverability.

A wing's angle of attack is the angle between the direction of the airflow and the chord. If the air strikes the lower surface of the wing, the pressure below the wing is augmented, and there is a greater backward downwash of air, thereby enhancing lift. Increasing the angle of attack up to about 15 or 20° progressively improves the lifting capability of the wing, but above this value the airflow breaks down (see STALLING).

With the cantilever type of wing con-

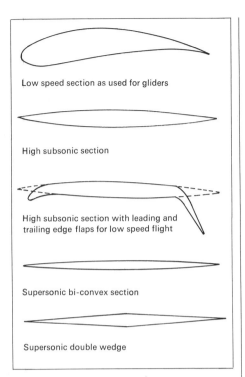

Typical wing sections: designs vary according to the performance required

struction the wing becomes gradually thinner from wing root to tip; this tapering is needed to preserve the proportions of the airfoil section. Braced wings, however, usually have a uniform thickness.

Wing sections vary according to the requirements of the aircraft. A thick section with a steep camber (or curvature) gives considerable lift at low speeds and thus reduces the speed at which the aircraft will stall. It generates too much drag for high-speed flight, however. This requires a thinner section, but as a result, fast aircraft tend to have high stalling speeds. For supersonic flight, symmetrical sections with similar upper and lower profiles are used, and various double-wedge-shaped airfoils are particularly efficient once the speed of sound has been exceeded.

By the 1970s it was possible to design an airfoil that would substantially delay the sudden increase in drag due to compression waves, which occurs above the critical Mach number (see SUPERSONIC FLIGHT). Wings constructed in this way are known as supercritical wings. Their characteristics are similar to those of airfoils used for most jets, but they have a better performance at low-lift coefficients (low-level high-speed flight).

The use of an oblique wing (with or without a pivot) has been investigated by Boeing and NASA. For transonic commercial use such a design would seem to offer savings in weight and fuel

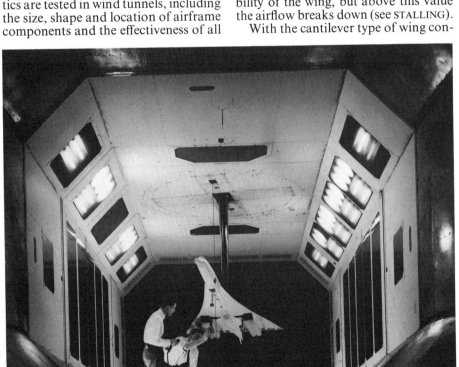

Preparing a scale-model of the Concorde for wind tunnel tests

consumption, together with a reduction in noise levels. Airelastic instability would require the use of additional stiffening if such a wing were made of aluminum, but construction in graphite-epoxy is also being considered. (See also AERODYNAMICS.)

Wolfpack

The name given to the U.S. Army Air Force's 56th Fighter Group in World War II and the U.S. Air Force's 8th Tactical Fighter Wing during the Vietnam War.

The 56th Fighter Group was the first USAAF organization to be equipped with the P-47 THUNDERBOLT. Early in 1943 the Group was established in England and in April began operations with this aircraft. At first Luftwaffe fighters were usually able to get the better of the P-47s in combat, but largely due to the tactics developed by the 56th's commander, Hubert ZEMKE, the group soon had a string of successful engagements to its credit. Basically these tactics exploited the P-47's "zoom" power and its armament—features in which it was superior to enemy fighters. Throughout the autumn and winter of 1943 the 56th Group led other P-47 units in combat victories and its tenacity earned it the nickname "Zemke's Wolfpack." At the end of hostilities this group was the highest scoring fighter organization in the EIGHTH AIR FORCE in air combat, with 675 victories.

The more contemporary 8th Tactical Fighter Wing, which operated from Ubon in Thailand, was given this nickname after its repeated successes in combat with MiG fighters encountered over North Vietnam during the period 1966–1968. The second wing to operate the F-4 PHANTOM in southeast Asia and the first to operate out of Thailand, the 8th Wing was led by Colonel Robin Olds for much of this period. Olds, a distinguished World War II fighter ace, destroyed four MiGs while serving with the 8th Wing. The Wing's total MiG "kills" during the Vietnam war was 39.

World Cruiser

A biplane design with a 450-hp Liberty engine, the Douglas World Cruiser was produced to a U.S. Army Air Service specification for a 1924 round-the-world flight. After prototype tests in 1923, four World Cruisers were built, each carrying two crew and 465 gallons (1,760 liters) of gasoline. Wingspan was 50 ft. (15.24 m) and length 35 ft. 2 in. (10.72 m). The aircraft *Seattle* (Maj. F. L. Martin), *Chicago* (Lt. L. Smith), *Boston* (Lt. L. Wade), and *New*

Douglas World Cruiser Chicago *photographed during its historic round-the-world flight, April–September 1924*

Orleans (Lt. E. Nelson), all fitted with floats, left Seattle on April 6, 1924, and flew to Alaska, where *Seattle* crashed. The flight continued via the Aleutians to Japan, China, Hong Kong, Bangkok, and Calcutta, where wheels were fitted for the overland section to Brough, England. The three aircraft took-off from England on July 30, 1924, for Iceland, but *Boston* sank off the Faeroe Islands on August 2. The remaining two aircraft were joined in Nova Scotia by the prototype, now named *Boston II*, and wheels were refitted on arrival at Boston. They returned to Seattle on September 13, 1924, having covered 26,345 mi. (42,389 km) in 363 hours' flying time. *Chicago* is now at the National Air and Space Museum in Washington, and *New Orleans* at the Air Force Museum, Dayton, Ohio.

World War I

The major air operations of World War I took place over the trenches, emphasizing the contrast between the static battles of vast armies locked together on the ground and the changeable, rapidly moving warfare of frail machines maneuvering in an altogether larger dimension. Clearly, for all the fury and heroism of the dogfights and the daring of the bombing raids, the air war made no decisive difference to the outcome of the larger conflict on the ground. But many who watched the development of aircraft during the course of the war came to see their strategic potential. By 1918 they had fought on all fronts, sometimes in very sizeable armadas, and powerful

weapons, in the form of bombs, torpedoes, and guns, had been developed for their use. The RAF had already become a separate branch of the armed forces, and other air forces were soon to gain the same recognition as independent fighting units.

At the outset of the war the airplane was seen merely as an extension of the cavalry's scouting role, and, indeed, the value of air reconnaissance soon became apparent to field commanders. In September 1914 detailed aerial reconnaissance of German positions and movements helped the Allies win the critical First Battle of the Marne. Shortly after this, aerial photography was introduced by Britain's Royal Flying Corps, and with increased use of radio communications scout airplanes were able to circle the field of battle and report directly to ground artillery crews (correcting their aim and pinpointing targets).

It was originally as observation platforms that lighter-than-air craft came into prominence. AIRSHIPS such as Zeppelins provided the German fleet with reconnaissance in advance of surface craft, and though they later posed a bombing threat to distant targets, their vulnerability precluded their use over the battlefront and limited their activities. The Allies concentrated on smaller nonrigid airships (blimps), mainly for antisubmarine patrolling. All the major belligerents used tethered observation balloons in the field, and unmanned tethered balloons were employed as aerial barrages.

With the recognition of the value of reconnaissance, the development of

FIGHTER aircraft to shoot down enemy scouts became a matter of urgency. Once they began to be armed, it became clear that aircraft could be fitted for bombing, though the machines available for all these tasks were at first mostly two-seat biplanes with engines of 80–150 hp giving speeds of some 80–130 mph (129–209 km/h) and three hours' endurance. Single-seaters were also used, and the first notable strategic bombing raid was in fact made by two small single-seater 80-hp Sopwith Tabloids, carrying 20-lb. (9-kg) bombs. Toward the end of the war came the development of large BOMBERS such as the Italian CAPRONIS, the British HANDLEY PAGE 0/400s and the German GOTHAS and R-PLANES.

By 1916 roles had become diversified into military observation, bombing, and fighting, and specialized units were formed for these separate functions, each operating aircraft of only one type to simplify maintenance and facilitate formation flying (which became necessary to concentrate firepower).

Scouting and bombing incursions over enemy fronts led to fast single-seat fighter aircraft, and in turn to even faster fighters to combat them. Replacing the rifles carried aloft by observers in 1914, when most aircraft were unarmed, the machine gun of rifle caliber became standard.

A PUSHER configuration gave early fighters such as the Vickers Experimental Fighting Biplane No. 1 (and its successor the F.B.5 "Gunbus") a good field of frontal fire. But since speed and maneuverability were vital to a fighter, and these qualities were most easily found in aircraft with front-mounted propellers, a way had to be found of firing past, or through, the propeller arc.

The first fighter fitted with an effective INTERRUPTER GEAR, the German Fokker EINDECKER, entered service in the second half of 1915 and immediately gave German pilots almost total air superiority. The "Fokker Scourge" lasted until May 1916, and for the rest of the war the standard fighters on both sides were single-seater tractor aircraft armed with twin synchronized machine guns firing between the spinning propeller blades. Interspersed smoke rounds enabled pilots to "trace" their fire.

The evolution of COMBAT MANEUVERS led to attacks from height out of the sun, gaining speed by diving to avoid early detection and then coming upon the enemy from below and behind—an area blind to pilot and observer. To meet this threat, observers also became gunners, with traversing gun mounts and in some cases yoked guns to increase firepower. At times, particularly in the

A hastily improvised RAF airfield in France, 1918, during the climax of air combat in World War I

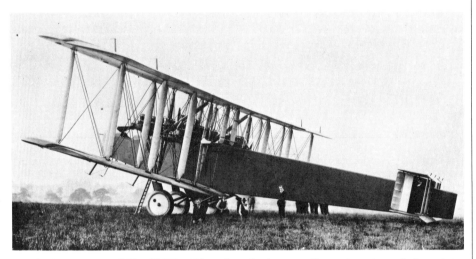

By the last stages of World War I bombers had grown from tiny aircraft dropping 20 lb. (9 kg) bombs to giants like this Handley Page V/1500, which was being prepared to bomb Berlin in November 1918

final battles of the war, fighters were adapted for dropping light bombs and for strafing troops. Multiengined bombers increased bombing range and posed a threat to industrial centers, forcing a diversion of war effort to defensive measures.

The British concentrated their bombing raids on German airfields and U-boat bases, but the main targets of German bombers were London and other cities in southeast England. In one raid, on June 13, 1917, 14 bombers dropped 118 bombs on London in daylight, and although 90 British fighters were sent against them all 14 escaped—clearly illustrating the British need for better organization, and better aircraft.

The message was not lost, and early in 1918 the world's first independent air arm, the Royal Air Force, was formed.

Floatplanes and flying boats, which could only operate off relatively calm seas, were being replaced by landplanes for maritime patrol duties at the end of hostilities. From the outset there were ships carrying seaplanes, but not until almost the close of 1918 did the world's first "flush-deck" aircraft carrier appear. By the end of the war, however, some 100 warships had facilities for launching spotting aircraft. The air-launched torpedo claimed its first ship in 1915, and carrier-based torpedo-bombers were being built in quantity during 1918.

During the course of World War I,

the development of military aviation had made air power a force to be reckoned with. For the first time in history warfare was no longer restricted to the movements of armies and navies. The scale of aerial activity between 1914 and 1918 was truly prodigious. Germany and Austria-Hungary built over 58,000 aircraft during the war and the Allies almost three times that number.

World War II

World War I had been preeminently a war of cannon, bayonet and machine guns, of fixed positions and infantry maneuvering—a static war in which aircraft fought their duels high over the muddy front and captured the imagination without materially affecting the outcome. By the time of World War II DOUHET, MITCHELL, the German General Walter Wever, the Soviet General V. B. Khripin, and others, had done their best to convince their respective General Staffs of the value of air power. Control of the skies for tactical support and strategic bombing was a principal factor on almost all fronts, though it did not completely destroy the enemies' capacity to wage war—as the most enthusiastic theorists had promised.

When the Germans struck at Poland on September 1, 1939, they used Blitzkrieg tactics to eliminate the Polish air force and facilitate the advance of their troops. The Luftwaffe had a numerical superiority of about 9 to 1 and the outmoded PZL P.11 fighters that equipped the Polish fighter squadrons were no match for the BF 109.

Eight months later similar tactics carried the Wehrmacht to the Channel coast. The French air force was in the process of re-equipping with modern types (DEWOITINE 520, BLOCH 174, Amiot 351, LeO 451) but nonetheless acquitted itself well against the Luftwaffe; the British HURRICANES proved just able to hold their own with the Bf 109, but SPITFIRES were committed to the fighting only at a late stage over Dunkirk.

After the Battle of BRITAIN, it was apparent that the ZERSTÖRER was a failure against single-engined fighters, and the Bf 109 and the Spitfire were evenly matched (the Spitfire was more maneuverable but not so fast, and had a carburetor that ceased to function under negative G).

Unescorted daylight bombing raids were apparently impractical, as heavy RAF losses over the North Sea in December 1939 (12 WELLINGTONS out of 22 shot down on December 18) and German experience in the Battle of Britain amply demonstrated. In the winter of 1940–41, night bombing offen-

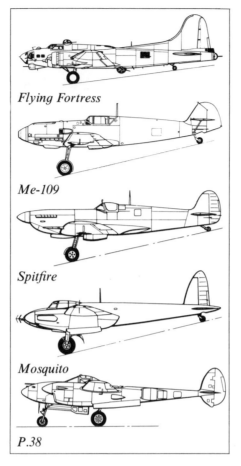

Flying Fortress

Me-109

Spitfire

Mosquito

P.38

sives were therefore embarked on by both the RAF and the Luftwaffe (the BLITZ), and in the spring of 1941 British fighters began to undertake offensive sweeps over occupied Europe in increasing strength. The introduction of the FOCKE WULF 190 in German squadrons temporarily outclassed the RAF's Spitfire Vs, but new Spitfire IXs partially redressed the balance in mid-1942.

Many Luftwaffe units were transferred to the newly-opened eastern front and initially outclassed their Soviet opponents, but the Russian air force quickly acquired combat experience and also began to receive American and British aircraft under Lend/Lease to supplement its new LAVOCHKIN LaGG 3 and MIG 3 fighters.

A number of British second-line aircraft (Blenheim Is, Gladiators, Tomahawks) were initially pitted against the Italians in the Middle East, who flew principally C.200 fighters and SPARVIERO bombers, but German help made it necessary to send first-line machines out from England (including American Maryland and Baltimore bombers).

On December 7, 1941, when the Japanese struck at PEARL HARBOR, the Americans found that their AIRACOBRAS and Brewster F2A Buffalos were outclassed by the Japanese ZERO (only the

Curtiss P-40 WARHAWK was competitive) and not until the battles of the CORAL SEA and MIDWAY was the Japanese advance checked. Carrier-borne fighters and strike planes were of greater importance than land-based planes in these actions, which established the preeminence of the aircraft carrier.

The RAF had begun introducing four-engined heavy bombers on night raids early in 1941 (STIRLINGS) and by 1942 LANCASTERS were available. American 8th Air Force bombers commenced daylight operations from Britain in the fall of 1942 and it was apparent that the absence of a strategic heavy bomber from the Luftwaffe's inventory was a significant omission. By the end of 1942 the Axis powers were withdrawing from Africa, and on the Russian front the Wehrmacht suffered a major reverse at Stalingrad, while increasingly heavy Anglo-American bombing raids required the retention of large German fighter forces for home defense; there was, as a result, a shortage of single-engined fighters—exacerbated by Hitler's demands to maintain bomber production for an offensive air campaign. B-24 LIBERATORS from north African bases raided the Ploesti oil fields for the first time on August 1, 1943.

An island-hopping campaign had begun in the Pacific to wrest back from the Japanese the territory they had seized in 1941 and 1942. American B-17 FLYING FORTRESSES and B-24s were being supplemented by medium bombers, but fighters capable of adequately combating the Zeros were still lacking and there was a continual demand from USAAF commanders for more long-range LIGHTNINGS. Air activity against a Japanese garrison in the Aleutians was limited by perpetual bad weather.

In 1943 the Russians were steadily assuming the initiative on the German eastern front. The Luftwaffe was now preoccupied with defense, and by the end of the year fighter production had been given priority to compensate for the heavy wastage in aircraft being suffered during combat with Anglo-American raids—particularly the U.S. daylight attacks, which were now escorted by THUNDERBOLTS and Lightnings for part of the journey. Deep penetration raids that went beyond escort range suffered heavy casualties, however (96 Fortresses lost in August 1943 during two raids on Schweinfurt).

The appearance of HELLCATS and CORSAIRS in the Pacific at last checked the Zero as the Japanese successively withdrew from the Gilberts, the Marshalls and the Carolines between Sep-

Classic fighters of World War II: Spitfires, seen here preparing to attack

tember 1943 and June 1944.

The Russian front continued to absorb vast numbers of German tactical bombers and transport aircraft as the Red Army pressed steadily westward, with the SHTURMOVIK acquiring a formidable reputation for ground attack work and new Russian fighters (La 7, YAKOVLEV Yak 3) helping to keep the Luftwaffe's Jagdstaffeln at bay.

The Italians capitulated in September 1943 as the Allied armies first occupied Sicily and then drove up Italy.

Even at this stage of the war, if they had been properly employed and available in quantity, the new German jet and rocket designs had the potential to wrest air superiority from the Allies. But Hitler's insistence on the development of the ME 262 jet fighter as a bomber delayed its deployment as a fighter and both the ARADO 234 jet bomber and Me 163 KOMET rocket fighter were too late and too few to have a material effect. After the Allied invasion of France in 1944 the Anglo-American air forces exercised command of the air through overwhelming numerical superiority, RAF Typhoons and USAAF Thunderbolts being assigned primarily to the ground attack role. The Luftwaffe still relied on the Bf 109 and Fw 190, but American MUSTANGS could now escort daylight raids to Berlin and back.

The British area-bombing raids had burned out Hamburg, Dresden and other German cities, and U.S. strategic bombing had devastated many aircraft production centers, ball bearing plants, and oil refineries. The Luftwaffe was therefore short of pilots, aircraft and fuel, and by April 1945 was largely impotent, although a few units (including jet fighters) fought bravely until the very end of the European war (May 8).

The introduction of SUPERFORTRESSES in the Pacific enabled regular bombing raids to be launched against Japan, using the Marianas as a base from the end of 1944. During the capture of these islands in June 1944, 402 Japanese planes were destroyed by U.S. Navy fighters and antiaircraft fire (the Marianas "Turkey Shoot").

The Japanese introduced several new fighters to supplement the now obsolescent Zero (the NAKAJIMA Ki.84 "Frank" and Ki.44 "Tojo," the MITSUBISHI J2M "Jack," the KAWASAKI Ki.100 and N1K "George") but the B-29s continued to devastate Japanese cities, going in at low altitude under cover of darkness with their guns stripped out to save weight. On the night of March 9/10, 1945, an incendiary raid on Tokyo devastated 15.8 square miles (40.9 km^2) of the Japanese capital and killed 83,783 people.

The dropping of atom bombs on Hiroshima (August 6, 1945) and Nagasaki (August 9, 1945) by Superfortresses precipitated the Japanese surrender on August 14, 1945. The war ended for America as it had begun—by an air strike.

Wright Brothers
The brothers Wilbur (1867–1912) and Orville (1871–1948) Wright made the world's first powered flight near Kitty Hawk, North Carolina, on December 17, 1903. The Wright Brothers had been interested in flying since boyhood. Wilbur extended his reading in 1899 by approaching the Smithsonian Institution for a list of aeronautical books and articles, after being impressed with the achievements and tragic death of Otto LILIENTHAL, builder

of man-carrying gliders. That same year they built their first aircraft, a biplane kite of 5-ft. (1.5 m) wingspan with a fixed stabilizer and wings that could be staggered. Originally bicycle salesmen in Dayton, Ohio, they eventually expanded to bicycle manufacturing, which provided a workshop. Correspondence and meetings with Octave CHANUTE of Chicago acted as a spur.

A glider built by the brothers at Dayton in 1900 was taken to the lonely sand dunes near Kitty Hawk where, as Weather Bureau records showed, there were strong and constant winds. This *No. 1* glider, based on the earlier kite, had fixed wings of 17-ft. (5.2 m) span with a warping control. The glider had a prone pilot position, but it was flown mainly as a kite. Unlike other pioneers who strove to find a perfectly stable craft, the Wrights accepted instability and concentrated on control. The Wright *No. 2* glider, also a biplane, but of a 22-ft (6.7-m) span, had drooping wings with warping wires operated by hip movement. First flown July 27, 1901, it made glides as long as 389 ft. (118 m). The following year the Wrights made nearly a thousand glides with their third glider, the result of wind-tunnel tests and featuring coordinated wing-warp and rudder control, and the important addition of a tail. Wright-type gliders built in Europe after lectures in Paris by Chanute in 1903 led to a renewal of interest there in the possibilities of powered flight.

In 1903, after their intense work with gliders, the Wrights confidently built their powered FLYER, the first successful airplane. A 12-hp engine drove two pusher propellers by crossed chains so that they counter-rotated. On December 17 the Wrights made their epic four flights, the last of which was of 59 seconds' duration. Orville had made the first flight and the brothers alternated. More flights might have been made that day but for damage to an elevator on the fourth flight, which covered 852 ft. (260 m) against a stiff breeze (equal to three times that much in the air).

The following spring the Wrights flew their *Flyer II*, with an engine of increased power, using a site near Dayton which became the world's first airfield. On September 20, 1904, Wilbur flew the first complete circle by an airplane, and followed this on November 9 by the first flight of over five minutes' duration.

In June 1905, with their *Flyer III*, the Wrights had the first fully practical airplane, and with this, in October, Orville made the first flight exceeding half an hour in duration. A flight later that month was the last the brothers made until 1908; meanwhile the Wright patents were published. Back at Kitty Hawk in the spring of 1908, the Wrights started using upright seating, and then began preparations for showing their aircraft to the world.

In the summer of 1908 Wilbur went to France and made a series of public demonstration flights beginning on August 8; a month later Orville was making public flights in the United States, achieving the first flight of an hour's duration on September 9. But eight days later Orville crashed and was seriously injured, and aviation suffered its first fatality when his passenger, Lt. Thomas E. Selfridge, was killed. Wilbur, still in France, continued to set records: the first flight of over 1½ hours and first of over an hour with a passenger. During 1909 the Wrights were feted in Europe, and Wright-type biplanes were being built in Britain, France, and the United States. The Wrights returned to Kitty Hawk for further experiments, but Wilbur died of typhoid in May 1912. Orville's work was subsequently largely concerned with legal aspects of the Wright patents.

Wright Corporation

U.S. aircraft engine manufacturer. The series of radial engines produced by Wright began with the 7- and 9-cylinder Whirlwinds (235–450 hp). The Cyclone Nine was introduced in 1929, the year Wright joined with CURTISS to form Curtiss-Wright Corporation. The Cyclone Nine (F, G, G100, and G200 series; 745–1,200 hp) powered the B-17 FLYING FORTRESS, STRATOLINER, DAUNTLESS and SBC HELLDIVER. The two-row 14-cylinder R-2600 Cyclone (1936), which developed upward of 1,500 hp,

The Wright R-3350 Cyclone, one of the most powerful piston engines ever built. A two-row 18-cylinder radial, it produced 3,250 hp in the version pictured here powering the postwar Lockheed P-2 Neptune

was used in the BOEING 314 and the B-25 MITCHELL, SB2C HELLDIVER, AVENGER and MARINER, among others, with the output eventually boosted to 1,700 hp. The two-row 18-cylinder R-3350 Cyclone (1941), initially of 2,200 hp, equipped the B-29 SUPERFORTRESS, the CONSTELLATION and the MARS.

After the war a turbocompound engine was produced, which had three exhaust-powered turbines driving through a gear chain to the back of the crankshaft. Later, Sapphire (J65) and Olympus (J67) jet engines were built under license.

X

X-1

The first aircraft to exceed the speed of sound in level flight was the rocket-propelled Bell X-1, which made its record-breaking flight on October 14, 1947, in the hands of "Chuck" Yeager. First flown on December 6, 1946, the straight-winged X-1 had a wingspan of 28 ft. (8.53 m) and a length of 31 ft. (9.45 m). It was originally carried aloft by a modified Superfortress, but in 1949 made its first takeoff under its own power. Duration of powered flight was only about 2½ minutes at full throttle.

A development of the X-1, the X-1A, had a fuselage 35 ft. 7 in. (10.85 m) in length. In 1953 it attained Mach 2.42 flown by Yeager. The similar X-1B was for research into high-speed thermal problems.

The two Bell X-2s, which had stainless steel swept-back wings and tail, were both lost in crashes (in 1954 and 1956), although the second X-2 apparently reached a speed of Mach 3.2 before it exploded.

Bell also constructed two X-5 swing-wing experimental aircraft, based on the design of the Messerschmitt P.1101. The first X-5, which made its maiden flight in 1951, crashed in 1953; the second continued flying until 1955. The sweep of the X-5's wings could be changed from 20° to 60° while in flight; the experience gained was vital for such later swing-wing aircraft as the F-14 TOMCAT and F-111.

X-15

A rocket-powered aircraft for research into flight at the uppermost limits of the atmosphere; the X-15 has achieved the highest speeds and greatest altitude of any airplane capable of being controlled in normal flight. During a series of flights from 1959 to 1968 the X-15, launched from a modified B-52 mother plane, reached an altitude of 67.08 mi. (107.93 km) and speed of Mach 6.72, or 4,534 mph (7,297 km/h).

The X-15, designed by North American Aviation, was built around a liquid-fuel rocket motor capable of 60,000 lb. (27,215 kg) thrust for a short period of time. More than half the X-15's weight at launch was accounted for by the liquid oxygen and anhydrous ammonia propellants. Wingspan of the X-15 was 22 ft. (6.70 m) and length 50 ft. (15.24 m).

Y

Yak-9

The Yak-9 and its immediate predecessors made up one of the Soviet Union's most successful family of World War II fighters. Designed by A. S. YAKOVLEV, the Yak-9 appeared in 1942 and was derived from the Yak-1 of 1940 via the Yak-7 fighter and advanced trainer. The basic Yak-9 single-seat fighter was powered by a 1,240-hp VK-105PF engine and had a wingspan of 32 ft. 10 in. (10.00 m) and length of 28 ft. 1 in. (8.55 m); maximum speed was 376 mph (605 km/h) at 14,110 ft. (4,300 m). A number of Yak-9 variants appeared. The Yak-9T had a heavy caliber (23–37-mm) cannon in the nose; the Yak-9B carried an internal load of four 220-lb. (100-kg) bombs; the Yak-9R was used for photo-reconnaissance; the Yak-9D had a range extended to 870 mi. (1,400 km); and the Yak-9DD's maximum range was 1,367 mi. (2,200 km). The Yak-9U and Yak-9P, which saw postwar service, were redesigns in which all-metal construction replaced the mixed construction of earlier versions, and a 1,700-hp VK-107A supplanted the VK-105. Maximum speed of the Yak-9P was 434 mph (698 km/h) at 18,045 ft. (5,500 m).

Total production of the Yakovlev series of wartime aircraft was 36,732, of which 16,769 were Yak-9s.

Yakovlev, Alexander Sergeevich (1906–)

A Soviet aircraft designer who first came to prominence during World War II, Yakovlev heads a design bureau that has been responsible for a wide range of aircraft. Yakovlev entered the Zhukovski Air Force Academy in 1927, after building two successful gliders and a prize-winning two-seat light biplane (the AIR-1). He graduated in 1931 and joined the Central Design Bureau as a member of one of POLIKARPOV's design teams. His preoccupation with building sportsplanes led to his expulsion in 1933, after which he established a lightplane factory that was officially recognized in 1935. By 1940 he had designed twenty sportsplanes and trainer aircraft, including the UT-2 trainer, the two-seater AIR-7 mailplane of 1932 (which had a maximum speed of 206 mph, 331 km/h, and was faster than the air force's standard fighter), a twin-engined fighter-bomber (the Yak-4), and a single-engined fighter (the Yak-1). From the Yak-1 was developed a highly successful

Yak-9P captured during the Korean War

Yakovlev Yak-28 "Brewer" tactical strike aircraft

series of wartime fighters, most prominent of which was the YAK-9.

At the end of the war, Yakovlev produced the Yak-16 small transport (derived from wartime light transports), the Yak-12 high-wing utility aircraft (a development of prewar cabin monoplanes), the Yak-11 and Yak-18 trainers (with origins in the Yak-7 and the prewar UT-2) and the Yak-14 military glider. The first of several Yakovlev helicopters was completed in 1947.

The Yak-15, which ranks together with the MiG-9 as the first Soviet jet fighter, made its maiden flight on April 24, 1946. The Yak-15 was developed from the airframe design of the Yak-3, and there were initial problems with the tailwheel undercarriage and the nose-mounted engine, a copy of a captured German Jumo 004, which had its exhaust under the fuselage. Maximum speed was 488 mph (786 km/h) at 16,400 ft. (5,000 m). The Yak-17 and Yak-23 (1947) were further developments of the Yak-15 design.

A series of twin-engined fighter-bombers dating from 1955 culminated in the Yak-28 of 1961. Among later Yakovlev military aircraft is the VTOL fighter, carried aboard the *Kiev*. Known

in the West as "Forger," this is a single-seat straight-winged aircraft with direct-thrust and vectored-thrust engines, which has been developed from the experimental "Freehand" shown at the 1967 Soviet Aviation Day display.

In civil aviation, the 24–32 passenger Yak-40 "Codling" feederliner, which first flew in 1966, has been very successful in the Soviet Union, though attempts to sell it to foreign customers have had little success outside the Communist countries. The Yak-42 is essentially a scaled-up version of the Yak-40 and is capable of carrying 120 passengers.

Z

Zemke, Hubert (1914–)

American World War II fighter leader and ace. "Hub" Zemke's grasp of technical matters led to his association with test programs, and in 1941 he was sent to Russia to instruct on the assembly and flying of fighters supplied under Lend Lease. Returning to America he was given command of the first U.S. fighter group equipped with the

P-47 THUNDERBOLT, the 56th, taking this unit to England late in 1942. Zemke developed new attack techniques with the P-47 and led his group to become the most successful of the 8th Air Force in aerial combat (see WOLFPACK). He was awarded the DSC and credited with some 18 air victories and 8½ by strafing, but became a POW on October 30, 1944, after his P-51 MUSTANG broke up in a storm. He retired from the Air Force in 1957.

Zeppelin, Count Ferdinand von
(1838–1917)

German AIRSHIP pioneer. A career soldier, he fought in the Civil War with the Union forces, where he saw balloons used for reconnaissance. He began designing airships in 1873, and after retiring from the German army in 1891 devoted himself fully to this work. He was the first to see the potential of very large rigid airships, and the first to produce a practical design.

Zeppelin completed plans for his first airship in 1893. Three years later the Association of German Engineers backed it, and construction of the *LZ-1* began. The airship was cigar-shaped, with an aluminum frame 420 ft. (128 m) long and 38 ft. (11.6 m) in diameter, covered with fabric. Lift was provided by internal rubberized bags filled with hydrogen. The *LZ-1* made its first flight in July 1900, but was wrecked soon afterward.

The success of the *LZ-3* in 1906 won government support for Zeppelin, and his factory began constructing airships on a large scale. In 1910 he founded Delag (Deutsche Luftschiffahrts Aktien Gesellschaft) to exploit commercial airship travel, and during the years 1910–1914 the company's five airships carried a total of 34,228 passengers without a single mishap. During World War I a great many Zeppelins were built for maritime reconnaissance duties, and they also carried out bombing raids over France and Britain, but the vulnerability of Zeppelins to antiaircraft fire and fighter airplanes soon took a heavy toll. After his death, Zeppelin's role as the champion of the rigid airship was filled by Hugo ECKENER.

Zero

More MITSUBISHI A6M Zero (Navy Type 0) fighters were built during World War II than any other Japanese warplane (10,499). This lightweight and highly maneuverable navy interceptor, with an armament of two machine guns and two 20-mm cannon, first flew on April 1, 1939, and appeared over China

Imperial German Navy Zeppelin L.48, shot down over England less than a week after being commissioned in 1917

in August 1940 as the A6M2. Despite warnings from General Chennault of the FLYING TIGERS, the presence of the Zero took American pilots by surprise in December 1941. The A6M3 of 1941 was easily able to out-turn and out-climb the P-39 AIRACOBRA and early P-40s, and until the advent of the F6F HELLCAT and the F4U CORSAIR in 1943, the Zero was the most effective fighter in the Pacific. Its early versions, however, lacked self-sealing fuel tanks or protection for the pilot, and its lightweight construction made it incapable of absorbing battle damage.

The A6M2 had a 925-hp Sakae 14-cylinder radial engine that gave a maximum speed of 332 mph (534 km/h) and a ceiling of 33,790 ft. (10,300 m). Wingspan was 39 ft. 4 in. (11.98 m) and length 29 ft. 9 in. (9.06 m). Later models (A6M5 of 1943, A6M6 of 1944, A6M7 of 1945) were heavier due to the addition of armor protection, self-sealing tanks, and other improvements, but the A6M6, with its 1,130-hp engine, could still attain 346 mph (557 km/h). Although clearly obsolescent by 1945, the Zero remained in production until the end of the war; the A6M8 (1,560-hp engine) never reached service units.

Zerstörer

One of many attempts to produce a "heavy fighter" capable of escorting long-range bombers was the Luftwaffe's Zerstörer ("Destroyer") program, which led to the twin-engined Messerschmitt Bf 110.

First flown on May 12, 1936, and powered by DB 600 engines, the initial production aircraft (Bf 110B) reached Luftwaffe squadrons in 1938, followed by the DB 601-powered Bf 110C in

1939. Wingspan of the Bf 110C was 53 ft. 5 in. (16.25 m) and length 39 ft. 7 in. (12.07 m).

With four machine guns and two 20-mm cannon in the nose, and a swiveling machine gun in the rear cockpit, the Zerstörer proved successful in Poland as a close-support aircraft (the Bf 110C-4 could carry a bomb load of 1,100 lb., 500 kg). Against the RAF in the Battle of BRITAIN, however, it was handicapped by its poor maneuverability and suffered heavy losses despite a top speed of nearly 350 mph (530 km/h).

The Zerstörer was also used as a fighter-bomber (Bf 110E-2), a night fighter equipped with radar, a photo-reconnaissance aircraft, and a daytime bomber destroyer. Armament was increased to include up to four cannon in the nose and twin cannon in the rear cockpit (Bf 110G-4 night fighter) or a 37-mm gun below the fuselage in place of two of the 20-mm cannon (Bf 110G-2 and H-2). The G and H variants both had 1,475-hp DB 605B engines.

Because of the failure of the Me 210, manufacture of the Zerstörer continued into 1945; total production was some 6,100 machines.

Zhukovski, Nikolai Yegorovich
(1847–1921)

Officially honored as the "Father of Russian Aviation," Zhukovski was the author of over 200 scientific papers and books, of which some 50 were pioneering works on aeronautical subjects. His first such paper, "The Theory of Flight," was read in January 1890, and it was followed in 1891 by "The Soaring of Birds," delivered to two meetings of the Mathematical Society. "On Aero-

Bf 110C-5 Zerstörer

nautics," delivered in 1898, was a far-sighted discussion of the future development of heavier-than-air flight. Zhukovski was in correspondence with Otto LILIENTHAL and had been given one of Lilienthal's 1894-type (No. 11) gliders. He began writing about propeller theory in 1898, and Zhukovski (NYeZh) propellers were in fact used on some ILYA MOUROMETZ bombers during World War I.

Zhukovski's first wind tunnel was built at Moscow University in 1902; two years later he became director of the Kuchino Aerodynamic Institute, which was equipped with two wind tunnels of his design (one a vertical tunnel) and a hydrodynamic tank. He left Kuchino in 1909 to become professor of aeronautics at Moscow Higher Technical School, where he organized the famous Aeronautical Circle, of which several future Soviet designers were young members.

During World War I Zhukovski was involved in producing training programs for Ilya Mourometz crews, and in 1915 he wrote several papers on aerial bombing. In 1917 he was the first to propose the creation of the Central Aero and Hydrodynamic Institute (TsAGI)—today the Soviet equivalent of NASA—and became its first director in 1918. He also became the first principal of the Moscow Aviation Technical School and the Air Force Academy, which were both named in his honor after his death.

Zlin

The Czech aircraft manufacturer Moravan builds a series of light aircraft under the name Zlin. Among them are a number of well-known aerobatic machines, notably the Z 526 AFS Akrobat, which has been highly successful in international competitions. The Akrobat is a single-seat development of the Z 526 F two-seat trainer (powered by a 180-hp Avia engine), which traces its origin through the Z-26, Z-126, Z-226, and Z 326. These aircraft have achieved notable successes in aerobatic championships since 1957, and the two-seat series has continued through the Z 726 (1973), with shorter-span wings and metal- (instead of fabric) covered rudder and elevators.

Moravan also builds the Zlin 42 M two-seat light trainer and tourer (180-hp Avia engine) and its derivative, the Zlin 43 two/four-seater with a 210-hp Avia engine.

Index

RICHARD K. SPIELMAN
19200 S. MAIN STREET
GARDENA, CA. 90248

RICHARD K. SPIELMAN
19200 S. MAIN STREET
GARDENA, CA. 90248

RICHARD K. SPIELMAN
19200 S. MAIN STREET
GARDENA, CA. 90248